T0233745

Springer Optimization and Its Applications

VOLUME 98

Aims and Scope

Optimization has been expanding in all directions at an astonishing rate during the last few decades. New algorithmic and theoretical techniques have been developed, the diffusion into other disciplines has proceeded at a rapid pace, and our knowledge of all aspects of the field has grown even more profound. At the same time, one of the most striking trends in optimization is the constantly increasing emphasis on the interdisciplinary nature of the field. Optimization has been a basic tool in all areas of applied mathematics, engineering, medicine, economics, and other sciences.

The series *Springer Optimization and Its Applications* publishes undergraduate and graduate textbooks, monographs and state-of-the-art expository work that focus on algorithms for solving optimization problems and also study applications involving such problems. Some of the topics covered include nonlinear optimization (convex and nonconvex), network flow problems, stochastic optimization, optimal control, discrete optimization, multi-objective programming, description of software packages, approximation techniques and heuristic approaches.

More information about this series at http://www.springer.com/series/7393

Khanh D. Pham

Resilient Controls for Ordering Uncertain Prospects

Change and Response

Springer

Khanh D. Pham
The Air Force Research Laboratory
Space Vehicles Directorate
Kirtland Air Force Base, NM, USA

ISSN 1931-6828 ISSN 1931-6836 (electronic)
ISBN 978-3-319-38300-2 ISBN 978-3-319-08705-4 (eBook)
DOI 10.1007/978-3-319-08705-4
Springer Cham Heidelberg New York Dordrecht London

Mathematics Subject Classification (2010): 93E20, 91A80, 62P30, 49J15, 49J20, 60G15

Printed on acid-free paper

Springer is part of Springer Science+Business Media (www.springer.com)

This monograph is dedicated to
increased awareness and understanding
through stochastic dominance
as exemplified by the growth
of performance-measure statistics
in statistical optimal control theory
during the past decade

and to Huong Nguyen, An and Duc,
my wife and children
who have encouraged much in this ideal

Preface

Thoroughly engaged in systematically exploring mathematical statistics which are conducive to feedback and stability analysis associated with stochastic regulators and performance uncertainty, I was an unlikely candidate for the task of chronicling and analyzing resilient controls from theoretical, algorithmic, and application perspectives described here. Such a project was far from my mind when this emerging field of resilient control research with far-reaching impact on a variety of applications has suddenly drawn my attention.

But as with any domain, resilient controls require the environments to be sensed in order to be able to have situational awareness. Performance uncertainty sensing seeks to exploit any part of the uncertainty spectrum in order to provide uncertain prospects necessary for that situational awareness so the integrity of performance assessment and the controlled systems that bind them can be better maintained and managed. Needless to say, *Resilient Controls for Ordering Uncertain Prospects* was written under the influence of this indignation and it suffers, no doubt, from the shortcomings of any work by subject matter experts.

The primary aim of this monograph is to give an account of resilient controls in risk-averse decision making for elective studies, research references, and graduate-level lectures in applied mathematics and electrical engineering with systems-theoretic concentration. Some of my arguments and supports for timely and responsive reforms on the use of asymmetry or skewness pertaining to the restrictive family of quadratic costs have appeared in various scholarly forums. Such theoretical constructs and design principles, especially when they underline the complexity of the resilient controls with risk-averse attitudes, are expected to have broad appeal.

Not to acknowledge the following would be an omission that detracts not only from their contribution to and ownership of parts of the monograph but also from its worth. I have also accumulated debts to individuals. It is no exaggeration to say that this particular monograph would not have been written without the encouragement and assistance of my wife, Huong, and my children, An and Duc, on the domestic front where I was allowed to use my spare time for writing this monograph.

Of course, Professors Stanley R. Liberty and Michael K. Sain were a part of this effort, as they were the research advisors with whom I had the pleasure of collaboration during the graduate studies at University of Nebraska and University of Notre Dame. They have some idea, I expect, of how grateful I am to them. In case of doubt, they should refer to this monograph's frontispiece.

Let me close the preliminaries of this monograph and claim sole responsibility for the work. *Resilient Controls for Ordering Uncertain Prospects* was written in anxious haste, and no one bears much responsibility for its final form. Certainly, no one with whom I am associated should be held accountable for the ideas or analyses expressed here.

Albuquerque, USA
January, 2014

Khanh D. Pham

Contents

List of Figures

Chapter 1
Introduction

1.1 Emerging Research, New Challenges

Rapid advances are taking place today in resilient control technologies. These are enabling the development of more capable controlled systems that are expected to have robust and improved performance with greater reliability and agility in degraded and/or contested operational environments. Operations in degraded environments and thereby raising concerns of mission assurance and performance riskiness, have created the need for large, expensive test programs. At this writing, it is necessary to provide research and development guidance and alignment between academia, industry and government organizations, and to raise awareness of emerging concepts, capabilities, testing, standards, and requirements for resilient control solutions.

Historically speaking, uncertainty analysis for probabilistic nature of performance uncertainty is relied on as part of the long range assessment of reliability. Some of the most widely used measures for performance reliability are mean and variance which attempt to summarize the underlying performance variations. However, other aspects of performance distributions, that do not appear in most of the existing progress, are skewness, flatness, etc. Over the years, there were of course signals – beginning as early as the 1970s – that all was not well in this paradigm of thinking [1] and [2]. Admittedly there are many reasons to believe that some performance with negative skewness appears riskier than performance with positive skewness, when expectation and variance are held constant. If skewness does, indeed, play an essential role in determining the perception of risks, then the range of applicability of the present theory for stochastic control and decision analysis should be restricted, e.g., to symmetric or equally skewed performance measures.

Having acquired a proper understanding of, and orientation toward, performance uncertainty, it has been known for some time [3] and [4] that the ubiquitous expected value approach and its variants are of limited generality and not realistic as they

K.D. Pham, *Resilient Controls for Ordering Uncertain Prospects*, Springer Optimization and Its Applications 98, DOI 10.1007/978-3-319-08705-4_1

rule out asymmetry or skewness in the probability distributions of the non-normal random costs. Thus, a control objective based on mean or expected value alone is indeed not justifiable on theoretical grounds.

Perhaps so motivated, this monograph will investigate the technical challenges associated with theoretical foundations and design principles pertaining to resilient controls of linear and bilinear stochastic systems with actuator faults, persistent disturbances, time-delay controls and observations, control rate and communication channel constraints. The admissible controls herein for ordering uncertain prospects are expected to make efficient use of the complete knowledge of entire chi-squared distributions associated with finite-horizon integral-quadratic-form costs; not just means and/or variances as often can be seen from the existing literature. Elsewhere on the related publications, it is possible to sample the way that the subject research by the author looks in real world applications such as seismic protection of buildings and bridges from earthquakes, winds and seas, by means of the IEEE and ASME conference proceedings, Journal of Optimization Theory and Applications, and book chapters by Birkhauser and Springer.

1.2 Monograph Ideas and Contributions

What are ultimately at stake are theoretical advances and potential techniques that would prove extremely valuable to those attempting to solve resilient control problems and post-design performance analyses using one-shot approaches. In this regard, it is the thesis of this monograph that resilient control designers are caught up in uncertain stochastic systems that link them to far-off expectations. Closed-loop performances have been altered by often unforeseen forces; e.g., actuator faults, exogenous disturbances, state and input time delays, network effects, etc. that must be brought into relief if quantification of performance uncertainty and resilient control optimization with performance risk aversion are to understand their lot. At least, any approaches that do not employ such an analysis will leave unanswered many of the questions central to an understanding of "resilient controls for ordering uncertain prospects". Specifically, left unanswered are the perennial "how" questions; e.g., How are uncertainties encoded on each of the crucial state variables? How to find the uncertainty in performance measure for each alternative control decisions? How to choose between two control strategies with different probability distributions on performance measures?

In order to address these questions, it is necessary, to abandon the formal methods and to leave the traditional mean-variance rules of the best alternative behind. To fill these explanatory lacunae, the work hereafter turns to other disciplines: applied decision theory, uncertainty analysis, and statistical optimal control. Cautious recourse to these disciplines is part of the primary aims of this monograph when risk preferences is established in the form of a utility function and then the best control strategy is the one whose performance index has the highest utility. Several very distinctive if not unique characteristics distinguish the present work.

Firstly, a general framework of the information-gathering scheme is obtained to transform performance uncertainty associated with various resilient control problem classes of linear-quadratic stochastic systems into the mathematical statistics for post-design analysis. Secondly, statistical measures of risk are pivotal, residual, and inherently fundamental to the modification of the probability distributions on the restrictive family of integral quadratic costs and thereby consequently affecting the control decision. Thirdly, risk preferences on performance uncertainty in accordance with the mean-risk aware performance indexes play crucial roles in the algorithmic development of resilient control strategies.

1.3 Methodology

Throughout the monograph, the methodology in each chapter/section is composed of five steps:

- Mathematical Modeling in which basic assumptions related to the state-space models are considered
- Mathematical Statistics for Performance Robustness whose backward differential equations are characterized by making use of both compactness from logics of state-space models and quantitativeness from a-priori knowledge of probabilistic processes
- Statements of Statistical Optimal Control which provide complete problem formulations composed of unique notations, terminologies, definitions and theorems to pave the way for subsequent analysis
- Existence of Control Solutions whereby information about accuracy of estimates for higher-order characteristics of performance-measure uncertainty is incorporated into risk-averse decision strategies. The results are provided most of the time in the form of theorems, lemmas, and corollaries, and
- Chapter Summaries which are given to shed some light of the relevance of the developed results.

For convenience, the relevant references are offered at the end of each chapter for the purpose of stimulating the reader. It is hoped that this way of articulating the information will attract the attention of a wide spectrum of readership.

1.4 Chapter Organization

As noted earlier, the use of expected values and variances as measures of risk for an entire class of non-symmetric distribution functions has been questioned by financial theorists [1] and [2]. It has been shown that the subclass that can be ordered by the mean-variance rule is indeed small. This popular approach and its variants, thus appear to be of limited generality. Of practical interest in engineering and resilient control applications hereafter, a finite linear combination of performance-measure statistics associated with the restrictive family of finite-horizon integral-quadratic-form costs is proposed to provide a strong rationale for using mean, variance, skewness, flatness, etc. to order uncertain prospects of the chi-squared random costs. As the result, this new paradigm is tempted to address both necessary and sufficient conditions as needed when ordering stochastic dominance for the class of probabilistic distributions with equal and unequal means. Thereafter, the monograph presents theoretical explorations on several fundamental problems for resilient controlled systems.

The monograph is primarily intended for researchers and engineers in the systems and control community. It can also serve as complementary reading for uncertainty analysis and control sciences at the postgraduate level. After the first introductory chapter, the core material of the monograph is organized as follows.

Chapter 2 seeks to address the development of modeling and technology to explore the emerging linear-quadratic class of stochastic fault-tolerant control systems and safely replicate critical actuator failures of the stuck-type. As the fidelities of model uncertainties and uncertain prospects of performance distributions moving toward closer to realities, new methodologies of performance information analysis and feedback control strategies with performance risk aversion that promote timely and cost effective development and analysis, are explored and developed.

Yet, as has been evident all along, there are crucial gaps between an operational reality of stochastic linear systems and a subset of its actuators be under the control of an external adversary. Such is the case where adversarial sources of uncertainty purposefully try to affect the functioning of the overall system; this behavior is denoted as a fault occurrence. As the result, the controlled dynamics can be seen as a zero-sum stochastic differential game where the system (via the remaining actuators) tries to counteract the negative influence of the external adversary (via the compromised actuators). Under this game-theoretic framework and with stochastic performance analysis, a quadratic performance criteria is considered and analyzed such that an optimal solution is found in Chap. 3.

Additionally, new problems have emerged within the broad areas of disturbance attenuation areas of unknown input time delay issues, adaptive and persistent disturbances, etc. that are difficult to tackle with conventional methods, since elusive and smart disturbances together with non-Gaussian performance distributions tend to be at the heart of these new types of problems. The purpose of Chap. 4 is to continue the discussion of current and ongoing efforts in using preferences of risk, dynamic game decision optimization, and disturbance mitigation techniques with output feedback measurements tailored toward the worst-case scenarios.

As noted, it is significant in resilient controls of bilinear stochastic systems that the expected value approach has undergone slow changes and there remain big and serious problems as they rule out the asymmetry or skewness in the probabilistic performance distributions of the chi-squared random costs. In order to do away with the fact that no distinct progress has been made in determining optimal selection rules for uncertain prospects beyond the expected values, Chap. 5 lays stress on the key aspects of the quadratic class of weakly coupled bilinear stochastic systems together with its characterization of performance-measure statistics so as to set up the corresponding statistical optimal control problem in which, from recently opened doors and also from windows, new perspectives on the procedural mechanism for resilient controller designs endowed with perfect state-feedback measurements and subject to performance risk aversion are revealed.

Continually to highlight the advances being made in resilient controls of bilinear stochastic systems, the main objective of Chap. 6 will be to promote synergistic exploitation of the ideas from the different areas of endeavor; e.g., weakly coupled bilinear stochastic systems, low sensitivity to small parameter variations, noisy measurements, decision analysis and feedback control optimization that together constitute another variant of statistical optimal control. In particular, the emphasis will be on the triplet: mathematical statistics associated with the restrictive family of finite-horizon integral-quadratic costs with sensitivity to small parameter variations, mean-risk aware performance index for ordering uncertain prospects of the chi-squared random costs, and closed-loop control strategies with state estimates and performance risk aversion.

Studies dealing with real-world issues, such as state and input time delays, norm-bounded and convex-bounded parameter uncertainties, stochastic exogenous disturbances, and the like are of particular interest in Chap. 7. On the theoretical front, recent advances in resilient controls with performance risk aversion provide new efficient tools to characterize the generalized chi-squared behaviors of performance distributions as well as deal with the design procedure of mean-risk control decision laws. In addition, the emphasis is to create a framework to foster the integration of decision theory and resilient control optimization that brings about closed-loop performance robustness against structural and stochastic uncertainties in presence of state and input time delay issues.

Chapter 8 will encompass all aspects of network time delays, communication channel constraints, etc. New breakthroughs as can be seen here, can help characterize how hard, at least on theoretical grounds, it is to design an admissible set of output feedback control strategies. In the case of time-delay measurements and communication channel constraints, such a feasible optimal control policy, all the time and random events can order uncertain prospects of the chi-squared random cost. Associated is the recently advocated mean-risk aware performance index, which then involves naturally two parameters, one of which is always the mean but the other parameter is the appropriate measures of performance dispersion. Ultimately, the results herein will provide a strong impetus for further research in the development of resilient controls for performance robustness while subject to networking effects.

The scope of Chap. 9 ranges from showcasing particular topics in basic research in the theoretical constructs and design principles for a class of networked stochastic systems, to the latest development in characterization and management of uncertain prospects pertaining to the restrictive family of finite-horizon integral quadratic costs. Given the enormous diversity of stochastic controlled systems, a class of time-invariant linear stochastic systems together with time-delay observations is considered and the admissible set of feedback control policies is subject to control rate constraints. In addition, it is anticipated that research advances here will foster cross-fertilization across different disciplines with participants being exposed not only to a broad range of financial and engineering problems associated with uncertainty quantification, but also the accompanying development of optimal rules for ordering uncertain prospects of the chi-squared random costs. Special emphasis is on performance uncertainty handling in resilient controls. Subsequently, the feasibility of time-varying linear memoryless control laws supported by Kalman-like estimator and subject to control rate constraints is finally put forward.

All in all, this monograph comes at a particular time when stochastic control is at a crucial crossroads in terms of both its selection rules for ordering uncertain prospects and foundational principles for resilient controls. The continuing task in Chap. 10 can be examined in two major dimensions: how to characterize and manage performance uncertainty for the restrictive family of chi-squared random costs; and how to use available resources of complete knowledge of the chi-squared probability distributions and stochastic controlled systems, in the most efficient and effective ways for the design principles for feedback control agility. This chapter sets out to give a brief account of opening the door of ordering uncertain prospects to a wider resilient control community, to provide through illustration and analysis a profile of current fault-tolerant/bilinear and networked control applications, and thereby to indicate some proposals that could lead to future research and development for resiliency.

References

1. Hirshleifer, J.: Investment, Interest and Capital. Prentice Hall, Englewood Cliffs (1970)
2. Markowitz, H.: Portfolio Selection: Efficient Diversification of Investments. Wiley, New York (1970)
3. Sain, M.K.: Control of linear systems according to the minimal variance criterion-a new approach to the disturbance problem. IEEE Trans. Autom. Control **11**, 118–122 (1966)
4. Sain, M.K., Liberty, S.R.: Performance measure densities for a class of LQG control systems. IEEE Trans. Autom. Control **16**, 431–439 (1971)

Chapter 2
Actuator Failure Accommodation in Risk-Averse Feedback Control

2.1 Introduction

Over the last three decades, there have been an increasing need for fault-tolerant control systems to continue operating acceptably to fulfill specified performance following faults in the system being controlled or in the controller. Many important issues in fault-tolerant control systems can be found in [1] and references therein. Motivated by the growing demand for reliability and maintainability, such efforts on different fault detection and diagnosis techniques have led to the development of many reconfigurable control systems [2] and [3]. However, little attention has been paid to the problem of graceful performance degradation with multiple risk considerations. More recently, risk-averse control feedback for stochastic control problems has begun to show promises and successes [4] and [5].

When the actuator failures of stuck-type ones occur, for example aircraft and robotic manipulators wherein the control surfaces or effectors would be stuck in certain positions, the command control inputs could not affect the actuator outputs. Hence, the total loss of control efficiency would result given the fact that there was very limited time and possibility to perform fault diagnosis. The main objective of this research is to show how to: (i) integrate an element of automatic allocation representation of apportioning performance robustness and reliability requirements into the multi-attribute requirement of qualitative characteristics of expected performance and performance risks, once the random performance measure of the generalized chi-squared type associated with the linear-quadratic class of stochastic fault-tolerant control systems has been recognized and (ii) adapt a risk-averse control strategy for the baseline controller in the stochastic fault-tolerant control system to retain some portion of its control integrity in the event of a specified set of possible severe actuator faults.

The structure of the chapter is as follows. Section 2.2 contains the problem description and the notion of admissible control for the linear-quadratic class of stochastic fault-tolerant systems. In addition, the development of all the

K.D. Pham, *Resilient Controls for Ordering Uncertain Prospects*, Springer Optimization and Its Applications 98, DOI 10.1007/978-3-319-08705-4_2

mathematical statistics for performance robustness is carefully discussed. The detailed problem statements and solution method in obtaining an optimal feedback controller with performance risk aversion are described in Sects. 2.3 and 2.4. In Sect. 2.5, some conclusions are also included.

2.2 Patterns of Performance Uncertainties

In this section, some preliminaries are in order. Some spaces of random variables and stochastic processes are introduced: A fixed probability space $(\Omega, \mathbb{F}, \{\mathbb{F}_{t_0,t} : t \in [t_0, t_f]\}, \mathbb{P})$ with filtration satisfying the usual conditions. All the filtrations are right continuous and complete and $\mathbb{F}_{t_f} \triangleq \{\mathbb{F}_{t_0,t} : t \in [t_0, t_f]\}$. In addition, let $\mathscr{L}^2_{\mathbb{F}_{t_f}}([t_0, t_f]; \mathbb{R}^n)$ denote the space of $\mathbb{F}_{t_f}$-adapted random processes $\{\hbar(t) : t \in [t_0, t_f]$ such that $E\{\int_{t_0}^{t_f} ||\hbar(t)||^2 \, dt\} < \infty\}$.

As for the setting, a class of stochastic fault-tolerant control problems with actuator failure accommodation is investigated as shown in Fig. 2.1. A typical description for the faulty system being controlled on the finite horizon $[t_0, t_f]$ makes use of the concept of "unknown inputs" acting upon a nominal linear model of the system described by

$$dx(t) = [(A + \Delta A(t))x(t) + (B + \Delta B(t))Lu(t) + (B + \Delta B(t))(I - L)f_a(t)]dt + G dw(t), \quad x(t_0) = x_0. \qquad (2.1)$$

With regards of the combined state and control effectiveness model (2.1), the nominal system coefficients $A \in \mathbb{R}^{n \times n}$, $B \in \mathbb{R}^{n \times m}$ and $G \in \mathbb{R}^{n \times p}$ are the constant matrices. The process disturbances $w(t)$ assume the form of a stationary Wiener process with the correlations of independent increments

$$E\left\{[w(\tau_1) - w(\tau_2)][w(\tau_1) - w(\tau_2)]^T\right\} = W|\tau_1 - \tau_2|, \quad \forall \tau_1, \tau_2 \in [t_0, t_f]$$

whose a-priori second-order statistic $W > 0$ is also known.

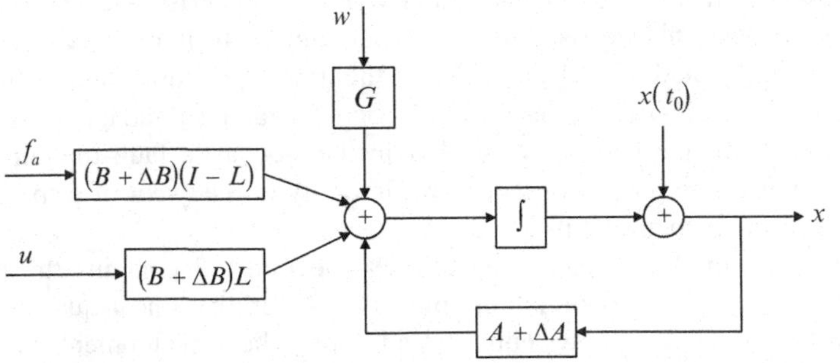

Fig. 2.1 Structure of stochastic fault-tolerant control with non-cognitive actuator failures

In addition, the parameter perturbations considered with robustness to minor model uncertainties are approximated as

$$\Delta A(t) = H_A F(t) E_A, \qquad \Delta B(t) = H_B F(t) E_B \tag{2.2}$$

where some a-priori knowledge about uncertainty are used for a direct derivation of known constant matrices H_A, H_B, E_A and E_B with appropriate dimensions. And $F(t)$ is an unknown real time-varying matrix with Lebesgue measurable elements satisfying the condition $F^T(t)F(t) \leq I$ for all $t \in [t_0, t_f]$.

In the discussion to follow $x(t)$ is the controlled state process valued in $\mathbb{R}^n$ and $u(t)$ is the normal control process valued in a closed convex subset of $\mathbb{R}^m$. Next the working condition of the actuators is further described by the distribution matrix $L \triangleq diag(l_1, \ldots, l_m)$, whereby $l_i = 1$ and $i \in \overline{L} \triangleq \{1, \ldots, m\}$ denotes a healthy ith actuator and $l_i = 0$ corresponds to the total failure of the ith actuator. In this case, the ith actuator is stuck at certain position and thus, its actual control input is only an offset value. Such an offset is denoted as $f_a(t)$ and is also considered as the disturbances added to the system through the failed actuators. In retrospect with a passive approach to control input failures, the robust feedback controller proposed herein is designed to tolerate control channel failures.

In most situations, the system (2.1) is assumed to possess $(m - l)$ degree of actuator redundancy if (†) the system is complete state controllable with all m actuators functional and (‡) if any l actuators out of m actuators fail, with the remaining $(m-l)$ actuators, the system is still completely state controllable. See [6].

Further let the σ-algebras

$$\mathbb{F}_{t_0,t} \triangleq \sigma\{w(\tau) : t_0 \leq \tau \leq t\}$$

$$\mathcal{G}^y_{t_0,t} \triangleq \sigma\{x(\tau) : t_0 \leq \tau \leq t\}, \qquad t \in [t_0, t_f].$$

Then, the admissible set of feedback control laws which makes efficient use of the information available $\mathcal{G}^y_{t_f} \triangleq \{\mathcal{G}^y_{t_0,t} : t \in [t_0, t_f]\} \subset \{\mathbb{F}_{t_0,t} : t \in [t_0, t_f]\}$, is defined as

$$\mathbb{U}^y[t_0, t_f] \triangleq \{u \in \mathcal{L}^2_{\mathcal{G}^y_{t_f}}([t_0, t_f], \mathbb{R}^m), \mathbb{P} - a.s.\}$$

where $\mathbb{U}^y[t_0, t_f]$ is a closed convex subset of $\mathcal{L}^2_{\mathbb{F}_{t_f}}([t_0, t_f], \mathbb{R}^m)$.

Associated with any admissible 2-tuple $(x(\cdot), u(\cdot))$ is an integral-quadratic form cost $J : \mathbb{R}^n \times \mathbb{U}^y[t_0, t_f] \mapsto \mathbb{R}^+$ on $[t_0, t_f]$, for which the fault-tolerant control reconfiguration attempts not only to stabilize the fault system with multiple degrees of performance robustness but also to closely track the reference trajectory (e.g., for additional details, refer to [7] and [8]) $r \in \mathcal{C}(t_0, t_f; \mathbb{R}^n)$, in presence of the fault-induced signal $f_a \neq 0$ added to the fault system; e.g.,

$$J(x_0, u(\cdot)) = (x(t_f) - r(t_f))^T Q_f (x(t_f) - r(t_f)) + \int_{t_0}^{t_f} [x^T(\tau) Q_1 x(\tau)$$
$$+ (x(\tau) - r(\tau))^T Q_2 (x(\tau) - r(\tau)) + u^T(\tau) R_1 u(\tau) - f_a^T(\tau) R_2 f_a(\tau)] d\tau \tag{2.3}$$

where the terminal penalty weighting $Q_f \in \mathbb{R}^{n\times n}$, the state weighting $Q_1 \in \mathbb{R}^{n\times n}$, the reference tracking weighting $Q_2 \in \mathbb{R}^{n\times n}$, control weighting $R_1 \in \mathbb{R}^{m\times m}$ and fault effect reduction weighting $R_2 \in \mathbb{R}^{m\times m}$ are continuous-time matrix functions with the properties of symmetry and positive semi-definiteness. In addition, R_1 is invertible.

The notion of admissible controls is next discussed. In the case of state-feedback measurements, an admissible control is of the form, for some $\gamma(\cdot,\cdot)$

$$u(t) = \gamma(t, x(t)), \quad t \in [t_0, t]. \tag{2.4}$$

As evidenced in the linear structure of (2.1) and (2.3) and the information structure $\mathcal{G}^y_{t_f}$, the admissible control (2.4) is further restricted to the linear mapping $\gamma : [t_0, t_f] \times \mathbb{R}^n \mapsto \mathbb{U}^y[t_0, t_f]$ with the accessible states $x(t)$; e.g.,

$$u(t) = K(t)x(t) + \ell(t), \quad t \in [t_0, t_f] \tag{2.5}$$

with $K \in \mathscr{C}([t_0, t_f]; \mathbb{R}^{m\times n})$ and $\ell \in \mathscr{C}([t_0, t_f]; \mathbb{R}^m)$ an admissible feedback gain and an affine vector whose further defining properties will be stated shortly.

Then, for the admissible $(K(\cdot), \ell(\cdot))$ and pair (t_0, x_0), it gives a sufficient condition for the existence of $x(t)$ in (2.1). In view of the control law (2.5), the controlled system (2.1) is rewritten as follows

$$\begin{aligned} dx(t) &= ((A + \Delta A(t)) + (B + \Delta B(t))LK(t))x(t)dt \\ &\quad + (B + \Delta B(t))L\ell(t)dt + Gdw(t), \quad x(t_0) = x_0 \end{aligned} \tag{2.6}$$

subject to the random performance (2.3) of the chi-squared type

$$\begin{aligned} J(K(\cdot), \ell(\cdot)) &= x^T(t_f)Q_f x(t_f) - 2x^T(t_f)Q_f r(t_f) + r^T(t_f)Q_f r(t_f) \\ &\quad + \int_{t_0}^{t_f} [x^T(\tau)(Q_1 + Q_2 + K^T(\tau)R_1K(\tau))x(\tau) \\ &\quad + 2x^T(\tau)(K^T(\tau)R_1\ell(\tau) - Q_2 r(\tau)) \\ &\quad + r^T(\tau)Q_2 r(\tau) + \ell^T(\tau)R_1\ell(\tau) - f_a^T(\tau)R_2 f_a(\tau)]d\tau. \end{aligned} \tag{2.7}$$

So far there are two types of information, i.e., process information (2.6) and goal information (2.7) have been given in advance to the controller (2.5). Since there is an external disturbance $w(\cdot)$ affecting the closed-loop performance, the controller now needs additional information about performance variations. This is coupling information and thus also known as performance information.

Understandably, given the linear-quadratic structure of (2.6) and (2.7), the path-wise performance measure (2.7) with which the control designer is risk averse is clearly a random variable of the generalized chi-squared type. Hence, the degree of uncertainty of the path-wise performance measure (2.7) must be assessed via a complete set of higher-order statistics beyond the statistical mean or average.

The essence of information about these higher-order performance-measure statistics in an attempt to describe or model performance uncertainty, is now considered as a source of information flow, which will affect perception of the problem and the environment at the risk-averse control designer. Next the question of how to characterize performance information is answered by further elaboration of modeling and management of cost cumulants associated with (2.7) as can be seen below.

Theorem 2.2.1 (Cumulant-Generating Function). *Consider the stochastic fault-tolerant system governed by (2.6)–(2.7) with the initial-value condition $x(\tau) \equiv x_\tau$ and $\tau \in [t_0, t_f]$. Further let the moment-generating function be defined by*

$$\varphi(\tau, x_\tau; \theta) \triangleq \varrho(\tau, \theta) \exp\left\{x_\tau^T \Upsilon(\tau, \theta) x_\tau + 2x_\tau^T \eta(\tau, \theta)\right\} \tag{2.8}$$

$$\upsilon(\tau, \theta) \triangleq \ln\{\varrho(\tau, \theta)\}, \qquad \theta \in \mathbb{R}^+. \tag{2.9}$$

Then, the cumulant-generating function has the form of quadratic affine

$$\psi(\tau, x_\tau; \theta) = x_\tau^T \Upsilon(\tau, \theta) x_\tau + 2x_\tau^T \eta(\tau, \theta) + \upsilon(\tau, \theta) \tag{2.10}$$

where the scalar-valued solution $\upsilon(\tau, \theta)$ solves the backward-in-time differential equation with the terminal-value condition $\upsilon(t_f, \theta) = r^T(t_f) Q_f r(t_f)$

$$\begin{aligned}\frac{d}{d\tau}\upsilon(\tau, \theta) = &- Tr\left\{\Upsilon(\tau, \theta) G W G^T\right\} - 2\eta^T(\tau, \theta)(B + \Delta B(\tau)) L \ell(\tau) \\ &- \theta(r^T(\tau) Q_2 r(\tau) + \ell^T(\tau) R_1 \ell(\tau) - f_a^T(\tau) R_2 f_a(\tau))\end{aligned} \tag{2.11}$$

and the matrix and vector-valued solutions $\Upsilon(\tau, \theta)$ and $\eta(\tau, \theta)$ satisfy the time-backward differential equations with the terminal-value conditions $\Upsilon(t_f, \theta) = \theta Q_f$ and $\eta(t_f, \theta) = -\theta Q_f r(t_f)$

$$\begin{aligned}\frac{d}{d\tau}\Upsilon(\tau, \theta) = &- [A + \Delta A(\tau) + (B + \Delta B(\tau)) L K(\tau)]^T \Upsilon(\tau, \theta) \\ &- \Upsilon(\tau, \theta)[A + \Delta A(\tau) + (B + \Delta B(\tau)) L K(\tau)] \\ &- 2\Upsilon(\tau, \theta) G W G^T \Upsilon(\tau, \theta) - \theta[Q_1 + Q_2 + K^T(\tau) R_1 K(\tau)]\end{aligned} \tag{2.12}$$

$$\begin{aligned}\frac{d}{d\tau}\eta(\tau, \theta) = &-[(A + \Delta A(\tau)) + (B + \Delta B(\tau)) L K(\tau)]^T \eta(\tau, \theta) \\ &- \Upsilon(\tau, \theta)(B + \Delta B(\tau)) L \ell(\tau) - \theta[K^T(\tau) R_1 \ell(\tau) - Q_2 r(\tau)].\end{aligned} \tag{2.13}$$

In addition, the scalar-valued solution $\varrho(\tau,\theta)$ satisfies the backward-in-time differential equation with the terminal-value condition $\varrho(t_f,\theta) = \exp\{\theta r^T(t_f) Q_f r(t_f)\}$

$$\frac{d}{d\tau}\varrho(\tau,\theta) = -\varrho(\tau,\theta)\,[Tr\{\Upsilon(\tau,\theta)GWG^T\} + 2\eta^T(\tau,\theta)(B+\Delta B(\tau))L\ell(\tau) + \theta(r^T(\tau)Q_2 r(\tau) + \ell^T(\tau)R_1\ell(\tau) - f_a^T(\tau)R_2 f_a(\tau))] \tag{2.14}$$

Proof. For notional simplicity, it is convenient to have $\varpi(\tau,x_\tau;\theta) \triangleq \exp\{\theta J(\tau,x_\tau)\}$ in which the performance measure (2.7) is rewritten as the cost-to-go function from an arbitrary state x_τ at a running time $\tau \in [t_0,t_f]$, that is,

$$\begin{aligned} J(\tau,x_\tau) &= x^T(t_f)Q_f x(t_f) - 2x^T(t_f)Q_f r(t_f) + r^T(t_f)Q_f r(t_f) \\ &\quad + \int_\tau^{t_f} [x^T(t)(Q_1+Q_2+K^T(t)R_1K(t))x(t) \\ &\quad +2x^T(t)(K^T(t)R_1\ell(t) - Q_2 r(t)) \\ &\quad +r^T(t)Q_2 r(t) + \ell^T(t)R_1\ell(t) - f_a^T(t)R_2 f_a(t)]dt \end{aligned} \tag{2.15}$$

subject to

$$\begin{aligned} dx(t) &= ((A+\Delta A(t)) + (B+\Delta B(t))LK(t))x(t)dt \\ &\quad +(B+\Delta B(t))L\ell(t)dt + Gdw(t), \quad x(\tau) = x_\tau . \end{aligned} \tag{2.16}$$

So defined, the moment-generating function is

$$\varphi(\tau,x_\tau;\theta) \triangleq E\{\varpi(\tau,x_\tau;\theta)\}.$$

The following result features the total time derivative of $\varphi(\tau,x_\tau;\theta)$; e.g.,

$$\begin{aligned} \frac{d}{d\tau}\varphi(\tau,x_\tau;\theta) &= -\theta[x_\tau^T(Q_1+Q_2+K^T(\tau)R_1K(\tau))x_\tau + 2x_\tau^T(K^T(\tau)R_1\ell(\tau) - Q_2 r(\tau)) \\ &\quad +r^T(\tau)Q_2 r(\tau) + \ell^T(\tau)R_1\ell(\tau) - f_a^T(\tau)R_2 f_a(\tau)]\varphi(\tau,x_\tau;\theta). \end{aligned}$$

Accordingly, the standard Ito's stochastic differential formula further results in

$$\begin{aligned} d\varphi(\tau,x_\tau;\theta) &= E\{d\varpi(\tau,x_\tau;\theta)\} \\ &= E\Big\{\varpi_\tau(\tau,x_\tau;\theta)\,d\tau + \varpi_{x_\tau}(\tau,x_\tau;\theta)\,dx_\tau + \frac{1}{2}\mathrm{Tr}\{\varpi_{x_\tau x_\tau}(\tau,x_\tau;\theta)GWG^T\}\,d\tau\Big\} \end{aligned}$$

Notably, in view of the controlled diffusion process (2.16), it is shown that

$$d\varphi(\tau,x_\tau;\theta) = \varphi_\tau(\tau,x_\tau;\theta)d\tau + \varphi_{x_\tau}(\tau,x_\tau;\theta)[((A+\Delta A(\tau)) + (B+\Delta B(t))LK(\tau))x_\tau$$
$$+(B+\Delta B(\tau))L\ell(\tau)]d\tau + \frac{1}{2}\mathrm{Tr}\left\{\varphi_{x_\tau x_\tau}(\tau,x_\tau;\theta)GWG^T\right\}d\tau$$

which, under the assumption of $\varphi(\tau,x_\tau;\theta) = \varrho(\tau,\theta)\exp\{x_\tau^T\Upsilon(\tau,\theta)x_\tau + 2x_\tau^T\eta(\tau,\theta)\}$ and its partial derivatives, leads to the result

$$d\varphi(\tau,x_\tau;\theta) = \left\{\frac{\frac{d}{d\tau}\varrho(\tau,\theta)}{\varrho(\tau,\theta)} + x_\tau^T\frac{d}{d\tau}\Upsilon(\tau,\theta)x_\tau + 2x_\tau^T\frac{d}{d\tau}\eta(\tau,\theta) + \mathrm{Tr}\{\Upsilon(\tau,\theta)GWG^T\}\right.$$
$$+ x_\tau^T((A+\Delta A(\tau)) + (B+\Delta B(\tau))LK(\tau))^T\Upsilon(\tau,\theta)x_\tau$$
$$+ x_\tau^T\Upsilon(\tau,\theta)((A+\Delta A(\tau)) + (B+\Delta B(\tau))LK(\tau))x_\tau$$
$$+2x_\tau^T\Upsilon(\tau,\theta)(B+\Delta B(\tau))L\ell(\tau) + 2x_\tau^T((A+\Delta A(\tau)) + (B+\Delta B(\tau))LK(\tau))^T\eta(\tau,\theta)$$
$$\left. + 2\eta^T(\tau,\theta)(B+\Delta B(\tau))L\ell(\tau) + 2x_\tau^T\Upsilon(\tau,\theta)GWG^T\Upsilon(\tau,\theta)x_\tau\right\}\varphi(\tau,x_\tau;\theta)d\tau.$$

As a consequence, to have the constant and quadratic terms being independent of arbitrary x_τ, one therefore obtains the following results

$$\frac{d}{d\tau}\Upsilon(\tau,\theta) = -[A+\Delta A(\tau) + (B+\Delta B(\tau))LK(\tau)]^T\Upsilon(\tau,\theta)$$
$$- \Upsilon(\tau,\theta)[A+\Delta A(\tau) + (B+\Delta B(\tau))LK(\tau)]$$
$$-2\Upsilon(\tau,\theta)GWG^T\Upsilon(\tau,\theta) - \theta[Q_1 + Q_2 + K^T(\tau)R_1K(\tau)], \quad \Upsilon(t_f,\theta) = \theta Q_f$$

$$\frac{d}{d\tau}\eta(\tau,\theta) = -[(A+\Delta A(\tau)) + (B+\Delta B(\tau))LK(\tau)]^T\eta(\tau,\theta)$$
$$-\Upsilon(\tau,\theta)(B+\Delta B(\tau))L\ell(\tau) - \theta[K^T(\tau)R_1\ell(\tau) - Q_2r(\tau)], \quad \eta(t_f,\theta) = -\theta Q_fr(t_f)$$

$$\frac{d}{d\tau}\varrho(\tau,\theta) = -\varrho(\tau,\theta)\left[\mathrm{Tr}\{\Upsilon(\tau,\theta)GWG^T\} + 2\eta^T(\tau,\theta)(B+\Delta B(\tau))L\ell(\tau)\right.$$
$$\left.+\theta(r^T(\tau)Q_2r(\tau) + \ell^T(\tau)R_1\ell(\tau) - f_a^T(\tau)R_2f_a(\tau))\right], \quad \varrho(t_f,\theta) = e^{\{\theta r^T(t_f)Q_fr(t_f)\}}.$$

More is promised and much more is needed; e.g., the backward-in-time differential equation satisfied by $\upsilon(\tau,\theta)$ is obtained

$$\frac{d}{d\tau}\upsilon(\tau,\theta) = -\mathrm{Tr}\{\Upsilon(\tau,\theta)GWG^T\} - 2\eta^T(\tau,\theta)(B+\Delta B(\tau))L\ell(\tau)$$
$$-\theta(r^T(\tau)Q_2r(\tau) + \ell^T(\tau)R_1\ell(\tau) - f_a^T(\tau)R_2f_a(\tau)), \quad \upsilon(t_f,\theta) = \theta r^T(t_f)Q_fr(t_f)$$

which completes the proof.

As will be seen in the following development, the shape and functional form of a utility function tell a great deal about the basic attitudes of control designers toward the uncertain outcomes or performance risks. Of particular, the new utility function or the so-called the generalized performance index, which is being proposed herein as a linear manifold defined by a finite number of higher-order statistics associated with (2.7) will provide a convenient allocation representation of apportioning performance robustness and reliability requirements into the multi-attribute requirement of qualitative characteristics of expected performance and performance risks. Subsequently, higher-order statistics that encapsulate the uncertain nature of (2.7) can now be generated via a Maclaurin series expansion of (2.10)

$$\psi(\tau, x_\tau; \theta) = \sum_{r=1}^{\infty} \frac{\partial^{(r)}}{\partial \theta^{(r)}} \psi(\tau, x_\tau; \theta)\bigg|_{\theta=0} \frac{\theta^r}{r!} \tag{2.17}$$

from which all $\kappa_r \triangleq \frac{\partial^{(r)}}{\partial \theta^{(r)}} \psi(\tau, x_\tau; \theta)\Big|_{\theta=0}$ are called the rth-order mathematical statistics of (2.15). Moreover, the series expansion coefficients are computed by using the cumulant-generating function (2.10)

$$\begin{aligned}\frac{\partial^{(r)}}{\partial \theta^{(r)}} \psi(\tau, x_\tau; \theta)\bigg|_{\theta=0} &= x_\tau^T \frac{\partial^{(r)}}{\partial \theta^{(r)}} \Upsilon(\tau, \theta)\bigg|_{\theta=0} x_\tau \\ &\quad + 2x_\tau^T \frac{\partial^{(r)}}{\partial \theta^{(r)}} \eta(\tau, \theta)\bigg|_{\theta=0} + \frac{\partial^{(r)}}{\partial \theta^{(r)}} \upsilon(\tau, \theta)\bigg|_{\theta=0}.\end{aligned}$$

In view of (2.17), the rth-order performance-measure statistic therefore follows

$$\kappa_r = x_\tau^T \frac{\partial^{(r)}}{\partial \theta^{(r)}} \Upsilon(\tau, \theta)\bigg|_{\theta=0} x_\tau + 2x_\tau^T \frac{\partial^{(r)}}{\partial \theta^{(r)}} \eta(\tau, \theta)\bigg|_{\theta=0} + \frac{\partial^{(r)}}{\partial \theta^{(r)}} \upsilon(\tau, \theta)\bigg|_{\theta=0}.$$

for any finite $1 \le r < \infty$. For notational convenience, the change of notations

$$H_r(\tau) \triangleq \frac{\partial^{(r)} \Upsilon(\tau, \theta)}{\partial \theta^{(r)}}\bigg|_{\theta=0}; \quad \breve{D}_r(\tau) \triangleq \frac{\partial^{(r)} \eta(\tau, \theta)}{\partial \theta^{(r)}}\bigg|_{\theta=0}; \quad D_r(\tau) \triangleq \frac{\partial^{(r)} \upsilon(\tau, \theta)}{\partial \theta^{(r)}}\bigg|_{\theta=0}$$

is introduced so that the next result provides an effective capability for forecasting all the higher-order characteristics associated with performance uncertainty.

Theorem 2.2.2 (Performance-Measure Statistics). *Let the stochastic fault-tolerant system be described by (2.6)–(2.7) whereby it is assumed to possess $(m-l)$ degree of actuator redundancy. The kth-order cumulant of performance measure (2.7) is*

$$\kappa_k = x_0^T H_k(t_0) x_0 + 2x_0^T \breve{D}_k(t_0) + D_k(t_0), \quad k \in \mathbb{Z}^+ \tag{2.18}$$

where the supporting variables $\{H_r(\tau)\}_{r=1}^k$, $\{\breve{D}_r(\tau)\}_{r=1}^k$ *and* $\{D_r(\tau)\}_{r=1}^k$ *evaluated at* $\tau = t_0$ *satisfy the matrix/vector/scalar-valued differential equations (with the dependence of* $H_r(\tau)$, $\breve{D}_r(\tau)$ *and* $D_r(\tau)$ *upon* $K(\tau)$ *and* $\ell(\tau)$ *suppressed)*

$$\frac{d}{d\tau}H_1(\tau) = -[A + \Delta A(\tau) + (B + \Delta B(\tau))LK(\tau)]^T H_1(\tau) - H_1(\tau)[A + \Delta A(\tau) + (B + \Delta B(\tau))LK(\tau)] - [Q_1 + Q_2 + K^T(\tau)R_1K(\tau)], \quad H_1(t_f) = Q_f \tag{2.19}$$

$$\frac{d}{d\tau}H_r(\tau) = -[A + \Delta A(\tau) + (B + \Delta B(\tau))LK(\tau)]^T H_r(\tau) - H_r(\tau)[A + \Delta A(\tau) + (B + \Delta B(\tau))LK(\tau)] - \sum_{s=1}^{r-1}\frac{2r!}{s!(r-s)!}H_s(\tau)GWG^T H_{r-s}(\tau), \quad H_r(t_f) = 0, \quad 2 \le r \le k \tag{2.20}$$

$$\frac{d}{d\tau}\breve{D}_1(\tau) = -[A + \Delta A(\tau) + (B + \Delta B(\tau))LK(\tau)]^T \breve{D}_1(\tau) - H_1(\tau)(B + \Delta B(\tau))L\ell(\tau) - K^T(\tau)R_1\ell(\tau) + Q_2 r(\tau), \quad \breve{D}_1(t_f) = -Q_f r(t_f) \tag{2.21}$$

$$\frac{d}{d\tau}\breve{D}_r(\tau) = -[A + \Delta A(\tau) + (B + \Delta B(\tau))LK(\tau)]^T \breve{D}_r(\tau) - H_r(\tau)(B + \Delta B(\tau))L\ell(\tau), \quad \breve{D}_r(t_f) = 0, \quad 2 \le r \le k \tag{2.22}$$

$$\frac{d}{d\tau}D_1(\tau) = -\operatorname{Tr}\{H_1(\tau)GWG^T\} - 2\breve{D}_1^T(\tau)(B + \Delta B(\tau))L\ell(\tau) - r^T(\tau)Q_2 r(\tau) - \ell^T(\tau)R_1\ell(\tau) + f_a^T(\tau)R_2 f_a(\tau), \quad D_1(t_f) = r^T(t_f)Q_f r(t_f) \tag{2.23}$$

$$\frac{d}{d\tau}D_r(\tau) = -\operatorname{Tr}\{H_r(\tau)GWG^T\} - 2\breve{D}_r^T(\tau)(B + \Delta B(\tau))L\ell(\tau), \; D_r(t_f) = 0. \tag{2.24}$$

Proof. The expression of performance-measure statistics described in (2.18) is readily justified by the coefficients of the Maclaurin series. What remains is to show that the solutions $H_r(\tau)$, $\breve{D}_r(\tau)$ and $D_r(\tau)$ for $1 \le r \le k$ indeed satisfy the backward-in-time differential equations (2.19)–(2.24). Notice that these equations (2.19)–(2.24) are satisfied by the solutions $H_r(\tau)$, $\breve{D}_r(\tau)$ and $D_r(\tau)$ and can be obtained by successively taking derivatives with respect to θ of the supporting equations (2.11)–(2.13) together with the assumptions of $(m - l)$ degree of actuator redundancy on $[t_0, t_f]$.

2.3 Moving Toward Problem Statements

Granted the insight of the roles played by performance-measure statistics on the generalized chi-squared performance measure (2.7), a basic shift in the development of a procedural mechanism of risk-averse feedback control policies with actuator failure accommodation is contemplated, a shift away from the traditional emphasis on depending the system states $x(t)$ vis-$\grave{a}$-vis one another and toward the new imperative of recognizing the time evolution trajectories by the time-backward differential equations (2.19)–(2.24), for the way they affect all the mathematical statistics (2.18).

For such problems it is important to have a compact statement of statistical optimal control so as to aid mathematical manipulation. To make this more precise, one may think of the k-tuple state variables

$$\mathscr{H}(\cdot) \triangleq (\mathscr{H}_1(\cdot), \ldots, \mathscr{H}_k(\cdot)),$$
$$\breve{\mathscr{D}}(\cdot) \triangleq (\breve{\mathscr{D}}_1(\cdot), \ldots, \breve{\mathscr{D}}_k(\cdot)),$$
$$\mathscr{D}(\cdot) \triangleq (\mathscr{D}_1(\cdot), \ldots, \mathscr{D}_k(\cdot))$$

whose continuously differentiable states $\mathscr{H}_r \in \mathscr{C}^1([t_0, t_f]; \mathbb{R}^{n \times n})$, $\breve{\mathscr{D}}_r \in \mathscr{C}^1([t_0, t_f]; \mathbb{R}^n)$ and $\mathscr{D}_r \in \mathscr{C}^1([t_0, t_f]; \mathbb{R})$ having the representations $\mathscr{H}_r(\cdot) \triangleq H_r(\cdot)$, $\breve{\mathscr{D}}_r(\cdot) \triangleq \breve{D}_r(\cdot)$ and $\mathscr{D}_r(\cdot) \triangleq D_r(\cdot)$ with the right members satisfying the dynamics (2.19)–(2.24) are defined on $[t_0, t_f]$.

In the remainder of the development, the bounded and Lipschitz continuous mappings are introduced as

$$\mathscr{F}_r : [t_0, t_f] \times (\mathbb{R}^{n \times n})^k \times \mathbb{R}^{m \times n} \mapsto \mathbb{R}^{n \times n}$$
$$\breve{\mathscr{G}}_r : [t_0, t_f] \times (\mathbb{R}^{n \times n})^k \times (\mathbb{R}^n)^k \times \mathbb{R}^{m \times n} \times \mathbb{R}^m \mapsto \mathbb{R}^n$$
$$\mathscr{G}_r : [t_0, t_f] \times (\mathbb{R}^{n \times n})^k \times (\mathbb{R}^n)^k \times \mathbb{R}^m \mapsto \mathbb{R}$$

where the rules of action are given by

$$\begin{aligned}
\mathscr{F}_1(\tau, \mathscr{H}, K) \triangleq & -[A + \Delta A(\tau) + (B + \Delta B(\tau))LK(\tau)]^T \mathscr{H}_1(\tau) \\
& - \mathscr{H}_1(\tau)[A + \Delta A(\tau) + (B + \Delta B(\tau))LK(\tau)] \\
& - [Q_1 + Q_2 + K^T(\tau) R_1 K(\tau)] \\
\mathscr{F}_r(\tau, \mathscr{H}, K) \triangleq & -[A + \Delta A(\tau) + (B + \Delta B(\tau))LK(\tau)]^T \mathscr{H}_r(\tau) \\
& - \mathscr{H}_r(\tau)[A + \Delta A(\tau) + (B + \Delta B(\tau))LK(\tau)] \\
& - \sum_{s=1}^{r-1} \frac{2r!}{s!(r-s)!} \mathscr{H}_s(\tau) G W G^T \mathscr{H}_{r-s}(\tau), \quad 2 \le r \le k
\end{aligned}$$

$$\breve{\mathscr{G}}_1(\tau,\mathscr{H},\breve{\mathscr{D}},K,\ell) \triangleq -\left[A+\Delta A(\tau)+(B+\Delta B(\tau))LK(\tau)\right]^T\breve{\mathscr{D}}_1(\tau) - \mathscr{H}_1(\tau)(B+\Delta B(\tau))L\ell(\tau) - K^T(\tau)R_1\ell(\tau) + Q_2 r(\tau)$$

$$\breve{\mathscr{G}}_r(\tau,\mathscr{H},\breve{\mathscr{D}},K,\ell) \triangleq -\left[A+\Delta A(\tau)+(B+\Delta B(\tau))LK(\tau)\right]^T\breve{\mathscr{D}}_r(\tau) - \mathscr{H}_r(\tau)(B+\Delta B(\tau))L\ell(\tau), \quad 2\le r\le k$$

$$\mathscr{G}_1(\tau,\mathscr{H},\breve{\mathscr{D}},\ell) \triangleq -\operatorname{Tr}\{\mathscr{H}_1(\tau)GWG^T\} - 2\breve{\mathscr{D}}_1^T(\tau)(B+\Delta B(\tau))L\ell(\tau) - r^T(\tau)Q_2 r(\tau) - \ell^T(\tau)R_1\ell(\tau) + f_a^T(\tau)R_2 f_a(\tau)$$

$$\mathscr{G}_r(\tau,\mathscr{H},\breve{\mathscr{D}},\ell) \triangleq -\operatorname{Tr}\{\mathscr{H}_r(\tau)GWG^T\} - 2\breve{\mathscr{D}}_r^T(\tau)(B+\Delta B(\tau))L\ell(\tau), \quad 2\le r\le k.$$

This is to be expected, for a compact formulation are the product mappings

$$\mathscr{F}:[t_0,t_f]\times(\mathbb{R}^{n\times n})^k\times\mathbb{R}^{m\times n}\mapsto(\mathbb{R}^{n\times n})^k$$
$$\breve{\mathscr{G}}:[t_0,t_f]\times(\mathbb{R}^{n\times n})^k\times(\mathbb{R}^{n})^k\times\mathbb{R}^{m\times n}\times\mathbb{R}^m\mapsto(\mathbb{R}^{n})^k$$
$$\mathscr{G}:[t_0,t_f]\times(\mathbb{R}^{n\times n})^k\times(\mathbb{R}^{n})^k\times\mathbb{R}^m\mapsto\mathbb{R}^k$$

where the corresponding notations

$$\mathscr{F}\triangleq\mathscr{F}_1\times\cdots\times\mathscr{F}_k,$$
$$\breve{\mathscr{G}}\triangleq\breve{\mathscr{G}}_1\times\cdots\times\breve{\mathscr{G}}_k,$$
$$\mathscr{G}\triangleq\mathscr{G}_1\times\cdots\times\mathscr{G}_k$$

are used. Thus, the dynamic equations (2.19)–(2.24) can be rewritten

$$\frac{d}{d\tau}\mathscr{H}(\tau)=\mathscr{F}(\tau,\mathscr{H}(\tau),K(\tau)),\quad \mathscr{H}(t_f) \tag{2.25}$$

$$\frac{d}{d\tau}\breve{\mathscr{D}}(\tau)=\breve{\mathscr{G}}(\tau,\mathscr{H}(\tau),\breve{\mathscr{D}}(\tau),K(\tau),\ell(\tau)),\quad \breve{\mathscr{D}}(t_f) \tag{2.26}$$

$$\frac{d}{d\tau}\mathscr{D}(\tau)=\mathscr{G}(\tau,\mathscr{H}(\tau),\breve{\mathscr{D}}(\tau),\ell(\tau)),\quad \mathscr{D}(t_f) \tag{2.27}$$

where the k-tuple terminal-value conditions

$$\mathscr{H}(t_f)\triangleq\mathscr{H}_f=(Q_f,0,\ldots,0),$$
$$\breve{\mathscr{D}}(t_f)\triangleq\breve{\mathscr{D}}_f=(-Q_f r(t_f),0,\ldots,0),$$
$$\mathscr{D}(t_f)\triangleq\mathscr{D}_f=(r^T(t_f)Q_f r(t_f),\ldots,0).$$

Notice that the product system uniquely determines the state matrices $\mathscr{H}$, $\breve{\mathscr{D}}$ and $\mathscr{D}$ once the admissible feedback K and ℓ are specified. Henceforth, these state variables will be considered as $\mathscr{H} \equiv \mathscr{H}(\cdot, K)$, $\breve{\mathscr{D}} \equiv \breve{\mathscr{D}}(\cdot, K, \ell)$ and $\mathscr{D} \equiv \mathscr{D}(\cdot, \ell)$.

For the given terminal data $(t_f, \mathscr{H}_f, \breve{\mathscr{D}}_f, \mathscr{D}_f)$, the classes of admissible feedback parameters are next defined.

Definition 2.3.1 (Admissible Feedback Parameters). Let compact subsets $\overline{L} \subset \mathbb{R}^m$ and $\overline{K} \subset \mathbb{R}^{m\times n}$ be the sets of allowable affine inputs and feedback gain values. For the given $k \in \mathbb{Z}^+$ and sequence $\mu = \{\mu_r \geq 0\}_{r=1}^k$ with $\mu_1 > 0$, the set of admissible affine inputs $\mathscr{L}_{t_f,\mathscr{H}_f,\breve{\mathscr{D}},\mathscr{D}_f;\mu}$ and feedback gains $\mathscr{K}_{t_f,\mathscr{H}_f,\breve{\mathscr{D}},\mathscr{D}_f;\mu}$ are assumed to be the class of $\mathscr{C}([t_0, t_f]; \mathbb{R}^m)$ and $\mathscr{C}([t_0, t_f]; \mathbb{R}^{m\times n})$ with values $\ell(\cdot) \in \overline{L}$ and $K(\cdot) \in \overline{K}$ for which solutions to the dynamic equations (2.25)–(2.27) with the terminal-value conditions $\mathscr{H}(t_f) = \mathscr{H}_f$, $\breve{\mathscr{D}}(t_f) = \breve{\mathscr{D}}_f$ and $\mathscr{D}(t_f) = \mathscr{D}_f$ exist on $[t_0, t_f]$ of optimization.

Beyond this, it is crucial to plan for robust decisions and performance reliability from the start because it is going to be much more difficult and expensive to add reliability to the process later. To be used in the design process, performance-based reliability requirements must be verifiable by analysis; in particular, they must be measurable, like all higher-order performance-measure statistics, as evidenced in the previous section. These higher-order performance-measure statistics become the test criteria for the requirement of performance-based reliability. What follows is the mean-risk aware performance index in statistical optimal control. It naturally contains some tradeoffs between performance values and risks for the subject class of stochastic control problems.

Definition 2.3.2 (Mean-Risk Aware Performance Index). Fix $k \in \mathbb{N}$ and the sequence of scalar coefficients $\mu = \{\mu_r \geq 0\}_{r=1}^k$ with $\mu_1 > 0$. Then for the given x_0, the risk-value aware performance index

$$\phi_0 : \{t_0\} \times (\mathbb{R}^{n\times n})^k \times (\mathbb{R}^n)^k \times \mathbb{R}^k \mapsto \mathbb{R}^+$$

pertaining to statistical optimal control of the stochastic fault-tolerant system with actuator failure accommodation over $[t_0, t_f]$ is defined by

$$\phi_0(t_0, \mathscr{H}(t_0), \breve{\mathscr{D}}(t_0), \mathscr{D}(t_0)) \triangleq \underbrace{\mu_1\kappa_1}_{\text{Mean Measure}} + \underbrace{\mu_2\kappa_2 + \cdots + \mu_k\kappa_k}_{\text{Risk Measures}}$$

$$= \sum_{r=1}^{k} \mu_r [x_0^T \mathscr{H}_r(t_0) x_0 + 2x_0^T \breve{\mathscr{D}}_r(t_0) + \mathscr{D}_r(t_0)] \qquad (2.28)$$

where additional design of freedom by means of μ_r's utilized by the control designer with risk-averse attitudes are sufficient to meet and exceed different levels of performance-based reliability requirements, for instance, mean (i.e., the average of performance measure), variance (i.e., the dispersion of values of performance measure around its mean), skewness (i.e., the anti-symmetry of the probability density of performance measure), kurtosis (i.e., the heaviness in the probability density tails of performance measure), etc., pertaining to closed-loop performance variations and uncertainties while the supporting solutions $\{\mathscr{H}_r(\tau)\}_{r=1}^k$, $\{\breve{\mathscr{D}}_r(\tau)\}_{r=1}^k$ and $\{\mathscr{D}_r(\tau)\}_{r=1}^k$ evaluated at $\tau = t_0$ satisfy the dynamical equations (2.25)–(2.27).

Given that the terminal time t_f and states $(\mathscr{H}_f, \breve{\mathscr{D}}_f, \mathscr{D}_f)$, the other end condition involved the initial time t_0 and state pair $(\mathscr{H}_0, \breve{\mathscr{D}}_0, \mathscr{D}_0)$ are specified by a target set requirement.

Definition 2.3.3 (Target Set). $(t_0, \mathscr{H}_0, \breve{\mathscr{D}}_0, \mathscr{D}_0) \in \mathscr{M}$, where the *target set* $\mathscr{M}$ is a closed subset of $[t_0, t_f] \times (\mathbb{R}^{n \times n})^k \times (\mathbb{R}^n)^k \times \mathbb{R}^k$.

Now, the optimization problem is to minimize the risk-value aware performance index (2.28) over all admissible feedback gains $K = K(\cdot)$ and affine inputs $\ell = \ell(\cdot)$ in $\mathscr{K}_{t_f, \mathscr{H}_f, \breve{\mathscr{D}}_f, \mathscr{D}_f; \mu}$ and $\mathscr{L}_{t_f, \mathscr{H}_f, \breve{\mathscr{D}}_f, \mathscr{D}_f; \mu}$.

Definition 2.3.4 (Optimization Problem of Mayer Type). Fix $k \in \mathbb{Z}^+$ and the sequence of scalar coefficients $\mu = \{\mu_r \geq 0\}_{r=1}^k$ with $\mu_1 > 0$. The optimization problem of the stochastic fault-tolerant system with actuator failure accommodation is defined by the minimization of the risk-value aware performance index (2.28) over $\mathscr{K}_{t_f, \mathscr{H}_f, \breve{\mathscr{D}}_f, \mathscr{D}; \mu} \times \mathscr{L}_{t_f, \mathscr{H}_f, \breve{\mathscr{D}}_f, \mathscr{D}_f; \mu}$ and subject to the dynamical equations (2.25)–(2.27) on $[t_0, t_f]$.

At best, it is important to recognize that the optimization considered here is in Mayer form and is solved by applying an adaptation of the Mayer-form verification theorem of dynamic programming as given in [9]. To embed the aforementioned optimization into a larger optimal control problem, the terminal time and states $(t_f, \mathscr{H}_f, \breve{\mathscr{D}}_f, \mathscr{D}_f)$ are parameterized as $(\varepsilon, \mathscr{Y}, \breve{\mathscr{Z}}, \mathscr{Z})$. Thus, the value function for this optimization problem now depends on the terminal condition parameterizations.

Definition 2.3.5 (Value Function). Suppose that $(\varepsilon, \mathscr{Y}, \breve{\mathscr{Z}}, \mathscr{Z}) \in [t_0, t_f] \times (\mathbb{R}^{n \times n})^k \times (\mathbb{R}^n)^k \times \mathbb{R}^k$ is given and fixed. Then, the value function $\mathscr{V}(\varepsilon, \mathscr{Y}, \breve{\mathscr{Z}}, \mathscr{Z})$ is defined by the greatest lower bound of $\phi_0(t_0, \mathscr{H}(t_0, K), \breve{\mathscr{D}}_r(t_0, K, \ell), \mathscr{D}(t_0, \ell))$ for all $K = K(\cdot)$ and $\ell = \ell(\cdot)$ in $\mathscr{K}_{\varepsilon, \mathscr{Y}, \breve{\mathscr{Z}}, \mathscr{Z}; \mu}$ and $\mathscr{L}_{\varepsilon, \mathscr{Y}, \breve{\mathscr{Z}}, \mathscr{Z}; \mu}$.

For convention, $\mathscr{V}(\varepsilon, \mathscr{Y}, \breve{\mathscr{Z}}, \mathscr{Z}) \triangleq \infty$ when $\mathscr{K}_{\varepsilon, \mathscr{Y}, \breve{\mathscr{Z}}, \mathscr{Z}; \mu} \times \mathscr{L}_{\varepsilon, \mathscr{Y}, \breve{\mathscr{Z}}, \mathscr{Z}; \mu}$ is empty. To avoid cumbersome notation, the dependence of trajectory solutions on $K(\cdot)$ and $\ell(\cdot)$ is suppressed. Next some candidates for the value function are constructed with the help of the concept of reachable set.

Definition 2.3.6 (Reachable Set). Let the reachable set $\mathscr{Q} \triangleq \{(\varepsilon, \mathscr{Y}, \breve{\mathscr{Z}}, \mathscr{Z}) \in [t_0, t_f] \times (\mathbb{R}^{n \times n})^k \times (\mathbb{R}^n)^k \times \mathbb{R}^k\}$ such that $\mathscr{K}_{\varepsilon, \mathscr{Y}, \breve{\mathscr{Z}}, \mathscr{Z}; \mu} \times \mathscr{L}_{\varepsilon, \mathscr{Y}, \breve{\mathscr{Z}}, \mathscr{Z}; \mu}$ is not empty.

Notice that $\mathscr{Q}$ contains a set of points $(\varepsilon, \mathscr{Y}, \breve{\mathscr{Z}}, \mathscr{Z})$, from which it is possible to reach the target set $\mathscr{M}$ with some trajectory pairs corresponding to a continuous feedback policy. Furthermore, the value function must satisfy both a partial differential inequality and an equation at each interior point of the reachable set, at which it is differentiable.

Theorem 2.3.1 (Hamilton-Jacobi-Bellman (HJB) Equation for Mayer Problem). *Let $(\varepsilon, \mathscr{Y}, \breve{\mathscr{Z}}, \mathscr{Z})$ be any interior point of the reachable set $\mathscr{Q}$, at which the scalar-valued function $\mathscr{V}(\varepsilon, \mathscr{Y}, \breve{\mathscr{Z}}, \mathscr{Z})$ is differentiable. Then $\mathscr{V}(\varepsilon, \mathscr{Y}, \breve{\mathscr{Z}}, \mathscr{Z})$ satisfies the partial differential inequality, for all $K \in \overline{K}$ and $\ell \in \overline{L}$*

$$0 \geq \frac{\partial}{\partial \varepsilon}\mathscr{V}(\varepsilon, \mathscr{Y}, \breve{\mathscr{Z}}, \mathscr{Z}) + \frac{\partial}{\partial \operatorname{vec}(\mathscr{Y})}\mathscr{V}(\varepsilon, \mathscr{Y}, \breve{\mathscr{Z}}, \mathscr{Z})\operatorname{vec}(\mathscr{F}(\varepsilon, \mathscr{Y}, K))$$
$$+ \frac{\partial}{\partial \operatorname{vec}(\breve{\mathscr{Z}})}\mathscr{V}(\varepsilon, \mathscr{Y}, \breve{\mathscr{Z}}, \mathscr{Z})\operatorname{vec}(\breve{\mathscr{G}}(\varepsilon, \mathscr{Y}, \breve{\mathscr{Z}}, K, \ell))$$
$$+ \frac{\partial}{\partial \operatorname{vec}(\mathscr{Z})}\mathscr{V}(\varepsilon, \mathscr{Y}, \breve{\mathscr{Z}}, \mathscr{Z})\operatorname{vec}(\mathscr{G}(\varepsilon, \mathscr{Y}, \breve{\mathscr{Z}}, \ell)) \tag{2.29}$$

wherein $\operatorname{vec}(\cdot)$ *the vectorizing operator of enclosed entities.*

If there is optimal feedback K^* in $\mathscr{K}_{\varepsilon,\mathscr{Y},\breve{\mathscr{Z}},\mathscr{Z};\mu}$ and feedforward ℓ^* in $\mathscr{L}_{\varepsilon,\mathscr{Y},\breve{\mathscr{Z}},\mathscr{Z};\mu}$, then the partial differential equation of dynamic programming

$$0 = \min_{K\in\overline{K},\ell\in\overline{L}} \Big\{\frac{\partial}{\partial \varepsilon}\mathscr{V}(\varepsilon, \mathscr{Y}, \breve{\mathscr{Z}}, \mathscr{Z}) + \frac{\partial}{\partial \operatorname{vec}(\mathscr{Y})}\mathscr{V}(\varepsilon, \mathscr{Y}, \breve{\mathscr{Z}}, \mathscr{Z}) \operatorname{vec}(\mathscr{F}(\varepsilon, \mathscr{Y}, K))$$
$$+ \frac{\partial}{\partial \operatorname{vec}(\breve{\mathscr{Z}})}\mathscr{V}(\varepsilon, \mathscr{Y}, \breve{\mathscr{Z}}, \mathscr{Z})\operatorname{vec}(\breve{\mathscr{G}}(\varepsilon, \mathscr{Y}, \breve{\mathscr{Z}}, K, \ell))$$
$$+ \frac{\partial}{\partial \operatorname{vec}(\mathscr{Z})}\mathscr{V}(\varepsilon, \mathscr{Y}, \breve{\mathscr{Z}}, \mathscr{Z}) \operatorname{vec}(\mathscr{G}(\varepsilon, \mathscr{Y}, \breve{\mathscr{Z}}, \ell))\Big\} \tag{2.30}$$

is satisfied. The minimum in (2.30) is achieved by the optimal feedback gain $K^*(\varepsilon)$ and feedforward input $\ell^*(\varepsilon)$ at ε.

Proof. Interested readers are referred to the mathematical details in [10].

Closely related to the continuing quest for an effective procedure of dynamic programming for optimality is a reliable process of testing the sufficient condition that is at once effective.

Theorem 2.3.2 (Verification Theorem). *Fix $k \in \mathbb{N}$ and let $\mathscr{W}(\varepsilon, \mathscr{Y}, \breve{\mathscr{Z}}, \mathscr{Z})$ be a continuously differentiable solution of the HJB equation (2.30), which satisfies the boundary $\mathscr{W}(t_0, \mathscr{H}(t_0), \breve{\mathscr{D}}(t_0), \mathscr{D}(t_0)) = \phi_0(t_0, \mathscr{H}(t_0), \breve{\mathscr{D}}(t_0), \mathscr{D}(t_0))$ for some $(\varepsilon, \mathscr{Y}, \breve{\mathscr{Z}}, \mathscr{Z}) \in \mathscr{M}$. Let $(t_f, \mathscr{H}_f, \breve{\mathscr{D}}_f, \mathscr{D}_f)$ be in $\mathscr{Q}$, (K, ℓ) be in $\mathscr{K}_{t_f,\mathscr{H}_f,\breve{\mathscr{D}}_f,\mathscr{D}_f;\mu} \times \mathscr{L}_{t_f,\mathscr{H}_f,\breve{\mathscr{D}}_f,\mathscr{D}_f;\mu}$, and let $\mathscr{H}(\cdot)$, $\breve{\mathscr{D}}(\cdot)$ and $\mathscr{D}(\cdot)$ be the corresponding solutions of Eqs. (2.25)–(2.27). Then, $\mathscr{W}(\tau, \mathscr{H}(\tau), \breve{\mathscr{D}}(\tau), \mathscr{D}(\tau))$ is a non-increasing time-backward function of τ.*

If (K^*, ℓ^*) is in $\mathscr{K}_{t_f,\mathscr{H}_f,\mathscr{D}_f;\mu} \times \mathscr{L}_{t_f,\mathscr{H}_f,\breve{\mathscr{D}}_f,\mathscr{D}_f;\mu}$ with the corresponding solutions $\mathscr{H}^*(\cdot)$, $\breve{\mathscr{D}}^*(\cdot)$ and $\mathscr{D}^*(\cdot)$ of Eqs. (2.25)–(2.27) such that, for $\tau \in [t_0, t_f]$,

$$
\begin{aligned}
0 = &\frac{\partial}{\partial \varepsilon}\mathscr{W}(\tau, \mathscr{H}^*(\tau), \breve{\mathscr{D}}^*(\tau), \mathscr{D}^*(\tau)) \\
&+ \frac{\partial}{\partial\, vec(\mathscr{Y})}\mathscr{W}(\tau, \mathscr{H}^*(\tau), \breve{\mathscr{D}}^*(\tau), \mathscr{D}^*(\tau)) vec(\mathscr{F}(\tau, \mathscr{H}^*(\tau), K^*(\tau))) \\
&+ \frac{\partial}{\partial vec(\breve{\mathscr{Z}})}\mathscr{W}(\tau, \mathscr{H}^*(\tau), \breve{\mathscr{D}}^*(\tau), \mathscr{D}^*(\tau))\; vec(\breve{\mathscr{G}}(\tau, \mathscr{H}^*(\tau), \breve{\mathscr{D}}^*(\tau), K^*(\tau), \ell^*(\tau))) \\
&+ \frac{\partial}{\partial\, vec(\mathscr{Z})}\mathscr{W}(\tau, \mathscr{H}^*(\tau), \breve{\mathscr{D}}^*(\tau), \mathscr{D}^*(\tau))\; vec(\mathscr{G}(\tau, \mathscr{H}^*(\tau), \breve{\mathscr{D}}^*(\tau), \ell^*(\tau)))
\end{aligned}
\tag{2.31}
$$

then both K^* and ℓ^* in $\mathscr{K}_{t_f, \mathscr{H}_f, \breve{\mathscr{D}}_f, \mathscr{D}_f; \mu} \times \mathscr{L}_{t_f, \mathscr{H}_f, \breve{\mathscr{D}}_f, \mathscr{D}_f; \mu}$ are optimal. Moreover, it follows that $\mathscr{W}(\varepsilon, \mathscr{Y}, \breve{\mathscr{Z}}, \mathscr{Z}) = \mathscr{V}(\varepsilon, \mathscr{Y}, \breve{\mathscr{Z}}, \mathscr{Z})$, where $\mathscr{V}(\varepsilon, \mathscr{Y}, \breve{\mathscr{Z}}, \mathscr{Z})$ is the value function.

Proof. Immediate from the detailed analysis found in the work [10].

2.4 Conceptualizing Risk-Averse Control Strategy

The aim of the present section is to recognize the optimization problem of Mayer form, which can therefore be solved by an adaptation of the Mayer-form verification theorem. To this end the terminal time and states $(t_f, \mathscr{H}_f, \breve{\mathscr{D}}_f, \mathscr{D}_f)$ are parameterized as $(\varepsilon, \mathscr{Y}, \breve{\mathscr{Z}}, \mathscr{Z})$ for a family of optimization problems. In particular, the states (2.25)–(2.27) defined on the interval $[t_0, \varepsilon]$ now have terminal values denoted by $\mathscr{H}(\varepsilon) \equiv \mathscr{Y}$, $\breve{\mathscr{D}}(\varepsilon) \equiv \breve{\mathscr{Z}}$ and $\mathscr{D}(\varepsilon) \equiv \mathscr{Z}$, where $\varepsilon \in [t_0, t_f]$. For given $k \in \mathbb{Z}^+$ and $(\varepsilon, \mathscr{Y}, \breve{\mathscr{Z}}, \mathscr{Z})$ in $\mathscr{Q}$, elementary interpretations of (2.28) further suggest a real-valued candidate for the value function of the form:

$$
\begin{aligned}
\mathscr{W}(\varepsilon, \mathscr{Y}, \breve{\mathscr{Z}}, \mathscr{Z}) =& x_0^T \sum_{r=1}^{k} \mu_r(\mathscr{Y}_r + \mathscr{E}_r(\varepsilon))\, x_0 \\
&+ 2x_0^T \sum_{r=1}^{k} \mu_r(\breve{\mathscr{Z}}_r + \breve{\mathscr{T}}_r(\varepsilon)) + \sum_{r=1}^{k} \mu_r(\mathscr{Z}_r + \mathscr{T}_r(\varepsilon))
\end{aligned}
\tag{2.32}
$$

where the time-parametric functions $\mathscr{E}_r \in \mathscr{C}^1([t_0, t_f]; \mathbb{R}^{n\times n})$, $\breve{\mathscr{T}}_r \in \mathscr{C}^1([t_0, t_f]; \mathbb{R}^n)$ and $\mathscr{T}_r \in \mathscr{C}^1([t_0, t_f]; \mathbb{R})$ are yet to be determined.

As it turns out, the total time derivative of $\mathscr{W}(\varepsilon, \mathscr{Y}, \breve{\mathscr{Z}}, \mathscr{Z})$ with respect to ε is easily obtained as follows.

$$\frac{d}{d\varepsilon}\mathscr{W}(\varepsilon,\mathscr{Y},\breve{\mathscr{Z}},\mathscr{Z}) = x_0^T\sum_{r=1}^{k}\mu_r(\mathscr{F}_r(\varepsilon,\mathscr{Y},K) + \frac{d}{d\varepsilon}\mathscr{E}_r(\varepsilon))\,x_0$$
$$+ 2x_0^T\sum_{r=1}^{k}\mu_r(\breve{\mathscr{G}}_r(\varepsilon,\mathscr{Y},\breve{\mathscr{Z}},K,\ell) + \frac{d}{d\varepsilon}\breve{\mathscr{T}}_r(\varepsilon)) + \sum_{r=1}^{k}\mu_r(\mathscr{G}_r(\varepsilon,\mathscr{Y},\breve{\mathscr{Z}},\ell) + \frac{d}{d\varepsilon}\mathscr{T}_r(\varepsilon)) \tag{2.33}$$

provided that $\ell \in \overline{L}$ and $K \in \overline{K}$. Trying this candidate for the value function (2.32) into the HJB equation (2.30) yields

$$0 \equiv \min_{\ell\in\overline{L},\,K\in\overline{K}}\Big\{x_0^T\sum_{r=1}^{k}\mu_r(\mathscr{F}_r(\varepsilon,\mathscr{Y},K) + \frac{d}{d\varepsilon}\mathscr{E}_r(\varepsilon))\,x_0$$
$$+ 2x_0^T\sum_{r=1}^{k}\mu_r(\breve{\mathscr{G}}_r(\varepsilon,\mathscr{Y},\breve{\mathscr{Z}},K,\ell) + \frac{d}{d\varepsilon}\breve{\mathscr{T}}_r(\varepsilon)) + \sum_{r=1}^{k}\mu_r(\mathscr{G}_r(\varepsilon,\mathscr{Y},\breve{\mathscr{Z}},\ell) + \frac{d}{d\varepsilon}\mathscr{T}_r(\varepsilon))\Big\}. \tag{2.34}$$

Since the initial condition x_0 is an arbitrary vector, the necessary condition for an extremum of (2.30) on $[t_0, \varepsilon]$ is obtained by differentiating the expression within the bracket of (2.34) with respect to the control parameters ℓ and K as follows

$$\ell = -R_1^{-1}[(B + \Delta B(\varepsilon))L]^T\sum_{r=1}^{k}\hat{\mu}_r\breve{\mathscr{Z}}_r \tag{2.35}$$

$$K = -R_1^{-1}[(B + \Delta B(\varepsilon))L]^T\sum_{r=1}^{k}\hat{\mu}_r\mathscr{Y}_r, \quad \hat{\mu}_r \triangleq \frac{\mu_i}{\mu_1} \tag{2.36}$$

where the convex bounded parametric uncertainties $\Delta A(\varepsilon)$ and $\Delta B(\varepsilon)$ as well as the actuator channels L are known whenever model mismatches and control input outages occur.

In the background is the substitution of (2.35) and (2.36) into the HJB equation (2.34) in a setting shaped by a potential value of the minimum. On the one side, it is possible to exhibit the time-parametric functions for the candidate function $\mathscr{W}(\varepsilon,\mathscr{Y},\breve{\mathscr{Z}},\mathscr{Z})$ of the value function, i.e., $\{\mathscr{E}_r(\cdot)\}_{r=1}^k$, $\{\breve{\mathscr{T}}_r(\cdot)\}_{r=1}^k$, and $\{\mathscr{T}_r(\cdot)\}_{r=1}^k$ which yield a sufficient condition to have the left-hand side of (2.31) being zero for any $\varepsilon \in [t_0, t_f]$, when $\{\mathscr{Y}_r\}_{r=1}^k$, $\{\breve{\mathscr{Z}}_r\}_{r=1}^k$ and $\{\mathscr{D}_r\}_{r=1}^k$ are evaluated along the solutions of the dynamical equations (2.25)–(2.27). But on the other side, with a careful examination of (2.34), one can infer that $\{\mathscr{E}_r(\cdot)\}_{r=1}^k$, $\{\breve{\mathscr{T}}_r(\cdot)\}_{r=1}^k$ and $\{\mathscr{T}_r(\cdot)\}_{r=1}^k$ can be indeed selected to satisfy the time-forward differential equations

$$\frac{d}{d\varepsilon}\mathscr{E}_1(\varepsilon) = [A + \Delta A(\varepsilon) + (B + \Delta B(\varepsilon))LK(\varepsilon)]^T\mathscr{H}_1(\varepsilon)$$
$$+ \mathscr{H}_1(\varepsilon)[A + \Delta A(\varepsilon) + (B + \Delta B(\varepsilon))LK(\varepsilon)]$$
$$+ [Q_1 + Q_2 + K^T(\varepsilon)R_1K(\varepsilon)], \quad \mathscr{E}_1(t_0) = 0 \tag{2.37}$$

$$\frac{d}{d\varepsilon}\mathscr{E}_r(\varepsilon) = [A + \Delta A(\varepsilon) + (B + \Delta B(\varepsilon))LK(\varepsilon)]^T \mathscr{H}_r(\varepsilon) + \mathscr{H}_r(\varepsilon)[A + \Delta A(\varepsilon) + (B + \Delta B(\varepsilon))LK(\varepsilon)] + \sum_{s=1}^{r-1} \frac{2r!}{s!(r-s)!} \mathscr{H}_s(\varepsilon) G W G^T \mathscr{H}_{r-s}(\varepsilon), \quad \mathscr{E}_r(t_0) = 0, \quad 2 \le r \le k \tag{2.38}$$

$$\frac{d}{d\varepsilon}\breve{T}_1(\varepsilon) = [A + \Delta A(\varepsilon) + (B + \Delta B(\varepsilon))LK(\varepsilon)]^T \breve{\mathscr{D}}_1(\varepsilon) + \mathscr{H}_1(\varepsilon)(B + \Delta B(\varepsilon))L\ell(\varepsilon) + K^T(\varepsilon)R_1\ell(\varepsilon) - Q_2 r(\varepsilon), \quad \breve{\mathscr{T}}_1(t_0) = 0 \tag{2.39}$$

$$\frac{d}{d\varepsilon}\breve{\mathscr{T}}_r(\varepsilon) = [A + \Delta A(\varepsilon) + (B + \Delta B(\varepsilon))LK(\varepsilon)]^T \breve{\mathscr{D}}_r(\varepsilon) + \mathscr{H}_r(\varepsilon)(B + \Delta B(\varepsilon))L\ell(\varepsilon), \quad \breve{\mathscr{T}}_r(t_0) = 0, \quad 2 \le r \le k \tag{2.40}$$

$$\frac{d}{d\varepsilon}\mathscr{T}_1(\varepsilon) = \mathrm{Tr}\{\mathscr{H}_1(\varepsilon) G W G^T\} + 2\breve{\mathscr{D}}_1^T(\varepsilon)(B + \Delta B(\varepsilon))L\ell(\varepsilon) + r^T(\varepsilon)Q_2 r(\varepsilon) + \ell^T(\varepsilon)R_1\ell(\varepsilon) - f_a^T(\varepsilon)R_2 f_a(\varepsilon), \quad \mathscr{T}_1(t_0) = 0 \tag{2.41}$$

$$\frac{d}{d\varepsilon}\mathscr{T}_r(\varepsilon) = \mathrm{Tr}\{\mathscr{H}_r(\varepsilon) G W G^T\} + 2\breve{\mathscr{D}}_r^T(\varepsilon)(B + \Delta B(\varepsilon))L\ell(\varepsilon), \quad \mathscr{T}_r(t_0) = 0 \quad 2 \le r \le k. \tag{2.42}$$

As often be seen, it is noteworthy that the replacement of the feedforward and feedback parameters (2.35) and (2.36) along the solution trajectories of the time-backward Riccati-type equations (2.25)–(2.27) enables the sufficient condition of (2.31) of the verification theorem satisfied. Henceforth, the optimal linear input (2.35) and feedback gain (2.36) minimizing (2.28) become optimal

$$\ell^*(\varepsilon) = -R_1^{-1}[(B + \Delta B(\varepsilon))L]^T \sum_{r=1}^{k} \hat{\mu}_r \breve{\mathscr{D}}_r^*(\varepsilon)$$

$$K^*(\varepsilon) = -R_1^{-1}[(B + \Delta B(\varepsilon))L]^T \sum_{r=1}^{k} \hat{\mu}_r \mathscr{H}_r^*(\varepsilon).$$

And yet it is not too soon to offer some assessment of the development here. As shown in Fig. 2.2, it acknowledges that the baseline control strategy, as in relation to resilient controls, is now capable of not only confronting a certain actuator failure of stuck type but also endorsing a particular claim of performance risk aversion.

Theorem 2.4.1 (Risk-Averse Strategy for Actuator Failures). *Let the stochastic fault-tolerant system be described by (2.6) and (2.7), whereby it is assumed to*

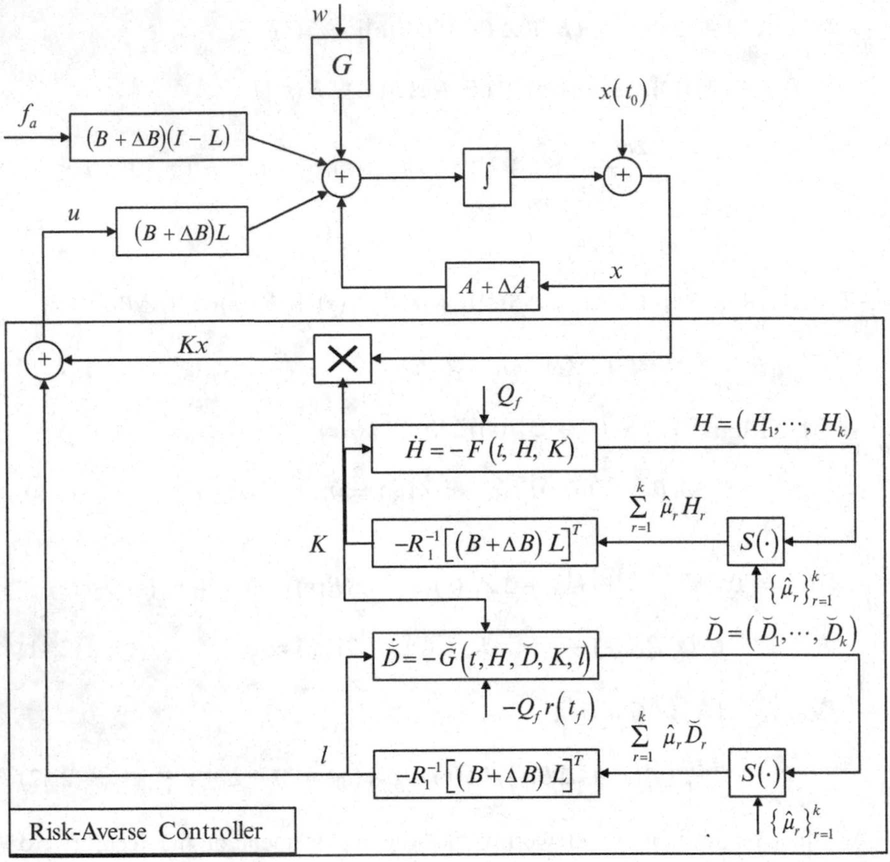

Fig. 2.2 Risk-averse feedback system with statistical measures of risks

possess $(m-l)$ *degree of actuator redundancy. Assume* $k \in \mathbb{N}$ *and the sequence* $\mu = \{\mu_i \geq 0\}_{i=1}^{k}$ *with* $\mu_1 > 0$ *fixed. The baseline control strategy* $u(t)$ *will tolerate to certain actuator failures while robustly tracking the desired trajectory* $r(t)$ *with statistical measures of risk, is given by*

$$u^*(t) = K^*(t)x^*(t) + \ell^*(t), \quad t = t_0 + t_f - \tau \tag{2.43}$$

$$K^*(\tau) = -R_1^{-1}[(B + \Delta B(\tau))L]^T \sum_{r=1}^{k} \hat{\mu}_r \mathcal{H}_r^*(\tau) \tag{2.44}$$

$$\ell^*(\tau) = -R_1^{-1}[(B + \Delta B(\tau))L]^T \sum_{r=1}^{k} \hat{\mu}_r \breve{\mathcal{D}}_r^*(\tau) \tag{2.45}$$

where the normalized weightings $\hat{\mu}_r \triangleq \mu_i/\mu_1$ emphasize on different design of freedom to shape the asymmetry or skewness of the chi-squared random cost (2.7).

The optimal control design solutions $\{\mathscr{H}_r^*(\tau)\}_{r=1}^k$, and $\{\breve{\mathscr{D}}_r^*(\tau)\}_{r=1}^k$ respectively satisfy the backward-in-time matrix-valued differential equations

$$\frac{d}{d\tau}\mathscr{H}_1^*(\tau) = -[A + \Delta A(\tau) + (B + \Delta B(\tau))LK^*(\tau)]^T \mathscr{H}_1^*(\tau) - \mathscr{H}_1^*(\tau)[A + \Delta A(\tau) + (B + \Delta B(\tau))LK^*(\tau)] - [Q_1 + Q_2 + K^{*T}(\tau)R_1K^*(\tau)], \quad \mathscr{H}_1^*(t_f) = Q_f \tag{2.46}$$

$$\frac{d}{d\tau}\mathscr{H}_r^*(\tau) = -[A + \Delta A(\tau) + (B + \Delta B(\tau))LK^*(\tau)]^T \mathscr{H}_r^*(\tau) - \mathscr{H}_r^*(\tau)[A + \Delta A(\tau) + (B + \Delta B(\tau))LK^*(\tau)] - \sum_{s=1}^{r-1}\frac{2r!}{s!(r-s)!}\mathscr{H}_s^*(\tau)GWG^T\mathscr{H}_{r-s}^*(\tau), \quad \mathscr{H}_r^*(t_f) = 0, \quad 2 \le r \le k \tag{2.47}$$

and the backward-in-time vector-valued differential equations

$$\frac{d}{d\tau}\breve{\mathscr{D}}_1^*(\tau) = -[A + \Delta A(\tau) + (B + \Delta B(\tau))LK^*(\tau)]^T \breve{\mathscr{D}}_1^*(\tau) - \mathscr{H}_1^*(\tau)(B + \Delta B(\tau))L\ell^*(\tau) - K^{*T}(\tau)R_1\ell^*(\tau) + Q_2 r(\tau), \quad \breve{\mathscr{D}}_1^*(t_f) = -Q_f r(t_f) \tag{2.48}$$

$$\frac{d}{d\tau}\breve{\mathscr{D}}_r^*(\tau) = -[A + \Delta A(\tau) + (B + \Delta B(\tau))LK^*(\tau)]^T \breve{\mathscr{D}}_r^*(\tau) - \mathscr{H}_r^*(\tau)(B + \Delta B(\tau))L\ell^*(\tau), \quad \breve{\mathscr{D}}_r^*(t_f) = 0, \quad 2 \le r \le k. \tag{2.49}$$

In the practitioner realm, a timely estimation for abrupt reduction of control effectiveness is a key to the implementation of stochastic fault-tolerant systems. As a consequence, a Kalman-like estimator remains an ever important capability to gain the knowledge of the amount of loss in the control effectiveness $f_a(\cdot)$. For example, such abrupt changes and rates of changes in the control effectiveness of the class of faulty systems here can be described by a continuous white noise acceleration model as given below

$$d\dot{f}_a(t) = V^{\frac{1}{2}}dv(t) \quad \text{and} \quad df_a(t) = \dot{f}_a(t)dt \tag{2.50}$$

where the time derivative of $f_a(t)$ is $\dot{f}_a(t)$ and the probability of fault occurrence in each control channel is governed by the additive stationary Wiener process $v(t) \equiv v(t,\omega)$ with the correlations of independent increments $\forall\, \tau_1, \tau_2 \in [t_0, t_f]$

$$E\left\{[v(\tau_1) - v(\tau_2)][v(\tau_1) - v(\tau_2)]^T\right\} = V|\tau_1 - \tau_2|$$

whose a-priori second-order statistic $V > 0$ should be designed appropriately for random actuator faults.

2.5 Chapter Summary

This chapter proposes a novel paradigm of designing feedback controls for the linear-quadratic class of stochastic fault-tolerant systems to not only track reference trajectories but also accommodate certain actuator failures, in accordance of the mean-risk aware performance index. This performance index with risk aversion consists of multiple selective performance-measure statistics beyond the traditional statistical average. On another front, the numerical procedure of calculating these mathematical statistics associated with the chi-squared performance measure is also developed. Yet the complexity of the risk-averse feedback controller may increase, depending on how many performance-measure statistics of the target probability density function are to be optimized.

References

1. Chen, J., Patton, R.J.: Robust Model-Based Fault Diagnosis for Dynamic Systems. Kluwer Academic, Norwell (1999)
2. Patton, R.J.: Fault-tolerant control: the 1997 situation survey. In: Proceedings of the IFAC Symposium on Fault Detection, Supervision and Safety for Technical Processes, Hull, pp. 1029–1052 (1997)
3. Zhou, D.H., Frank, P.M.: Fault diagnosis and fault tolerant control. IEEE Trans. Aerosp. Electron. Syst. **34**(2), 420–427 (1998)
4. Pham, K.D., Sain, M.K., Liberty, S.R.: Cost cumulant control: state-feedback, finite-horizon paradigm with application to seismic protection. In: Miele, A. (ed.) Spec. Issue J. Optim. Theory Appl. **115**(3), 685–710. Kluwer Academic/Plenum, New York (2002)
5. Pham, K.D.: New risk-averse control paradigm for stochastic two-time-scale systems and performance robustness. In: Miele, A. (ed.) J. Optim. Theory Appl. **146**(2), 511–537 (2010)
6. Zhao, Q., Jiang, J.: Reliable state feedback control system design against actuator failures. Automatica **34**(10), 1267–1272 (1998)
7. Pham, K.D.: Performance information in risk-averse control of model-following systems. In: Proceedings of the 18th International Federation of Automatic Control World Congress, Milano, pp. 12413–12420 (2011)
8. Pham, K.D.: Performance information in risk-averse control of model-following systems: output feedback compensation. In: Proceedings of the IEEE Multi-Conference on Systems and Control, Denver, pp. 593–600 (2011)
9. Fleming, W.H., Rishel, R.W.: Deterministic and Stochastic Optimal Control. Springer, New York (1975)
10. Pham, K.D.: Performance-reliability aided decision making in multiperson quadratic decision games against jamming and estimation confrontations. In: Giannessi, F. (ed.) J. Optim. Theory Appl. **149**(3), 559–629 (2011)

Chapter 3
Towards a Risk Sensitivity and Game-Theoretic Approach of Stochastic Fault-Tolerant Systems

3.1 Introduction

Modern research on auxiliary signal design for fault detection and robust control of dynamic systems subject to certain ranges of perturbations or disturbances are well under way [1] and [2]. The principles of many reconfigurable control systems through various fault detection and diagnosis are eloquently described in [3] and [4]. In addition, a resilient control and filtering framework for a class of uncertain dynamical systems has been proposed in [5] which encompasses inherent time-delay model, parametric uncertainties and external disturbances.

Until now, most of the research in this field was devoted to understanding the effects of controller uncertainties in the implementation of robust controllers which optimize the standard measure of average performance in dynamical systems. From a different perspective, it is crucial to design highly reliable and robust controlled systems that are fault tolerant and non-fragile to both structured uncertainties and unexpected extreme and rare events. Building such non-fragile and robust controlled systems requires a new science which needs to bridge the gap between mathematical statistics of performance uncertainties and control decision optimization for control feedback synthesis. There resulted from the work of [6] a set of mathematical statistics of probabilistic performance distributions is adept at predicting performance behavior in the simple but important class of linear-quadratic stochastic control problems. The robust and resilient control design pivoted on the inherent system tradeoff between robustness for actuator failure accommodation and resilience for performance risk considerations is initially summarized in the recent report of [7].

The research reported in this chapter represents an integrated, systematic attempt at applying a multi-sided decision making to the wide class of robust and non-fragile control situations in which: continuous-time stochastic dynamical systems are linear and subject to both norm-bounded and convex-bounded parameter uncertainties as well as actuator anomalies represent worst-case errors and/or physical attacks in

K.D. Pham, *Resilient Controls for Ordering Uncertain Prospects*, Springer Optimization and Its Applications 98, DOI 10.1007/978-3-319-08705-4_3

executing the intended control inputs. Hence, the total loss of control efficiency would result given the fact that there was very limited time and possibility to perform fault diagnosis.

When there are increasing concerns about closed-loop performance in response to various environmental and adversarial stimuli, several robust and resilient control design problems can be solved in an elegant manner if dynamic game-theoretic paradigms for competitive decision making and higher-order performance statistics for capturing the frequency of rare events are resorted. This will be accomplished in Sect. 3.2. The discussion of the problem statements will be occurred in Sect. 3.3. To complete this chapter's proposed solutions and final remarks, Sects. 3.4 and 3.5 are to overcome the norm nowadays through developing and utilizing: on the one hand, non-zero skewness and kurtosis analysis that proves inadequacies of average performances, is needed in order to properly characterize the generalized chi-squared type of performance behavior; and on the other hand, the synthesis of mean-risk control decision laws that bring about both robustness against structural and environmental uncertainties and resilience towards adversarial elements.

3.2 Revisiting Performance Statistics for Resiliency

In order to discuss the mathematical underpinnings, time t is modeled as continuous and the notation of a finite horizon interval is $[t_0, t_f]$. The formulation presupposes a fixed probability space $(\Omega, \mathbb{F}, \{\mathbb{F}_{t_0,t} : t \in [t_0, t_f]\}, \mathbb{P})$ with filtration satisfying the usual conditions. All the filtrations are right continuous and complete and $\mathbb{F}_{t_f} \triangleq \{\mathbb{F}_{t_0,t} : t \in [t_0, t_f]\}$. In addition, let $\mathcal{L}^2_{\mathbb{F}_{t_f}}([t_0, t_f]; \mathbb{R}^n)$ denote the space of $\mathbb{F}_{t_f}$-adapted random processes $\{\hbar(t) : t \in [t_0, t_f]$ such that $E\{\int_{t_0}^{t_f} ||\hbar(t)||^2 \, dt\} < \infty\}$.

For decades, the science of systems design has tacitly assumed that actuator failures can be modeled as random errors. This situation has to change and the major development in dynamic game theory developed specifically for processes characterized by a stochastic zero-sum game between a resilient controller or a defender and a cognitive actuator tamper or an attacker should become essential tools for designing future robust and non-fragile controlled systems.

In the initial consideration, the block diagram of a new formalism as illustrated in Fig. 3.1 is proposed that allows not only for the resilient control of actuator tampers but also incorporates the robust control of the stochastic uncertain dynamical systems on the time interval $[t_0, t_f]$

$$\begin{aligned} dx(t) = [&(A + \Delta A(t))x(t) + (B + \Delta B(t))Lu(t) \\ &+ (B + \Delta B(t))(I - L)f_a(t)]dt + G dw(t), \quad x(t_0) = x_0 \end{aligned} \tag{3.1}$$

where $x(t)$ is the controlled state process valued in $\mathbb{R}^n$ and $u(t)$ is the resilient control process valued in an action set $\mathbb{A}^u \subset \mathbb{R}^m$; and $w(t) \equiv w(t, \omega)$ is the

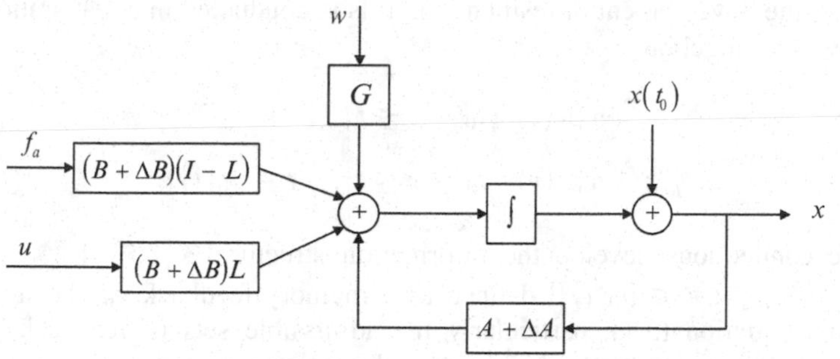

Fig. 3.1 Structure of resilient controlled systems with actuator tampers

process noise taking the form of a stationary Wiener process with the correlations of independent increments $E\left\{[w(\tau_1) - w(\tau_2)][w(\tau_1) - w(\tau_2)]^T\right\} = W|\tau_1 - \tau_2|$, for all $\tau_1, \tau_2 \in [t_0, t_f]$ and $W > 0$ known.

So the system established is consisted of the nominal system coefficients $A \in \mathbb{R}^{n\times n}$, $B \in \mathbb{R}^{n\times m}$ and $G \in \mathbb{R}^{n\times p}$ with the constant matrices. Moreover, it is proposed that the parameter perturbations $\Delta A(t)$ and $\Delta B(t)$ are real, time-varying matrix functions representing the norm-bounded parameter uncertainties and governed by

$$\Delta A(t) = H_A F(t) E_A, \qquad \Delta B(t) = H_B F(t) E_B \tag{3.2}$$

where a priori knowledge about uncertainty are used for a direct derivation of known constant matrices H_A, H_B, E_A and E_B with appropriate dimensions. And $F(t)$ is an unknown real time-varying matrix with Lebesgue measurable elements satisfying $F^T(t)F(t) \le I$ for all $t \in [t_0, t_f]$.

In fact, to such actuators involved in the system (3.1), the working condition of the actuators are described by the distribution matrix $L \triangleq diag(l_1, \ldots, l_m)$, where $l_i = 1$ and $i \in \overline{L} \triangleq \{1, \ldots, m\}$ denotes a healthy ith actuator and $l_i = 0$ corresponds to the total failure of the ith actuator. When the ith actuator is tampered and stuck at certain position, it is clear that its actual control input is only an offset value. Such an offset is presumably manipulated by a cognitive tamper or attack $f_a(t)$, which is the tamper process valued in an action set $\mathbb{A}^{f_a} \subset \mathbb{R}^m$ and thereby, impacting the system through the failed actuators. In the context of any active approaches to control input failures and stochastic environmental uncertainties, the feedback controller proposed here robustly fights through persistent control channel failures by a means of dynamic game-theoretic principles and resiliently guarantees performance against probabilities of rare events via high-order statistical analysis.

A proposal based on recent research [8] included the following recommendation: that the fault-tolerant system (3.1) be complete state controllable with all m actuators functional; that any l actuators out of m actuators be failed so that the system with the remaining $(m - l)$ actuators is still completely state controllable; and so that the system with actuator failures will be regarded to possess $(m - l)$ degree of actuator redundancy.

Most of the development of resilient controls is conducted in collaboration with the following σ-algebras

$$\mathbb{F}_{t_0,t} \triangleq \sigma\{w(\tau) : t_0 \le \tau \le t\}$$
$$\mathscr{G}^y_{t_0,t} \triangleq \sigma\{x(\tau) : t_0 \le \tau \le t\}, \quad t \in [t_0, t_f].$$

Such the connections develop the information structure $\mathscr{G}^y_{t_f} \triangleq \{\mathscr{G}^y_{t_0,t} : t \in [t_0, t_f]\} \subset \{\mathbb{F}_{t_0,t} : t \in [t_0, t_f]\}$ defined by a memory feedback via the stochastic differential equation (3.1), particularly the admissible sets of feedback control decisions, and they greatly assist in defining the action sets

$$\mathbb{U}^{y,u}[t_0, t_f] \triangleq \{u \in \mathscr{L}^2_{\mathscr{G}^y_{t_f}}([t_0, t_f], \mathbb{R}^m) : u(t) \in \mathbb{A}^u \subset \mathbb{R}^m, \; a.e. \quad t \in [t_0, t_f], \; \mathbb{P} - a.s.\}$$

$$\mathbb{U}^{y,f_a}[t_0, t_f] \triangleq \{d \in \mathscr{L}^2_{\mathscr{G}^y_{t_f}}([t_0, t_f], \mathbb{R}^m) : f_a(t) \in \mathbb{A}^{f_a} \subset \mathbb{R}^m, \; a.e. \quad t \in [t_0, t_f], \; \mathbb{P} - a.s.\}$$

whereupon $\mathbb{U}^{y,u}[t_0, t_f]$ and $\mathbb{U}^{y,f_a}[t_0, t_f]$ are closed convex subsets of $\mathscr{L}^2_{\mathbb{F}_{t_f}}([t_0, t_f], \mathbb{R}^m)$.

In essence, a 2-tuple of control strategies is by definition

$$(u, f_a) \in \mathbb{U}^{(2),y^{u,f_a}}[t_0, t_f] \triangleq \mathbb{U}^{y,u}[t_0, t_f] \times \mathbb{U}^{y,f_a}[t_0, t_f] \tag{3.3}$$

and hence it is a family of 2 functions which are non-anticipative with respect to the information structure $\mathscr{G}^y_{t_0,t}$.

Next the resilient controller maintains a commitment to ensure the state trajectory $x(t)$ following a target trajectory $r \in \mathscr{C}([t_0, t_f]; \mathbb{R}^n)$ and to keep the control effort $u(t)$ to be small. So a loss incurred by potential mismatch between $x(t)$ and $r(t)$ and excessive magnitude of $u(t)$ will be perceived by the controller. In order to do away with the efforts by the resilient controller, the actuator tamper lays stress on the common loss so as to set up false impressions with the negative effects. Assuming quadratic losses, the loss for both resilient controller and persistent actuator tamper $J : \mathbb{U}^{y,u}[t_0, t_f] \times \mathbb{U}^{y,f_a}[t_0, t_f] \mapsto \mathbb{R}^+$ is given by

$$\begin{aligned} J(u, f_a) &= [x(t_f) - r(t_f)]^T Q_f [x(t_f) - r(t_f)] \\ &\quad + \int_{t_0}^{t_f} [x^T(\tau) Q_1 x(\tau) + (x(\tau) - r(\tau))^T Q_2 (x(\tau) - r(\tau))] d\tau \\ &\quad + \int_{t_0}^{t_f} \left[u^T(\tau) R_1 u(\tau) - f_a^T(\tau) R_2 f_a(\tau)\right] d\tau, \end{aligned} \tag{3.4}$$

where the terminal penalty weighting $Q_f \in \mathbb{R}^{n \times n}$, the state weighting $Q_1 \in \mathbb{R}^{n \times n}$, the reference tracking weighting $Q_2 \in \mathbb{R}^{n \times n}$, control weighting $R_1 \in \mathbb{R}^{m \times m}$ and tamper weighting $R_2 \in \mathbb{R}^{m \times m}$ are continuous-time matrix functions with the properties of symmetry and positive semi-definiteness. In addition, R_1 and R_2 are invertible.

In the wake of state feedback measurements, both resilient controller and actuator tamper are assumed to act purely on the basis of $\mathcal{G}^y_{t_f}$, for some $\gamma_u(\cdot,\cdot)$ and $\gamma_a(\cdot,\cdot)$

$$u(t) = \gamma_u(t, x(t)); \quad f_a(t) = \gamma_a(t, x(t)), \quad t \in [t_0, t_f] \tag{3.5}$$

Given the least square error criterion (3.4) and Gaussian processes, an optimum linear control system produces results at least as good as any nonlinear control system. Henceforth, the search for optimal control solutions is productively restricted to linear time-varying feedback laws generated from the accessible state $x(t)$ by

$$u(t) = K(t)x(t) + \ell(t), \tag{3.6}$$

$$f_a(t) = K_a(t)x(t) + \ell_a(t), \quad t \in [t_0, t_f] \tag{3.7}$$

where $K, K_a \in \mathcal{C}([t_0, t_f]; \mathbb{R}^{m \times n})$ and $\ell, \ell_a \in \mathcal{C}([t_0, t_f]; \mathbb{R}^m)$ are admissible feedback gains and affine vectors whose further property refinements will be stated shortly.

Then, when such admissible pairs $(K(\cdot), \ell(\cdot))$, $(K_a(\cdot), \ell_a(\cdot))$ and (t_0, x_0) exist, a sufficient condition for the existence of $x(t)$ in (3.1) is ensured. In view of (3.6)–(3.7), the controlled system (3.1) is rewritten as follows

$$\begin{aligned} dx(t) = (((A + \Delta A(t)) + (B + \Delta B(t))LK(t) \\ + (B + \Delta B(t))(I - L)K_a(t))x(t) + (B + \Delta B(t))L\ell(t) \\ + (B + \Delta B(t))(I - L)\ell_a(t))dt + Gdw(t) \end{aligned} \tag{3.8}$$

subject to the sample-path performance measure (3.4)

$$\begin{aligned} J = x^T(t_f)Q_f x(t_f) - 2x^T(t_f)Q_f r(t_f) + r^T(t_f)Q_f r(t_f) \\ + \int_{t_0}^{t_f} x^T(\tau)[Q_1 + Q_2 + K^T(\tau)R_1 K(\tau) - K_a^T(\tau)R_2 K_a(\tau)]x(\tau)d\tau \\ + \int_{t_0}^{t_f} 2x^T(\tau)[K^T(\tau)R_1\ell(\tau) - K_a^T(\tau)R_2\ell_a(\tau) - Q_2 r(\tau)]d\tau \\ + \int_{t_0}^{t_f} [r^T(\tau)Q_2 r(\tau) + \ell^T(\tau)R_1\ell(\tau) - \ell_a^T(\tau)R_2\ell_a(\tau)]d\tau. \end{aligned} \tag{3.9}$$

Against the linear-quadratic structure and background of (3.8) and (3.9), it is enlightening to realize that the performance measure (3.9) is clearly a random variable of the generalized chi-squared type. This emphasis on upholding the accurate modeling of performance uncertainty proves to be beneficial to the design of resilient controllers in the face of both structured and stochastic uncertainties as well as the onset of persistent actuator tamper. As has been often noted, to mathematical statistics which have been successful in explaining nature of performance fluctuations, the performance statistics are now taken shape as follows.

Theorem 3.2.1 (Cumulant-Generating Function). *Let initial system states $x(\tau) \equiv x_\tau$ for $\tau \in [t_0, t_f]$; the moment-generating function (also known as the first-order characteristic function) $\varphi(\tau, x_\tau, \theta) = \varrho(\tau, \theta)\exp\{x_\tau^T \Upsilon(\tau,\theta)x_\tau + 2x_\tau^T \eta(\tau,\theta)\}$, the affine function $\upsilon(\tau,\theta) = \ln\{\varrho(\tau,\theta)\}$ and the risk sensitive parameter $\theta \in \mathbb{R}^+$. Then, the cumulant-generating function is given by*

$$\psi(\tau, x_\tau, \theta) = x_\tau^T \Upsilon(\tau,\theta)x_\tau + 2x_\tau^T \eta(\tau,\theta) + \upsilon(\tau,\theta) \tag{3.10}$$

where the backward-in-time scalar valued $\upsilon(\tau,\theta)$ is satisfying with the terminal-value condition $\upsilon(t_f,\theta) = r^T(t_f)Q_f r(t_f)$

$$\begin{aligned}\frac{d}{d\tau}\upsilon(\tau,\theta) = &-\operatorname{Tr}\{\Upsilon(\tau,\theta)GWG^T\}\\ &-2\eta^T(\tau,\theta)[(B+\Delta B(\tau))L\ell(\tau) + (B+\Delta B(\tau))(I-L)\ell_a(\tau)]\\ &-\theta[r^T(\tau)Q_2 r(\tau) + \ell^T(\tau)R_1\ell(\tau) - \ell_a^T(\tau)R_2\ell_a(\tau)]\end{aligned} \tag{3.11}$$

and the backward-in-time matrix valued $\Upsilon(\tau,\theta)$ and vector valued $\eta(\tau,\theta)$ solve

$$\begin{aligned}\frac{d}{d\tau}\Upsilon(\tau,\theta) = &-[A+\Delta A(\tau) + (B+\Delta B(\tau))LK(\tau)\\ &+(B+\Delta B(\tau))(I-L)K_a(\tau)]^T\Upsilon(\tau,\theta) - \Upsilon(\tau,\theta)[A+\Delta A(\tau)\\ &+(B+\Delta B(\tau))LK(\tau) + (B+\Delta B(\tau))(I-L)K_a(\tau)]\\ &-2\Upsilon(\tau,\theta)GWG^T\Upsilon(\tau,\theta) - \theta[Q_1+Q_2+K^T(\tau)R_1K(\tau)\\ &-K_a^T(\tau)R_2K_a(\tau)], \quad \Upsilon(t_f,\theta) = \theta Q_f\end{aligned} \tag{3.12}$$

and

$$\begin{aligned}\frac{d}{d\tau}\eta(\tau,\theta) = &-[A+\Delta A(\tau)+(B+\Delta B(\tau))LK(\tau) + (B+\Delta B(\tau))(I-L)K_a(\tau)]^T\eta(\tau,\theta)\\ &-\Upsilon(\tau,\theta)[(B+\Delta B(\tau))L\ell(\tau) + (B+\Delta B(\tau))(I-L)\ell_a(\tau)]\\ &-\theta[K^T(\tau)R_1\ell(\tau) - K_a^T(\tau)R_2\ell_a(\tau) - Q_2 r(\tau)], \quad \eta(t_f,\theta) = -\theta Q_f r(t_f).\end{aligned} \tag{3.13}$$

Proof. For notional simplicity, it is convenient to have $\varpi(\tau, x_\tau, \theta) \triangleq \exp\{\theta J(\tau, x_\tau)\}$, within which the performance measure (3.9) is rewritten as the cost-to-go function

$$\begin{aligned}J(\tau, x_\tau) = &\; x^T(t_f)Q_f x(t_f) - 2x^T(t_f)Q_f r(t_f) + r^T(t_f)Q_f r(t_f)\\ &+\int_\tau^{t_f} x^T(t)[Q_1+Q_2+K^T(t)R_1K(t) - K_a^T(t)R_2K_a(t)]x(t)dt\end{aligned}$$

$$+ \int_\tau^{t_f} 2x^T(t)[K^T(t)R_1\ell(t) - K_a^T(t)R_2\ell_a(t) - Q_2 r(\tau)]dt$$
$$+ \int_\tau^{t_f} [r^T(t)Q_2 r(t) + \ell^T(t)R_1\ell(t) - \ell_a^T(t)R_2\ell_a(t)]dt \tag{3.14}$$

subject to the stochastic dynamics (3.8) with the initial condition $x(\tau) = x_\tau$.

Also relevant is $\varphi(\tau, x_\tau, \theta) \triangleq E\{\varpi(\tau, x_\tau, \theta)\}$. Therefore, its total time derivative is of the form

$$\begin{aligned}\frac{d}{d\tau}\varphi(\tau, x_\tau, \theta) = &-\theta\Big\{x_\tau^T[Q_1 + Q_2 + K^T(\tau)R_1K(\tau)\\ &- K_a^T(\tau)R_2K_a(\tau)]x_\tau + 2x_\tau^T[K^T(\tau)R_1\ell(\tau) - K_a^T(\tau)R_2\ell_a(\tau)\\ &- Q_2 r(\tau)] + r^T(\tau)Q_2 r(\tau) + \ell^T(\tau)R_1\ell(\tau) - \ell_a^T(\tau)R_2\ell_a(\tau)\Big\}\varphi(\tau, x_\tau, \theta).\end{aligned}$$

Next it is now safe to say, in fact, that the standard Ito's formula results in

$$\begin{aligned}d\varphi(\tau, x_\tau, \theta) &= E\{d\varpi(\tau, x_\tau, \theta)\}\\ &= \varphi_\tau(\tau, x_\tau, \theta)\,d\tau + \varphi_{x_\tau}(\tau, x_\tau, \theta)\,dx_\tau + \frac{1}{2}\mathrm{Tr}\left\{\varphi_{x_\tau x_\tau}(\tau, x_\tau, \theta)GWG^T\right\}d\tau.\end{aligned}$$

What are needed are $\varphi(\tau, x_\tau, \theta) = \varrho(\tau, \theta)\exp\left\{x_\tau^T\Upsilon(\tau, \theta)x_\tau + 2x_\tau^T\eta(\tau, \theta)\right\}$ and its partial derivatives that in turn accommodate the following result

$$\begin{aligned}\frac{d}{d\tau}\varphi(\tau, x_\tau, \theta) = &\left\{\frac{\frac{d}{d\tau}\varrho(\tau, \theta)}{\varrho(\tau, \theta)} + x_\tau^T\frac{d}{d\tau}\Upsilon(\tau, \theta)x_\tau + 2x_\tau^T\frac{d}{d\tau}\eta(\tau, \theta)\right.\\ &+x_\tau^T[A+\Delta A(\tau)+(B+\Delta B(\tau))LK(\tau)+(B+\Delta B(\tau))(I-L)K_a(\tau)]^T\Upsilon(\tau, \theta)x_\tau\\ &+x_\tau^T\Upsilon(\tau, \theta)[A+\Delta A(\tau)+(B+\Delta B(\tau))LK(\tau)+(B+\Delta B(\tau))(I-L)K_a(\tau)]x_\tau\\ &+2x_\tau^T\Upsilon(\tau, \theta)[(B+\Delta B(\tau))L\ell(\tau)+(B+\Delta B(\tau))(I-L)\ell_a(\tau)]\\ &+2x_\tau^T[A+\Delta A(\tau)+(B+\Delta B(\tau))LK(\tau)+(B+\Delta B(\tau))(I-L)K_a(\tau)]^T\eta(\tau, \theta)\\ &+2\eta^T(\tau, \theta)[(B+\Delta B(\tau))L\ell(\tau)+(B+\Delta B(\tau))(I-L)\ell_a(\tau)]\\ &\left.+\mathrm{Tr}\{\Upsilon(\tau, \theta)GWG^T\}+2x_\tau^T\Upsilon(\tau, \theta)GWG^T\Upsilon(\tau, \theta)x_\tau\right\}\varphi(\tau, x_\tau, \theta).\end{aligned}$$

It is clear that the quest for the constant, linear and quadratic terms being independent of arbitrary x_τ is addressed by the validity of the results (3.11)–(3.13). At $\tau = t_f$, it follows that $\varphi(t_f, x(t_f), \theta) = \rho(t_f, \theta)\exp\{x^T(t_f)\Upsilon(t_f, \theta)x(t_f) + 2x^T(t_f)\eta(t_f, \theta)\} = \exp\{x^T(t_f)\theta Q_f x(t_f) - 2x^T(t_f)\theta Q_f r(t_f) + \theta r^T(t_f)Q_f r(t_f)\}$. In other words, the terminal-value conditions are $\Upsilon(t_f, \theta) = \theta Q_f$, $\eta(t_f, \theta) = -\theta Q_f r(t_f)$, $\rho(t_f, \theta) = \exp\{\theta r^T(t_f)Q_f r(t_f)\}$, and $\upsilon(t_f, \theta) = \theta r^T(t_f)Q_f r(t_f)$, which completes the proof.

In the background remains the intrinsic behavior of the generalized chi-squared random variable (3.9). Still the context of the mathematical statistics, e.g., mean, variance, skewness, kurtosis, etc. that essentially encapsulate the frequency of rare events associated with (3.9) calls for a Maclaurin series expansion of (3.10)

$$\psi(\tau, x_\tau, \theta) = \sum_{r=1}^{\infty} \frac{\partial^{(r)}}{\partial\theta^{(r)}} \psi(\tau, x_\tau, \theta)\bigg|_{\theta=0} \frac{\theta^r}{r!}, \tag{3.15}$$

where all $\kappa_r \triangleq \frac{\partial^{(r)}}{\partial\theta^{(r)}} \psi(\tau, x_\tau, \theta)\Big|_{\theta=0}$ are called the mathematical statistics of (3.9).

To maintain this goal, the series expansion coefficients are further determined by the cumulant-generating function or the second characteristic function (3.9)

$$\begin{aligned}\kappa_r \triangleq \frac{\partial^{(r)}}{\partial\theta^{(r)}} \psi(\tau, x_\tau, \theta)\bigg|_{\theta=0} &= x_\tau^T \frac{\partial^{(r)}}{\partial\theta^{(r)}} \Upsilon(\tau, \theta)\bigg|_{\theta=0} x_\tau \\ &\quad + 2x_\tau^T \frac{\partial^{(r)}}{\partial\theta^{(r)}} \eta(\tau, \theta)\bigg|_{\theta=0} + \frac{\partial^{(r)}}{\partial\theta^{(r)}} \upsilon(\tau, \theta)\bigg|_{\theta=0}.\end{aligned} \tag{3.16}$$

For the standpoint of notational convenience, it requires

$$\begin{aligned} H_r(\tau) &\triangleq \frac{\partial^{(r)}\Upsilon(\tau, \theta)}{\partial\theta^{(r)}}\bigg|_{\theta=0}; \quad \breve{D}_r(\tau) \triangleq \frac{\partial^{(r)}\eta(\tau, \theta)}{\partial\theta^{(r)}}\bigg|_{\theta=0} \\ D_r(\tau) &\triangleq \frac{\partial^{(r)}\upsilon(\tau, \theta)}{\partial\theta^{(r)}}\bigg|_{\theta=0} \end{aligned} \tag{3.17}$$

so that the inquiry into the powerful capability of effectively forecasting all the higher-order characteristics pertaining to (3.9) is obtained as follows.

Theorem 3.2.2 (Performance-Measure Statistics). *Let the stochastic system with a resilient controller subject to actuator tampering be described by (3.8) and (3.9) where it is assumed to possess $(m - l)$ degree of actuator redundancy. For $k \in \mathbb{N}$, the kth performance-measure statistics of (3.9) takes the form of*

$$\kappa_k = x_0^T H_k(t_0) x_0 + 2x_0^T \breve{D}_k(t_0) + D_k(t_0), \tag{3.18}$$

where the supporting variables $\{H_r(\tau)\}_{r=1}^k$, $\{\breve{D}_r(\tau)\}_{r=1}^k$ and $\{D_r(\tau)\}_{r=1}^k$ evaluated at $\tau = t_0$ satisfy the matrix/vector/scalar-valued differential equations (with the dependence of $H_r(\tau)$, $\breve{D}_r(\tau)$ and $D_r(\tau)$ upon $K(\tau)$, $K_a(\tau)$, $\ell(\tau)$ and $\ell_a(\tau)$ suppressed)

$$\begin{aligned}\frac{d}{d\tau} H_1(\tau) = &- [A + \Delta A(\tau) + (B + \Delta B(\tau))LK(\tau) + (B + \Delta B(\tau))(I - L)K_a(\tau)]^T H_r(\tau) \\ &- H_r(\tau)[A + \Delta A(\tau) + (B + \Delta B(\tau))LK(\tau) + (B + \Delta B(\tau))(I - L)K_a(\tau)] \\ &- [Q_1 + Q_2 + K^T(\tau)R_1K(\tau) - K_a^T(\tau)R_2K_a(\tau)], \quad H_1(t_f) = Q_f\end{aligned} \tag{3.19}$$

$$\frac{d}{d\tau}H_r(\tau) = -[A + \Delta A(\tau) + (B + \Delta B(\tau))LK(\tau) + (B + \Delta B(\tau))(I - L)K_a(\tau)]^T H_r(\tau)$$
$$- H_r(\tau)[A + \Delta A(\tau) + (B + \Delta B(\tau))LK(\tau) + (B + \Delta B(\tau))(I - L)K_a(\tau)]$$
$$- \sum_{s=1}^{r-1} \frac{2r!}{s!(r-s)!} H_s(\tau) G W G^T H_{r-s}(\tau), \quad H_r(t_f) = 0 \tag{3.20}$$

$$\frac{d}{d\tau}\breve{D}_1(\tau) = -[A + \Delta A(\tau) + (B + \Delta B(\tau))LK(\tau) + (B + \Delta B(\tau))(I - L)K_a(\tau)]^T \breve{D}_1(\tau)$$
$$- H_1(\tau)[(B + \Delta B(\tau))L\ell(\tau) + (B + \Delta B(\tau))(I - L)\ell_a(\tau)]$$
$$+ Q_2 r(\tau) - K^T(\tau) R_1 \ell(\tau) + K_a^T(\tau) R_2 \ell_a(\tau), \quad \breve{D}_1(t_f) = -Q_f r(t_f) \tag{3.21}$$

$$\frac{d}{d\tau}\breve{D}_r(\tau) = -[A + \Delta A(\tau) + (B + \Delta B(\tau))LK(\tau) + (B + \Delta B(\tau))(I - L)K_a(\tau)]^T \breve{D}_r(\tau)$$
$$- H_r(\tau)[(B + \Delta B(\tau))L\ell(\tau) + (B + \Delta B(\tau))(I - L)\ell_a(\tau)], \quad \breve{D}_r(t_f) = 0 \tag{3.22}$$

$$\frac{d}{d\tau}D_1(\tau) = -2\breve{D}_1^T(\tau)[(B + \Delta B(\tau))L\ell(\tau) + (B + \Delta B(\tau))(I - L)\ell_a(\tau)]$$
$$- \mathrm{Tr}\{H_1(\tau) G W G^T\} - r^T(\tau) Q_2 r(\tau)$$
$$- \ell^T(\tau) R_1 \ell(\tau) + \ell_a^T(\tau) R_2 \ell_a(\tau), \quad D_1(t_f) = r^T(t_f) Q_f r(t_f) \tag{3.23}$$

$$\frac{d}{d\tau}D_r(\tau) = -2\breve{D}_r^T(\tau)[(B + \Delta B(\tau))L\ell(\tau) + (B + \Delta B(\tau))(I - L)\ell_a(\tau)]$$
$$- \mathrm{Tr}\{H_r(\tau) G W G^T\}, \quad D_r(t_f) = 0, \quad 2 \le r \le k. \tag{3.24}$$

Proof. At this stage it is plausible to see that the expression of performance-measure statistics (3.18) is readily justified by using the result (3.17). The remaining problem is how to confer on the solutions $H_r(\tau)$, $\breve{D}_r(\tau)$ and $D_r(\tau)$ for $1 \le r \le k$ the backward-in-time differential equations (3.19)–(3.24). It is also worth noting these Eqs. (3.19)–(3.24) are satisfied by the solutions $H_r(\tau)$, $\breve{D}_r(\tau)$ and $D_r(\tau)$ and can be obtained by successively taking derivatives with respect to θ of the supporting equations (3.11)–(3.13) together with the assumptions of $(m - l)$ degree of actuator redundancy on $[t_0, t_f]$.

3.3 Framing the Problem Statements

From the previous section, the statistical performance processes (3.19)–(3.24) relevant to probabilistic distributions of (3.9) clearly exhibit a systematic relationship at different statistical scales and time. Such processes are now considered to induce complex behavior of closed-loop performance in response to various concurrent feedback actions (K, ℓ) and (K_a, ℓ_a) from the resilient controller and persistent actuator tampers on actuator while subject to structured uncertainties and stochastic environmental stimuli.

To better understand the advantages of the new formalism proposed for optimal statistical control, it is important to have a compact statement so as to aid mathematical manipulations. More precisely, one can think of the k-tuple state variables

$$\mathcal{H}(\cdot) \triangleq (\mathcal{H}_1(\cdot), \ldots, \mathcal{H}_k(\cdot))$$
$$\breve{\mathcal{D}}(\cdot) \triangleq (\breve{\mathcal{D}}_1(\cdot), \ldots, \breve{\mathcal{D}}_k(\cdot))$$
$$\mathcal{D}(\cdot) \triangleq (\mathcal{D}_1(\cdot), \ldots, \mathcal{D}_k(\cdot))$$

whose continuously differentiable states $\mathcal{H}_r \in \mathcal{C}^1([t_0, t_f]; \mathbb{R}^{n \times n})$, $\breve{\mathcal{D}}_r \in \mathcal{C}^1([t_0, t_f]; \mathbb{R}^n)$ and $\mathcal{D}_r \in \mathcal{C}^1([t_0, t_f]; \mathbb{R})$ having the representations

$$\mathcal{H}_r(\cdot) \triangleq H_r(\cdot)\,; \qquad \breve{\mathcal{D}}_r(\cdot) \triangleq \breve{D}_r(\cdot)\,; \qquad \mathcal{D}_r(\cdot) \triangleq D_r(\cdot)$$

with the right members satisfying the dynamics (3.19)–(3.24) are defined on $[t_0, t_f]$.

At the same time, there has been the corresponding mappings of convenient consciousness with respect to such the dynamics (3.19)–(3.24) introduced as follows

$$\mathcal{F}_r : [t_0, t_f] \times (\mathbb{R}^{n \times n})^k \times (\mathbb{R}^{m \times n})^2 \mapsto \mathbb{R}^{n \times n}$$
$$\breve{\mathcal{G}}_r : [t_0, t_f] \times (\mathbb{R}^{n \times n})^k \times (\mathbb{R}^n)^k \times (\mathbb{R}^{m \times n})^2 \times (\mathbb{R}^m)^2 \mapsto \mathbb{R}^n$$
$$\mathcal{G}_r : [t_0, t_f] \times (\mathbb{R}^{n \times n})^k \times (\mathbb{R}^n)^k \times (\mathbb{R}^m)^2 \mapsto \mathbb{R}$$

where the rules of action are given by

$$\begin{aligned}
\mathcal{F}_1(\tau, \mathcal{H}, K, K_a) \triangleq {} & -[A + \Delta A(\tau) + (B + \Delta B(\tau))LK(\tau) \\
& + (B + \Delta B(\tau))(I - L)K_a(\tau)]^T \mathcal{H}_r(\tau) - \mathcal{H}_r(\tau)[A + \Delta A(\tau) \\
& + (B + \Delta B(\tau))LK(\tau) + (B + \Delta B(\tau))(I - L)K_a(\tau)] \\
& - [Q_1 + Q_2 + K^T(\tau)R_1K(\tau) - K_a^T(\tau)R_2K_a(\tau)]
\end{aligned}$$

$$\begin{aligned}
\mathcal{F}_r(\tau, \mathcal{H}, K, K_a) \triangleq {} & -[A + \Delta A(\tau) + (B + \Delta B(\tau))LK(\tau) \\
& + (B + \Delta B(\tau))(I - L)K_a(\tau)]^T \mathcal{H}_r(\tau) - \mathcal{H}_r(\tau)[A + \Delta A(\tau) \\
& + (B + \Delta B(\tau))LK(\tau) + (B + \Delta B(\tau))(I - L)K_a(\tau)] \\
& - \sum_{s=1}^{r-1} \frac{2r!}{s!(r-s)!} \mathcal{H}_s(\tau) G W G^T \mathcal{H}_{r-s}(\tau)
\end{aligned}$$

$$\breve{\mathscr{G}}_1(\tau, \mathscr{H}, \breve{\mathscr{D}}, K, \ell, K_a, \ell_a) \triangleq -[A + \Delta A(\tau) + (B + \Delta B(\tau))LK(\tau) + (B + \Delta B(\tau))(I - L)K_a(\tau)]^T \breve{\mathscr{D}}_1(\tau) - \mathscr{H}_1(\tau)[(B + \Delta B(\tau))L\ell(\tau) + (B + \Delta B(\tau))(I - L)\ell_a(\tau)] + Q_2 r(\tau) - K^T(\tau)R_1\ell(\tau) + K_a^T(\tau)R_2\ell_a(\tau)$$

$$\breve{\mathscr{G}}_r(\tau, \mathscr{H}, \breve{\mathscr{D}}, K, \ell, K_a, \ell_a) \triangleq -[A + \Delta A(\tau) + (B + \Delta B(\tau))LK(\tau) + (B + \Delta B(\tau))(I - L)K_a(\tau)]^T \breve{\mathscr{D}}_r(\tau) - \mathscr{H}_r(\tau)[(B + \Delta B(\tau))L\ell(\tau) + (B + \Delta B(\tau))(I - L)\ell_a(\tau)]$$

$$\mathscr{G}_1(\tau, \mathscr{H}, \breve{\mathscr{D}}, \ell, \ell_a) \triangleq -\text{Tr}\{\mathscr{H}_1(\tau)GWG^T\} - 2\breve{\mathscr{D}}_1^T(\tau)[(B+\Delta B(\tau))L\ell(\tau)+(B+\Delta B(\tau))(I-L)\ell_a(\tau)] - r^T(\tau)Q_2 r(\tau) - \ell^T(\tau)R_1\ell(\tau) + \ell_a^T(\tau)R_2\ell_a(\tau)$$

$$\mathscr{G}_r(\tau, \mathscr{H}, \breve{\mathscr{D}}, \ell, \ell_a) \triangleq -\text{Tr}\{\mathscr{H}_r(\tau)GWG^T\} - 2\breve{\mathscr{D}}_r^T(\tau)[(B+\Delta B(\tau))L\ell(\tau)+(B+\Delta B(\tau))(I-L)\ell_a(\tau)].$$

Insofar as the compactness of the representation of the dynamics (3.19)–(3.24) is concerned, the product mappings are concentrated on

$$\mathscr{F} : [t_0, t_f] \times (\mathbb{R}^{n\times n})^k \times (\mathbb{R}^{m\times n})^2 \mapsto (\mathbb{R}^{n\times n})^k$$

$$\breve{\mathscr{G}} : [t_0, t_f] \times (\mathbb{R}^{n\times n})^k \times (\mathbb{R}^{n})^k \times (\mathbb{R}^{m\times n})^2 \times (\mathbb{R}^{m})^2 \mapsto (\mathbb{R}^{n})^k$$

$$\mathscr{G} : [t_0, t_f] \times (\mathbb{R}^{n\times n})^k \times (\mathbb{R}^{n})^k \times (\mathbb{R}^{m})^2 \mapsto \mathbb{R}^k$$

within which the settings

$$\mathscr{F} \triangleq \mathscr{F}_1 \times \cdots \times \mathscr{F}_k; \quad \breve{\mathscr{G}} \triangleq \breve{\mathscr{G}}_1 \times \cdots \times \breve{\mathscr{G}}_k; \quad \mathscr{G} \triangleq \mathscr{G}_1 \times \cdots \times \mathscr{G}_k$$

are posed. Therefore, the dynamic equations (3.19)–(3.24) can be rewritten as

$$\frac{d}{d\tau}\mathscr{H}(\tau) = \mathscr{F}(\tau, \mathscr{H}(\tau), K(\tau), K_a(\tau)) \tag{3.25}$$

$$\frac{d}{d\tau}\breve{\mathscr{D}}(\tau) = \breve{\mathscr{G}}(\tau, \mathscr{H}(\tau), \breve{\mathscr{D}}(\tau), K(\tau), \ell(\tau), K_a(\tau), \ell_a(\tau)) \tag{3.26}$$

$$\frac{d}{d\tau}\mathscr{D}(\tau) = \mathscr{G}(\tau, \mathscr{H}(\tau), \breve{\mathscr{D}}(\tau), \ell(\tau), \ell_a(\tau)) \tag{3.27}$$

together with k-tuple terminal-value conditions

$$\mathscr{H}(t_f) \triangleq \mathscr{H}_f = (Q_f, 0, \dots, 0)$$

$$\breve{\mathscr{D}}(t_f) \triangleq \breve{\mathscr{D}}_f = (-Q_f r(t_f), 0, \dots, 0)$$

$$\mathscr{D}(t_f) \triangleq \mathscr{D}_f = (r^T(t_f) Q_f r(t_f), \dots, 0).$$

It is now more critical to embark upon a fundamental rethinking of the product system (3.25)–(3.27). An important aspect of such a rethinking is to establish the connection between the state matrices $\mathscr{H}$, $\breve{\mathscr{D}}$ and $\mathscr{D}$ and the admissible feedback pairs (K, ℓ) and (K_a, ℓ_a). Henceforth, this interplay suggests the importance of $\mathscr{H} \equiv \mathscr{H}(\cdot, K, K_a)$, $\breve{\mathscr{D}} \equiv \breve{\mathscr{D}}(\cdot, K, \ell, K_a, \ell_a)$ and $\mathscr{D} \equiv \mathscr{D}(\cdot, \ell, \ell_a)$. In accordance of the dynamics (3.25)–(3.27) and the terminal data $(t_f, \mathscr{H}_f, \breve{\mathscr{D}}_f, \mathscr{D}_f)$, the non-anticipative classes of admissible feedback parameters characterizing the performance evolutions are defined below.

Definition 3.3.1 (Non-anticipative Feedback Strategies). Given $k \in \mathbb{N}$ and the set of scalars $\mu = \{\mu_r \geq 0\}_{r=1}^k$ with $\mu_1 > 0$, a non-anticipative feedback (K, ℓ) defined by: on the one hand, $K \in \mathscr{K}^u_{\varepsilon, \mathscr{Y}, \breve{\mathscr{Z}}, \mathscr{Z}; \mu} \triangleq \{\mathscr{C}([t_0, \varepsilon]; \mathbb{R}^{m \times n}) \times \mathscr{C}([t_0, \varepsilon]; \mathbb{R}^m)$ such that for any $K_a, \tilde{K}_a \in \mathscr{C}([t_0, \varepsilon]; \mathbb{R}^{m \times n})$ and $\ell_a, \tilde{\ell}_a \in \mathscr{C}([t_0, \varepsilon]; \mathbb{R}^m)$ and $K_a(\tau) \approx \tilde{K}_a(\tau)$ and $\ell_a(\tau) \approx \tilde{\ell}_a(\tau)$ for $t_0 \leq \epsilon \leq \tau \leq \varepsilon \leq t_f$ imply $K[K_a, \ell_a](\tau) \approx K[\tilde{K}_a, \tilde{\ell}_a](\tau)$ for $t_0 \leq \epsilon \leq \tau \leq \varepsilon \leq t_f\}$ where $K_a(\tau) \approx \tilde{K}_a(\tau)$ and $\ell_a(\tau) \approx \tilde{\ell}_a(\tau)$ on $[\epsilon, \varepsilon]$ if the probabilities $P(K_a(\tau) = \tilde{K}_a(\tau)$ and $P(\ell_a(\tau) = \tilde{\ell}_a(\tau)$ almost everywhere in $[\epsilon, \varepsilon]) = 1$ and on the other hand, $\ell \in \mathscr{L}^u_{\varepsilon, \mathscr{Y}, \breve{\mathscr{Z}}, \mathscr{Z}; \mu} \triangleq \{\mathscr{C}([t_0, \varepsilon]; \mathbb{R}^{m \times n}) \times \mathscr{C}([t_0, \varepsilon]; \mathbb{R}^m)$ such that for any $K_a, \tilde{K}_a \in \mathscr{C}([t_0, \varepsilon]; \mathbb{R}^{m \times n})$ and $\ell_a, \tilde{\ell}_a \in \mathscr{C}([t_0, \varepsilon]; \mathbb{R}^m)$ and $K_a(\tau) \approx \tilde{K}_a(\tau)$ and $\ell_a(\tau) \approx \tilde{\ell}_a(\tau)$ for $t_0 \leq \epsilon \leq \tau \leq \varepsilon \leq t_f$ imply $\ell[K_a, \ell_a](\tau) \approx \ell[\tilde{K}_a, \tilde{\ell}_a](\tau)$ for $t_0 \leq \epsilon \leq \tau \leq \varepsilon \leq t_f\}$.

A similar case in which the non-anticipative feedback pair (K_a, ℓ_a) are associated with persistent and actuator faults can be interpreted as $K_a \in \mathscr{K}^{f_a}_{\varepsilon, \mathscr{Y}, \breve{\mathscr{Z}}, \mathscr{Z}; \mu}$ and $\ell_a \in \mathscr{L}^{f_a}_{\varepsilon, \mathscr{Y}, \breve{\mathscr{Z}}, \mathscr{Z}; \mu}$. In addition, $\mathscr{K}^u_{\varepsilon, \mathscr{Y}, \breve{\mathscr{Z}}, \mathscr{Z}; \mu}$, $\mathscr{L}^u_{\varepsilon, \mathscr{Y}, \breve{\mathscr{Z}}, \mathscr{Z}; \mu}$, $\mathscr{K}^{f_a}_{\varepsilon, \mathscr{Y}, \breve{\mathscr{Z}}, \mathscr{Z}; \mu}$ and $\mathscr{L}^{f_a}_{\varepsilon, \mathscr{Y}, \breve{\mathscr{Z}}, \mathscr{Z}; \mu}$ assume the respective values in $\overline{K}_u \in \mathbb{R}^{m \times n}$, $\overline{L}_u \in \mathbb{R}^m$, $\overline{K}_{f_a} \in \mathbb{R}^{m \times n}$ and $\overline{L}_{f_a} \in \mathbb{R}^m$ for which the solutions to Eqs. (3.25)–(3.27) with the terminal-value conditions $\mathscr{H}(\varepsilon) = \mathscr{Y}$, $\breve{\mathscr{D}}(\varepsilon) = \breve{\mathscr{Z}}$ and $\mathscr{D}(\varepsilon) = \mathscr{Z}$ exist on $[t_0, \varepsilon]$.

The successful description of performance uncertainties via the new formalism as proposed in (3.25)–(3.27) should allow for better optimization and control of the performance behavior. For the highly nonlinear behavior of the generalized chi-squared type, it is therefore necessary to define multi-attribute utilities such that the deviations (or chances) of rare events (or risks) associated with (3.9) are optimized.

On the 4-tuple $\mathscr{K}^u_{\varepsilon, \mathscr{Y}, \breve{\mathscr{Z}}, \mathscr{Z}; \mu} \times \mathscr{L}^u_{\varepsilon, \mathscr{Y}, \breve{\mathscr{Z}}, \mathscr{Z}; \mu} \times \mathscr{K}^{f_a}_{\varepsilon, \mathscr{Y}, \breve{\mathscr{Z}}, \mathscr{Z}; \mu} \times \mathscr{L}^{f_a}_{\varepsilon, \mathscr{Y}, \breve{\mathscr{Z}}, \mathscr{Z}; \mu}$, a series of substantive concerns has given rise to the expected values and risks associated with the closed-loop performance herein. These substantive concerns relate to the closely linked performance index of mean and risk considerations as follows.

Definition 3.3.2 (Mean-Risk Aware Performance Index). Fix $k \in \mathbb{N}$ and the profile of risk-averse attitudes $\mu = \{\mu_r \geq 0\}_{r=1}^{k}$ with $\mu_1 > 0$. Then for the given x_0, the mean-risk aware performance index

$$\phi_0 : \{t_0\} \times (\mathbb{R}^{n\times n})^k \times (\mathbb{R}^n)^k \times \mathbb{R}^k \mapsto \mathbb{R}^+$$

pertaining to a risk-averse noncooperative game over $[t_0, t_f]$ is defined by

$$\begin{aligned}\phi_0(t_0, \mathscr{H}(t_0), \breve{\mathscr{D}}(t_0), \mathscr{D}(t_0)) &\triangleq \underbrace{\mu_1\kappa_1}_{\text{Mean}} + \underbrace{\mu_2\kappa_2 + \cdots + \mu_k\kappa_k}_{\text{Risk}} \\ &= \sum_{r=1}^{k} \mu_r[x_0^T \mathscr{H}_r(t_0)x_0 + 2x_0^T \breve{\mathscr{D}}_r(t_0) + \mathscr{D}_r(t_0)]\end{aligned} \tag{3.28}$$

where additional design freedom, μ_r's are sufficient to meet different levels of performance reliability; e.g., mean (i.e., the average of performance), variance (i.e., the dispersion of performance around its mean), skewness (i.e., the anti-symmetry of the performance density), kurtosis (i.e., the heaviness in the performance density tails), etc.; whereas the supporting solutions $\{\mathscr{H}_r(\tau)\}_{r=1}^{k}$, $\{\breve{\mathscr{D}}_r(\tau)\}_{r=1}^{k}$ and $\{\mathscr{D}_r(\tau)\}_{r=1}^{k}$ evaluated at $\tau = t_0$ satisfy the dynamical equations (3.25)–(3.27).

When necessary to indicate the dependence of the mean-risk aware performance index (3.28) expressed in Mayer form on the 4-tuple $\mathscr{K}^{u}_{\varepsilon,\mathscr{Y},\breve{\mathscr{Z}},\mathscr{Z};\mu} \times \mathscr{L}^{u}_{\varepsilon,\mathscr{Y},\breve{\mathscr{Z}},\mathscr{Z};\mu} \times \mathscr{K}^{f_a}_{\varepsilon,\mathscr{Y},\breve{\mathscr{Z}},\mathscr{Z};\mu} \times \mathscr{L}^{f_a}_{\varepsilon,\mathscr{Y},\breve{\mathscr{Z}},\mathscr{Z};\mu}$, it is then rewritten explicitly as $\phi_0(K, \ell, K_a, \ell_a)$. Notice that the research investigation herein is concerned with the application of multi-person zero-sum differential game wherein Player 1 (u or (K, ℓ)) is trying to minimize (3.28) and Player 2 (f_a or (K_a, ℓ_a)) is attempting to maximize (3.28). If the game admits a saddle-point equilibrium, there exist best responses (K^*, ℓ^*) and (K_a^*, ℓ_a^*) which are satisfying the following inequalities

$$\phi_0(K^*, \ell^*, K_a, \ell_a) \leq \phi_0(K^*, \ell^*, K_a^*, \ell_a^*) \leq \phi_0(K, \ell, K_a^*, \ell_a^*)$$

for all the non-anticipative feedback strategies $(K, \ell) \in \mathscr{K}^{u}_{\varepsilon,\mathscr{Y},\breve{\mathscr{Z}},\mathscr{Z};\mu} \times \mathscr{L}^{u}_{\varepsilon,\mathscr{Y},\breve{\mathscr{Z}},\mathscr{Z};\mu}$ and the counterparts $(K_a, \ell_a) \in \mathscr{K}^{f_a}_{\varepsilon,\mathscr{Y},\breve{\mathscr{Z}},\mathscr{Z};\mu} \times \mathscr{L}^{f_a}_{\varepsilon,\mathscr{Y},\breve{\mathscr{Z}},\mathscr{Z};\mu}$.

Theorem 3.3.1 (Existence of a Saddle-Point Equilibrium). *If $\mathscr{K}^{u}_{\varepsilon,\mathscr{Y},\breve{\mathscr{Z}},\mathscr{Z};\mu} \times \mathscr{L}^{u}_{\varepsilon,\mathscr{Y},\breve{\mathscr{Z}},\mathscr{Z};\mu} \times \mathscr{K}^{f_a}_{\varepsilon,\mathscr{Y},\breve{\mathscr{Z}},\mathscr{Z};\mu} \times \mathscr{L}^{f_a}_{\varepsilon,\mathscr{Y},\breve{\mathscr{Z}},\mathscr{Z};\mu}$ is nonempty, compact and convex and the continuous performance index (6.43) is strictly convex in (K, ℓ) and strictly concave in (K_a, ℓ_a), the zero-sum differential game admits a saddle-point equilibrium in pure strategies.*

Proof. The proof can be derived from [9] and [10].

As a tenet of transition from the principle of optimality, a family of games based on different starting points is now of concerned. With an intermission of time, ε in mid-play considered here, the path has reached some definitive point at its commencement. There exist potential trajectories $(\mathscr{Y}, \breve{\mathscr{Z}}, \mathscr{Z})$ which may be reached at the end of the intermission for all possible choices of (K, ℓ) and (K_a, ℓ_a). Hence, the concepts of *playable set* and *value function* are defined below.

Definition 3.3.3 (Playable Set). Let $\mathscr{Q} \triangleq \Big\{(\varepsilon, \mathscr{Y}, \breve{\mathscr{Z}}, \mathscr{Z}) \in [t_0, t_f] \times (\mathbb{R}^{n\times n})^k \times (\mathbb{R}^n)^k \times \mathbb{R}^k$ so that $\mathscr{K}^u_{\varepsilon,\mathscr{Y},\breve{\mathscr{Z}},\mathscr{Z};\mu} \times \mathscr{L}^u_{\varepsilon,\mathscr{Y},\breve{\mathscr{Z}},\mathscr{Z};\mu} \times \mathscr{K}^{f_a}_{\varepsilon,\mathscr{Y},\breve{\mathscr{Z}},\mathscr{Z};\mu} \times \mathscr{L}^{f_a}_{\varepsilon,\mathscr{Y},\breve{\mathscr{Z}},\mathscr{Z};\mu} \neq \emptyset\Big\}$.

Definition 3.3.4 (Value Function). Suppose $(\varepsilon, \mathscr{Y}, \breve{\mathscr{Z}}, \mathscr{Z}) \in [t_0, t_f] \times (\mathbb{R}^{n\times n})^k \times (\mathbb{R}^n)^k \times \mathbb{R}^k$ is given. A saddle-point equilibrium is obtained by solving two optimal decision problems for the lower and upper values of the game

$$\underline{\mathscr{V}}(\varepsilon, \mathscr{Y}, \breve{\mathscr{Z}}, \mathscr{Z}) \triangleq \inf_{(K,\ell)\in\mathscr{K}^u_{\varepsilon,\mathscr{Y},\breve{\mathscr{Z}},\mathscr{Z};\mu}\times\mathscr{L}^u_{\varepsilon,\mathscr{Y},\breve{\mathscr{Z}},\mathscr{Z};\mu}} \sup_{(K_a(\cdot),\ell_a(\cdot))\in\overline{K}_{f_a}\times\overline{L}_{f_a}}$$

$$\phi_0(t_0, \mathscr{H}(t_0, K[K_a, \ell_a], \ell[K_a, \ell_a], K_a, \ell_a), \mathscr{D}(t_0, K[K_a, \ell_a], \ell[K_a, \ell_a], K_a, \ell_a))$$

$$\overline{\mathscr{V}}(\varepsilon, \mathscr{Y}, \breve{\mathscr{Z}}, \mathscr{Z}) \triangleq \sup_{(K_a,\ell_a)\in\mathscr{K}^{f_a}_{\varepsilon,\mathscr{Y},\breve{\mathscr{Z}},\mathscr{Z};\mu}\times\mathscr{L}^{f_a}_{\varepsilon,\mathscr{Y},\breve{\mathscr{Z}},\mathscr{Z};\mu}} \inf_{(K(\cdot),\ell(\cdot))\in\overline{K}_u\times\overline{L}_u}$$

$$\phi_0(t_0, \mathscr{H}(t_0, K, \ell, K_a[K, \ell], \ell_a[K, \ell]), \mathscr{D}(t_0, K, \ell, K_a[K, \ell], \ell_a[K, \ell])).$$

Then, a saddle-point equilibrium exists when the following Isaacs condition holds

$$\mathscr{V}(\varepsilon, \mathscr{Y}, \breve{\mathscr{Z}}, \mathscr{Z}) \triangleq \overline{\mathscr{V}}(\varepsilon, \mathscr{Y}, \breve{\mathscr{Z}}, \mathscr{Z}) = \underline{\mathscr{V}}(\varepsilon, \mathscr{Y}, \mathscr{Z})$$

where $\mathscr{V}(\varepsilon, \mathscr{Y}, \breve{\mathscr{Z}}, \mathscr{Z})$ is thus called the value of the game.

In general, it is true that

$$\underline{\mathscr{V}}(\varepsilon, \mathscr{Y}, \breve{\mathscr{Z}}, \mathscr{Z}) \leq \overline{\mathscr{V}}(\varepsilon, \mathscr{Y}, \breve{\mathscr{Z}}, \mathscr{Z}).$$

For each endpoint in $\mathscr{Q}$, the game beginning there has already been solved. Then, the lower and upper values $\underline{\mathscr{V}}(\varepsilon, \mathscr{Y}, \breve{\mathscr{Z}}, \mathscr{Z})$ and $\overline{\mathscr{V}}(\varepsilon, \mathscr{Y}, \breve{\mathscr{Z}}, \mathscr{Z})$ resulted from each choice of (K, ℓ) and (K_a, ℓ_a) are focused.

Theorem 3.3.2 (HJI Equation-Mayer Problem). *Let $(\varepsilon, \mathscr{Y}, \breve{\mathscr{Z}}, \mathscr{Z})$ be any interior point of the playable set $\mathscr{Q}$ at which $\underline{\mathscr{V}}(\varepsilon, \mathscr{Y}, \breve{\mathscr{Z}}, \mathscr{Z})$ and $\overline{\mathscr{V}}(\varepsilon, \mathscr{Y}, \breve{\mathscr{Z}}, \mathscr{Z})$ are differentiable. If there exists a saddle point defined by $(K^*, \ell^*) \in \mathscr{K}^u_{\varepsilon,\mathscr{Y},\breve{\mathscr{Z}},\mathscr{Z};\mu} \times \mathscr{L}^u_{\varepsilon,\mathscr{Y},\breve{\mathscr{Z}},\mathscr{Z};\mu}$ and $(K^*_a, \ell^*_a) \in \mathscr{K}^{f_a}_{\varepsilon,\mathscr{Y},\breve{\mathscr{Z}},\mathscr{Z};\mu} \times \mathscr{L}^{f_a}_{\varepsilon,\mathscr{Y},\breve{\mathscr{Z}},\mathscr{Z};\mu}$, then the Isaacs equations of the zero-sum deterministic game*

$$0 = \min_{(K,\ell)\in\bar{K}_u\times\bar{L}_u} \max_{(K_a,\ell_a)\in\bar{K}_{fa}\times\bar{L}_{fa}} \Big\{\frac{\partial}{\partial\varepsilon}\underline{\mathscr{V}}(\varepsilon,\mathscr{Y},\breve{\mathscr{Z}},\mathscr{Z})$$

$$+ \frac{\partial}{\partial\,\mathrm{vec}(\mathscr{Y})}\underline{\mathscr{V}}(\varepsilon,\mathscr{Y},\breve{\mathscr{Z}},\mathscr{Z})\cdot\mathrm{vec}(\mathscr{F}(\varepsilon,\mathscr{Y},K,K_a))$$

$$+ \frac{\partial}{\partial\,\mathrm{vec}(\breve{\mathscr{Z}})}\underline{\mathscr{V}}(\varepsilon,\mathscr{Y},\breve{\mathscr{Z}},\mathscr{Z})\cdot\mathrm{vec}(\breve{\mathscr{G}}(\varepsilon,\mathscr{Y},\breve{\mathscr{Z}},K,\ell,K_a,\ell_a))$$

$$+ \frac{\partial}{\partial\,\mathrm{vec}(\mathscr{Z})}\underline{\mathscr{V}}(\varepsilon,\mathscr{Y},\breve{\mathscr{Z}},\mathscr{Z})\cdot\mathrm{vec}(\mathscr{G}(\varepsilon,\mathscr{Y},\breve{\mathscr{Z}},\ell,\ell_a))\Big\} \tag{3.29}$$

$$0 = \max_{(K_a,\ell_a)\in\bar{K}_{fa}\times\bar{L}_{fa}} \min_{(K,\ell)\in\bar{K}_u\times\bar{L}_u} \Big\{\frac{\partial}{\partial\varepsilon}\overline{\mathscr{V}}(\varepsilon,\mathscr{Y},\breve{\mathscr{Z}},\mathscr{Z})$$

$$+ \frac{\partial}{\partial\,\mathrm{vec}(\mathscr{Y})}\overline{\mathscr{V}}(\varepsilon,\mathscr{Y},\breve{\mathscr{Z}},\mathscr{Z})\cdot\mathrm{vec}(\mathscr{F}(\varepsilon,\mathscr{Y},K,K_a))$$

$$+ \frac{\partial}{\partial\,\mathrm{vec}(\breve{\mathscr{Z}})}\overline{\mathscr{V}}(\varepsilon,\mathscr{Y},\breve{\mathscr{Z}},\mathscr{Z})\cdot\mathrm{vec}(\breve{\mathscr{G}}(\varepsilon,\mathscr{Y},\breve{\mathscr{Z}},K,\ell,K_a,\ell_a))$$

$$+ \frac{\partial}{\partial\,\mathrm{vec}(\mathscr{Z})}\overline{\mathscr{V}}(\varepsilon,\mathscr{Y},\breve{\mathscr{Z}},\mathscr{Z})\cdot\mathrm{vec}(\mathscr{G}(\varepsilon,\mathscr{Y},\breve{\mathscr{Z}},\ell,\ell_a))\Big\} \tag{3.30}$$

are satisfied and subject to the boundary conditions

$$\underline{\mathscr{V}}(t_0,\mathscr{H}_0,\breve{\mathscr{D}}_0,\mathscr{D}_0) = \overline{\mathscr{V}}(t_0,\mathscr{H}_0,\breve{\mathscr{D}}_0,\mathscr{D}_0) \equiv \phi_0(t_0,\mathscr{H}_0,\breve{\mathscr{D}}_0,\mathscr{D}_0).$$

Of note, vec$(\cdot)$ is the vectorizing operator of enclosed entities.

Proof. The proof of the theorem here can be shown by extending the results Theorem 4.1 on page 159 in [11].

3.4 Saddle-Point Strategies with Risk Aversion

In this section, the existence of an explicit, closed-form solution for the game-theoretic approach to uncertain dynamical systems is now investigated. The two premises of the optimization problem defined in (3.29)–(3.30) are just a starting point, however, in determining admissible feedback strategies (K,ℓ) and (K_a,ℓ_a). It is time to realize that these feedback strategies and the resulting performance profiles are balanced by the framework of responsibilities (3.28).

Particularly revealing in the development that follows is the initial system states with arbitrary values x_0 directly affecting both linear and quadratic characteristics of the mean-risk aware performance index (3.28). It is evident that it may be possible, at least in principle, to find a candidate function which is physically appropriate to produce both linear and quadratic contributions; e.g.,

$$\mathscr{W}(\varepsilon, \mathscr{Y}, \breve{\mathscr{Z}}, \mathscr{Z}) = x_0^T \sum_{r=1}^{k} \mu_r (\mathscr{Y}_r + \mathscr{E}_r(\varepsilon))\, x_0 + 2x_0^T \sum_{r=1}^{k} \mu_r (\breve{\mathscr{Z}}_r + \breve{\mathscr{T}}_r(\varepsilon)) + \sum_{r=1}^{k} \mu_r (\mathscr{Z}_r + \mathscr{T}_r(\varepsilon)) \quad (3.31)$$

where the time parametric functions $\mathscr{E}_r \in \mathscr{C}^1([t_0, t_f]; \mathbb{R}^{n \times n})$, $\breve{\mathscr{T}}_r \in \mathscr{C}^1([t_0, t_f]; \mathbb{R}^n)$ and $\mathscr{T}_r \in \mathscr{C}^1([t_0, t_f]; \mathbb{R})$ are yet to be determined.

Of note, the states of the statistical performance processes (3.25)–(3.27) defined on the interval $[t_0, \varepsilon]$ now have terminal values denoted by $\mathscr{H}(\varepsilon) \equiv \mathscr{Y}$, $\breve{\mathscr{D}}(\varepsilon) \equiv \breve{\mathscr{Z}}$ and $\mathscr{D}(\varepsilon) \equiv \mathscr{Z}$ for all $\varepsilon \in [t_0, t_f]$. As shown in [12], the total time derivative of $\mathscr{W}(\varepsilon, \mathscr{Y}, \breve{\mathscr{Z}}, \mathscr{Z})$ with respect to ε is obtained as follows

$$\begin{aligned}\frac{d}{d\varepsilon}\mathscr{W}(\varepsilon, \mathscr{Y}, \breve{\mathscr{Z}}, \mathscr{Z}) &= x_0^T \sum_{r=1}^{k} \mu_r \Big[\mathscr{F}_r(\varepsilon, \mathscr{Y}, K, K_a) + \frac{d}{d\varepsilon}\mathscr{E}_r(\varepsilon)\Big] x_0 \\ &+ 2x_0^T \sum_{r=1}^{k} \mu_r \Big[\breve{\mathscr{G}}_r(\varepsilon, \mathscr{Y}, \breve{\mathscr{Z}}, K, \ell, K_a, \ell_a) + \frac{d}{d\varepsilon}\breve{\mathscr{T}}_r(\varepsilon)\Big] \\ &+ \sum_{r=1}^{k} \mu_r \Big[\mathscr{G}_r(\varepsilon, \mathscr{Y}, \breve{\mathscr{Z}}, \ell, \ell_a) + \frac{d}{d\varepsilon}\mathscr{T}_r(\varepsilon)\Big] \qquad (3.32)\end{aligned}$$

provided that $(K, \ell) \in \overline{K}_u \times \overline{L}_u$ and $(K_a, \ell_a) \in \overline{K}_{f_a} \times \overline{L}_{f_a}$.

In relation to the Isaacs equations (3.29)–(3.30) which render the imperative of competitiveness between two players, the results (3.31)–(3.32) contribute to the necessary conditions with separable structures

$$\begin{aligned}0 \equiv \min_{(K,\ell)\in\overline{K}_u\times\overline{L}_u} \max_{(K_a,\ell_a)\in\overline{K}_{fa}\times\overline{L}_{fa}} &\Big\{x_0^T \sum_{r=1}^{k} \mu_r \Big[\mathscr{F}_r(\varepsilon, \mathscr{Y}, K, K_a) + \frac{d}{d\varepsilon}\mathscr{E}_r(\varepsilon)\Big] x_0 \\ &+ 2x_0^T \sum_{r=1}^{k} \mu_r \Big[\breve{\mathscr{G}}_r(\varepsilon, \mathscr{Y}, \breve{\mathscr{Z}}, K, \ell, K_a, \ell_a) + \frac{d}{d\varepsilon}\breve{\mathscr{T}}_r(\varepsilon)\Big] \\ &+ \sum_{r=1}^{k} \mu_r \Big[\mathscr{G}_r(\varepsilon, \mathscr{Y}, \breve{\mathscr{Z}}, \ell, \ell_a) + \frac{d}{d\varepsilon}\mathscr{T}_r(\varepsilon)\Big]\Big\} \qquad (3.33)\end{aligned}$$

and

$$0 \equiv \max_{(K_a,\ell_a)\in\overline{K}_{fa}\times\overline{L}_{fa}} \min_{(K,\ell)\in\overline{K}_u\times\overline{L}_u} \Big\{x_0^T \sum_{r=1}^{k} \mu_r \Big[\mathscr{F}_r(\varepsilon,\mathscr{Y},K,K_a) + \frac{d}{d\varepsilon}\mathscr{E}_r(\varepsilon)\Big] x_0$$

$$+ 2x_0^T \sum_{r=1}^{k} \mu_r \Big[\breve{\mathscr{G}}_r(\varepsilon,\mathscr{Y},\breve{\mathscr{Z}},K,\ell,K_a,\ell_a) + \frac{d}{d\varepsilon}\breve{\mathscr{T}}_r(\varepsilon)\Big]$$

$$+ \sum_{r=1}^{k} \mu_r \Big[\mathscr{G}_r(\varepsilon,\mathscr{Y},\breve{\mathscr{Z}},\ell,\ell_a) + \frac{d}{d\varepsilon}\mathscr{T}_r(\varepsilon)\Big]\Big\}. \tag{3.34}$$

Also essential to this development is the fact that the initial condition x_0 is an arbitrary vector. On one level, the necessary conditions for a minimax and a maximin on the time interval $[t_0, \varepsilon]$ are obtained by taking the gradients with respect to (K, ℓ) and (K_a, ℓ_a) of the expressions within the brackets of (3.33)–(3.34)

$$\ell = -R_1^{-1}[(B + \Delta B(\varepsilon))L]^T \sum_{r=1}^{k} \hat{\mu}_r \breve{\mathscr{Z}}_r \tag{3.35}$$

$$K = -R_1^{-1}[(B + \Delta B(\varepsilon))L]^T \sum_{r=1}^{k} \hat{\mu}_r \mathscr{Y}_r \tag{3.36}$$

and

$$\ell_a = -R_2^{-1}[(B + \Delta B(\varepsilon))(I - L)]^T \sum_{r=1}^{k} \hat{\mu}_r \breve{\mathscr{Z}}_r \tag{3.37}$$

$$K_a = -R_2^{-1}[(B + \Delta B(\varepsilon))(I - L)]^T \sum_{r=1}^{k} \hat{\mu}_r \mathscr{Y}_r \tag{3.38}$$

where $\hat{\mu}_r \triangleq \frac{\mu_i}{\mu_1}$ and the convex bounded parametric uncertainties $\Delta A(\varepsilon)$ and $\Delta B(\varepsilon)$ as well as the actuator channels L are known whenever model mismatches and control input outages occur.

Next it is important to realize that respect of extremal feedback strategies (3.35)–(3.38) contributed significantly in the game value when the Isaacs equations (3.33)–(3.34) are evaluated. Thus being sensitive to the time parametric functions for the candidate function $\mathscr{W}(\varepsilon,\mathscr{Y},\breve{\mathscr{Z}},\mathscr{Z})$ of the value function, i.e., $\{\mathscr{E}_r(\cdot)\}_{r=1}^k$, $\{\breve{\mathscr{T}}_r(\cdot)\}_{r=1}^k$, and $\{\mathscr{T}_r(\cdot)\}_{r=1}^k$ is a prudential means to yield a sufficient condition to have the left-hand sides of (3.29)–(3.30) being zero for any $\varepsilon \in [t_0, t_f]$. Such sensitivity is particularly evident when the $\{\mathscr{Y}_r\}_{r=1}^k$, $\{\breve{\mathscr{Z}}_r\}_{r=1}^k$ and $\{\mathscr{D}_r\}_{r=1}^k$ are evaluated along the solutions of the dynamical equations (3.25)–(3.27).

In this regard, a careful examination of (3.33)–(3.34) can call for attention to the selection of the time parametric functions $\{\mathscr{E}_r(\cdot)\}_{r=1}^k$, $\{\breve{\mathscr{T}}_r(\cdot)\}_{r=1}^k$ and $\{\mathscr{T}_r(\cdot)\}_{r=1}^k$ to satisfy the time-forward differential equations; for example

$$\frac{d}{d\varepsilon}\mathscr{E}_1(\varepsilon) = [A+\Delta A(\varepsilon)+(B+\Delta B(\varepsilon))LK(\varepsilon)+(B+\Delta B(\varepsilon))(I-L)K_a(\varepsilon)]^T \mathscr{H}_r(\varepsilon)$$
$$+\mathscr{H}_r(\varepsilon)[A+\Delta A(\varepsilon)+(B+\Delta B(\varepsilon))LK(\varepsilon)+(B+\Delta B(\varepsilon))(I-L)K_a(\varepsilon)]$$
$$+ [Q_1 + Q_2 + K^T(\varepsilon)R_1K(\varepsilon) - K_a^T(\varepsilon)R_2K_a(\varepsilon)], \quad \mathscr{E}_1(t_0) = 0 \tag{3.39}$$

$$\frac{d}{d\varepsilon}\mathscr{E}_r(\varepsilon) = [A+\Delta A(\varepsilon)+(B+\Delta B(\varepsilon))LK(\varepsilon)+(B+\Delta B(\varepsilon))(I-L)K_a(\varepsilon)]^T \mathscr{H}_r(\varepsilon)$$
$$+\mathscr{H}_r(\varepsilon)[A+\Delta A(\varepsilon)+(B+\Delta B(\varepsilon))LK(\varepsilon)+(B+\Delta B(\varepsilon))(I-L)K_a(\varepsilon)]$$
$$+\sum_{s=1}^{r-1}\frac{2r!}{s!(r-s)!}\mathscr{H}_s(\varepsilon)GWG^T\mathscr{H}_{r-s}(\varepsilon), \quad \mathscr{E}_r(t_0) = 0, \quad 2 \le r \le k \tag{3.40}$$

$$\frac{d}{d\varepsilon}\breve{\mathscr{T}}_1(\varepsilon) = [A+\Delta A(\varepsilon)+(B+\Delta B(\varepsilon))LK(\varepsilon)+(B+\Delta B(\varepsilon))(I-L)K_a(\varepsilon)]^T \breve{\mathscr{D}}_1(\varepsilon)$$
$$+ \mathscr{H}_1(\varepsilon)[(B + \Delta B(\varepsilon))L\ell(\varepsilon) + (B + \Delta B(\varepsilon))(I - L)\ell_a(\varepsilon)]$$
$$- Q_2r(\varepsilon) + K^T(\varepsilon)R_1\ell(\varepsilon) - K_a^T(\varepsilon)R_2\ell_a(\varepsilon), \quad \breve{\mathscr{T}}_1(t_0) = 0 \tag{3.41}$$

$$\frac{d}{d\varepsilon}\breve{\mathscr{T}}_r(\varepsilon) = [A+\Delta A(\varepsilon)+(B+\Delta B(\varepsilon))LK(\varepsilon)+(B+\Delta B(\varepsilon))(I-L)K_a(\varepsilon)]^T \breve{\mathscr{D}}_r(\varepsilon)$$
$$+ \mathscr{H}_r(\varepsilon)[(B + \Delta B(\varepsilon))L\ell(\varepsilon) + (B + \Delta B(\varepsilon))(I - L)\ell_a(\varepsilon)], \quad \breve{\mathscr{T}}_r(t_0) = 0 \tag{3.42}$$

$$\frac{d}{d\varepsilon}\mathscr{T}_1(\varepsilon) = \mathrm{Tr}\{\mathscr{H}_1(\varepsilon)GWG^T\}$$
$$+ 2\breve{\mathscr{D}}_1^T(\varepsilon)[(B + \Delta B(\varepsilon))L\ell(\varepsilon) + (B + \Delta B(\varepsilon))(I - L)\ell_a(\varepsilon)]$$
$$+ r^T(\varepsilon)Q_2r(\varepsilon) + \ell^T(\varepsilon)R_1\ell(\varepsilon) - \ell_a^T(\varepsilon)R_2\ell_a(\varepsilon), \quad \mathscr{T}_1(t_0) = 0 \tag{3.43}$$

$$\frac{d}{d\varepsilon}\mathscr{T}_r(\varepsilon) = 2\breve{\mathscr{D}}_r^T(\varepsilon)[(B + \Delta B(\varepsilon))L\ell(\varepsilon) + (B + \Delta B(\varepsilon))(I - L)\ell_a(\varepsilon)]$$
$$+ \mathrm{Tr}\{\mathscr{H}_r(\varepsilon)GWG^T\}, \quad \mathscr{T}_r(t_0) = 0, \quad 2 \le r \le k. \tag{3.44}$$

At present, the feedback strategies (3.35)–(3.36) and (3.37)–(3.38) in the setting of the noncooperative game are now applied along the solution trajectories of the time-backward Riccati-type equations (3.25)–(3.27). For their parts, the Isaac conditions result in equal upper and lower values for the zero-sum differential game under consideration. Subsequently, these noncooperative feedback strategies optimizing the mean-risk aware performance index (3.28) guarantee a saddle-point equilibrium.

$$\ell^*(\varepsilon) = -R_1^{-1}[(B + \Delta B(\varepsilon))L]^T \sum_{r=1}^{k} \hat{\mu}_r \breve{\mathscr{D}}_r^*(\varepsilon)$$

$$K^*(\varepsilon) = -R_1^{-1}[(B + \Delta B(\varepsilon))L]^T \sum_{r=1}^{k} \hat{\mu}_r \mathscr{H}_r^*(\varepsilon)$$

$$\ell_a^*(\varepsilon) = -R_2^{-1}[(B + \Delta B(\varepsilon))(I - L)]^T \sum_{r=1}^{k} \hat{\mu}_r \breve{\mathscr{D}}_r^*(\varepsilon)$$

$$K_a^*(\varepsilon) = -R_2^{-1}[(B + \Delta B(\varepsilon))(I - L)]^T \sum_{r=1}^{k} \hat{\mu}_r \mathscr{H}_r^*(\varepsilon).$$

To carry forward Fig. 3.2 illustrated here is simply presenting the emergence of the opposing faces of the resilient controller and persistent actuator tamper which have both been exemplified by the game-theoretic approach.

Theorem 3.4.1 (Noncooperative Risk-Averse Strategies). *Let the stochastic fault-tolerant system be described by (3.8)–(3.9), whereby it is assumed to possess $(m - l)$ degree of actuator redundancy. Fix $k \in \mathbb{N}$ and the risk-averse profile*

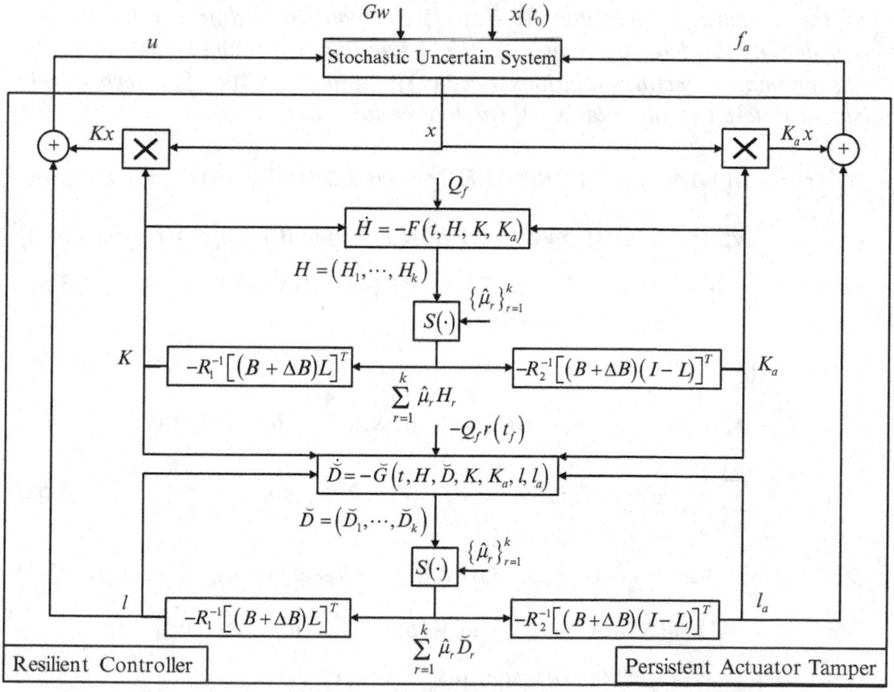

Fig. 3.2 Engagement structure between resilient controller and persistent actuator tamper

$\mu = \{\mu_i \geq 0\}_{i=1}^k$ with $\mu_1 > 0$. The baseline control strategy $u(t)$ which will not only robustly track the reference trajectory $r(t)$ but also resiliently guarantee multiple levels of performance reliability, is given by

$$u^*(t) = K^*(t)x^*(t) + \ell^*(t), \quad t = t_0 + t_f - \tau \tag{3.45}$$

$$K^*(\tau) = -R_1^{-1}[(B + \Delta B(\tau))L]^T \sum_{r=1}^{k} \hat{\mu}_r \mathscr{H}_r^*(\tau) \tag{3.46}$$

$$\ell^*(\tau) = -R_1^{-1}[(B + \Delta B(\tau))L]^T \sum_{r=1}^{k} \hat{\mu}_r \breve{\mathscr{D}}_r^*(\tau) \tag{3.47}$$

despite of persistent actuator tampering strategy $f_a(t)$ as described below

$$f_a^*(t) = K_a^*(t)x^*(t) + \ell_a^*(t), \quad t = t_0 + t_f - \tau \tag{3.48}$$

$$K_a^*(\tau) = -R_2^{-1}[(B + \Delta B(\tau))(I - L)]^T \sum_{r=1}^{k} \hat{\mu}_r \mathscr{H}_r^*(\tau) \tag{3.49}$$

$$\ell_a^*(\tau) = -R_2^{-1}[(B + \Delta B(\tau))(I - L)]^T \sum_{r=1}^{k} \hat{\mu}_r \breve{\mathscr{D}}_r^*(\tau) \tag{3.50}$$

where the normalized weightings $\hat{\mu}_r \triangleq \mu_i/\mu_1$ emphasize on different design freedom of shaping the density function of the chi-squared performance-measure (3.9).

The optimal trajectory solutions $\{\mathscr{H}_r^*(\tau)\}_{r=1}^k$, and $\{\breve{\mathscr{D}}_r^*(\tau)\}_{r=1}^k$ respectively satisfy the time-backward matrix-valued differential equations

$$\begin{aligned}\frac{d}{d\tau}\mathscr{H}_1^*(\tau) = &-[A+\Delta A(\tau)+(B+\Delta B(\tau))LK^*(\tau)+(B+\Delta B(\tau))(I-L)K_a^*(\tau)]^T\mathscr{H}_r^*(\tau)\\ &-\mathscr{H}_r^*(\tau)[A+\Delta A(\tau)+(B+\Delta B(\tau))LK^*(\tau)+(B+\Delta B(\tau))(I-L)K_a^*(\tau)]\\ &-[Q_1 + Q_2 + K^{*T}(\tau)R_1K^*(\tau) - K_a^{*T}(\tau)R_2K_a^*(\tau)]\end{aligned} \tag{3.51}$$

$$\begin{aligned}\frac{d}{d\tau}\mathscr{H}_r^*(\tau) = &-[A+\Delta A(\tau)+(B+\Delta B(\tau))LK^*(\tau)+(B+\Delta B(\tau))(I-L)K_a^*(\tau)]^T\mathscr{H}_r^*(\tau)\\ &-\mathscr{H}_r^*(\tau)[A+\Delta A(\tau)+(B+\Delta B(\tau))LK^*(\tau)+(B+\Delta B(\tau))(I-L)K_a^*(\tau)]\\ &-\sum_{s=1}^{r-1}\frac{2r!}{s!(r-s)!}\mathscr{H}_s^*(\tau)GWG^T\mathscr{H}_{r-s}^*(\tau), \quad 2 \leq r \leq k\end{aligned} \tag{3.52}$$

$$\begin{aligned}\frac{d}{d\tau}\breve{\mathscr{D}}_1^*(\tau) = &-[A+\Delta A(\tau)+(B+\Delta B(\tau))LK^*(\tau)+(B+\Delta B(\tau))(I-L)K_a^*(\tau)]^T\breve{\mathscr{D}}_1^*(\tau)\\ &-\mathscr{H}_1^*(\tau)[(B+\Delta B(\tau))L\ell^*(\tau)+(B+\Delta B(\tau))(I-L)\ell_a^*(\tau)]\\ &+Q_2r(\tau) - K^{*T}(\tau)R_1\ell^*(\tau) + K_a^{*T}(\tau)R_2\ell_a^*(\tau)\end{aligned} \tag{3.53}$$

$$\frac{d}{d\tau}\breve{\mathcal{D}}_r^*(\tau) = -[A+\Delta A(\tau)+(B+\Delta B(\tau))LK^*(\tau)+(B+\Delta B(\tau))(I-L)K_a^*(\tau)]^T\breve{\mathcal{D}}_r^*(\tau) - \mathcal{H}_r^*(\tau)[(B+\Delta B(\tau))L\ell^*(\tau)+(B+\Delta B(\tau))(I-L)\ell_a^*(\tau)],\ r \geq 2 \tag{3.54}$$

provided the terminal-value conditions $\mathcal{H}_1^*(t_f) = Q_f$, $\mathcal{H}_r^*(t_f) = 0$, $\breve{\mathcal{D}}_1^*(t_f) = -Q_f r(t_f)$, $\breve{\mathcal{D}}_r^*(t_f) = 0$, *and* $2 \leq r \leq k$.

3.5 Chapter Summary

In this chapter, a radically new view of the linear-quadratic design of stochastic fault-tolerant systems is presented. Complex characteristics of performance distributions in presence of cognitive actuator tampering, structured uncertainties, and stochastic environmental stimuli are modeled via the statistical performance equations. Upon identification of performance profiles through the statistical processes, the focus of robust resilient control design is placed on transient analysis of higher-order statistics for the target probability density of the generalized chi-squared performance measure.

Adopting the statistical performance approach not only allows the deterministic game-theoretic framework for a robust and resilient baseline controller in synchronization with a persistent and adaptive actuator tamper but also opens new pathways for active defensive countermeasures on agile feedback design and optimization algorithms for performance risk aversion. Last but not least, the complexity of resilient baseline controllers may increase, depending on how many performance-measure statistics of the desired probability density function are to be optimized. As for future work, output dynamical feedback algorithms are under way. Another interesting research direction could incorporate decentralized resilient controllers with common goals in the presence of (possibly) adversarial uncertainty. Thereby, the resulting problem would correspond to potential games with uncertainty. A relevant work along those lines can be found in [13].

References

1. Chen, J., Patton, R.J.: Robust Model-Based Fault Diagnosis for Dynamic Systems. Kluwer Academic, Boston (1999)
2. Campbell, S.L., Nikoukhah, R.: Auxiliary Signal Design for Failure Detection. Princeton Series in Applied Mathematics. Princeton University Press, Princeton (2004)
3. Patton, R.J.: Fault-tolerant control: the 1997 situation survey. In: Proceedings of the IFAC Symposium on Fault Detection, Supervision and Safety for Technical Processes: SAFEPROCESS'97, Pergamon 1998, University of Hull, pp. 1029–1052 (1997)
4. Zhou, D.H., Frank, P.M.: Fault diagnosis and fault tolerant control. IEEE Trans. Aerosp. Electron. Syst. **34**(2), 420–427 (1998)

5. Mahmoud, M.S.: Resilient Control of Uncertain Dynamical Systems. Springer Lecture Notes in Control and Information Sciences, vol. 303. Springer, Berlin/New York (2004)
6. Pham, K.D.: Linear-Quadratic Controls in Risk-Averse Decision Making: Performance-Measure Statistics and Control Decision Optimization. Springer Briefs in Optimization. Springer, New York (2012). ISBN:978-1-4614-5078-8
7. Pham, K.D.: Risk-averse feedback for stochastic fault-tolerant control systems with actuator failure accommodation. In: Proceedings of the 20th Mediterranean Conference on Control and Automation, Barcelona, pp. 504–511 (2012)
8. Zhao, Q., Jiang, J.: Reliable state feedback control system design against actuator failures. Automatica **34**(10), 1267–1272 (1998)
9. Basar, T., Olsder, G.J.: Dynamic Non-cooperative Game Theory, 2nd edn. Society for Industrial and Applied Mathematics, Philadelphia, PA (1998)
10. Pham, K.D.: Risk-averse based paradigms for uncertainty forecast and management in differential games of persistent disruptions and denials. In: Proceedings of American Control Conference, Baltimore, pp. 5526–5531 (2010)
11. Fleming, W.H., Rishel, R.W.: Deterministic and Stochastic Optimal Control. Springer, New York (1975)
12. Pham, K.D.: Statistical control paradigms for structural vibration suppression. Ph.D. dissertation (2004). Department of Electrical Engineering, University of Notre Dame, Indiana. Available via the http://etd.nd.edu/ETD-db/theses/available/etd-04152004-121926/unrestricted/PhamKD052004.pdf. Cited 24 Feb 2014
13. Piliouras, G., Nikolova, E., Shamma, J.S.: Risk sensitivity of price of anarchy under uncertainty. In: 14th ACM Conference on Electronic Commerce, Philadelphia (2013)

Chapter 4
Disturbance Attenuation Problems with Delayed Feedback Measurements

4.1 Introduction

Cyber-physical systems that are dynamically complex are difficult to predict and understand because their behavior is shaped by relationships in which feedbacks are separated from their effects in time and space. Implications of temporal features on delays, adaptive updating and cumulative effects have been explored in a variety of realms, including wide-area power systems [1]. In addition, research on robust decision and control synthesis of a finite-time horizon disturbance attenuation problem where the responsive controller interacts with persistent yet worst-case disturbances systematically overlook the effects of delays and cumulations [2]. The tendency of complex systems to resist attempts to improve performance has been linked to cognitive limitations in understanding and predicting dynamic behaviors.

To understand the evolution of stochastic systems, recognizing the probabilistic aspects of performance uncertainties [3] and [4] is as important as accounting for delayed feedback [5, 6] and references therein. In fact, performance uncertainties and temporal features are not separable, but intertwined. Henceforth, any system with feedback necessarily entails delays, adversarial disturbances and decision making with performance risk aversion that generate complex dynamics. In turn, delayed feedback, exogenous disturbances and risk quantification are the subjects of the research effort herein.

The remainder of this chapter investigates some key process representations of feedback lags, process and measurement disturbances, risk-averse decision making and cumulative effects operating within a class of disturbance attenuation problems with unknown time-delays and linear mutual influence dynamics subjected to process and measurement disturbances. Section 4.2 contains the problem formulation whereby unknown time-delays are estimated as auxiliary states and feedback delays are then modeled through 1st-order Pade approximations. In addition, all the mathematical statistics associated with the finite-horizon integral-quadratic-form (IQF) cost are developed to illustrate what risk-averse decision making and control

K.D. Pham, *Resilient Controls for Ordering Uncertain Prospects*, Springer Optimization and Its Applications 98, DOI 10.1007/978-3-319-08705-4_4

synthesis need when performance reliability accounts for its riskiness effects. Detailed problem statements and solution method of zero-sum game-theoretic optimization for conservative saddle-point strategies between risk-averse controller and cognitive disturbances are described in Sects. 4.3 and 4.4. In Sect. 4.5, some conclusions are also included.

4.2 A Digression on the Problem, States and Observables

In this section, time t is modeled as continuous and the notation of a finite horizon interval is $[t_0, t_f]$. The formulation presupposes a fixed probability space $(\Omega, \mathbb{F}, \{\mathbb{F}_{t_0,t} : t \in [t_0, t_f]\}, \mathbb{P})$ with filtration satisfying the usual conditions. All the filtrations are right continuous and complete and $\mathbb{F}_{t_f} \triangleq \{\mathbb{F}_{t_0,t} : t \in [t_0, t_f]\}$. In addition, let $\mathcal{L}^2_{\mathbb{F}_{t_f}}([t_0, t_f]; \mathbb{R}^n)$ denote the space of $\mathbb{F}_{t_f}$-adapted random processes $\{\hbar(t) : t \in [t_0, t_f]$ such that $E\{\int_{t_0}^{t_f} ||\hbar(t)||^2 \, dt\} < \infty\}$.

A starting point for a class of disturbance attenuation problems considered herein is a stochastic differential equation that explains the controlled systems adapt to external disturbances with the initial condition $x(t_0) = x_0$ known as follows

$$dx(t) = (A(t)x(t) + B_u(t)u(t) + B_d(t)d(t) + B_f(t)f(t))dt + G(t)dw(t) \tag{4.1}$$

to which, the control design has available measurements

$$dy(t) = (C(t)x(t) + D_u(t)u(t) + D_d(t)d(t) + D_f(t)f(t))dt + dv(t). \tag{4.2}$$

Against such a background and the block diagram of the disturbance attenuation investigated as in Fig. 4.1, $x(t)$ is the controlled state process valued in $\mathbb{R}^n$; $u(t)$ is the control process valued in an action space $\mathbb{A}^u \subset \mathbb{R}^{m_u}$; $d(t)$ is the control process valued in an action space $\mathbb{A}^d \subset \mathbb{R}^{m_d}$ representing finite energy disturbances due to exogenous signals or linearization uncertainties; and $f(t)$ is the control process valued in an action space $\mathbb{A}^f \subset \mathbb{R}^{m_f}$ considering possible process, sensor, or actuator faults.

As for a further refinement in the problem description by (4.1)–(4.2), the system coefficients $A(t) \equiv A(t,\omega) : [t_0, t_f] \times \Omega \mapsto \mathbb{R}^{n \times n}$, $B_u(t) \equiv B_u(t,\omega) : [t_0, t_f] \times \Omega \mapsto \mathbb{R}^{n \times m_u}$, $D_u(t) \equiv D_u(t,\omega) : [t_0, t_f] \times \Omega \mapsto \mathbb{R}^{q \times m_u}$ and $C(t) \equiv C(t,\omega) : [t_0, t_f] \times \Omega \mapsto \mathbb{R}^{q \times n}$ are the continuous-time matrix functions. The disturbance distribution coefficients $B_d(t) \equiv B_d(t,\omega) : [t_0, t_f] \times \Omega \mapsto \mathbb{R}^{n \times m_d}$, $B_f(t) \equiv B_f(t,\omega) : [t_0, t_f] \times \Omega \mapsto \mathbb{R}^{n \times m_f}$, $D_d(t) \equiv D_d(t,\omega) : [t_0, t_f] \times \Omega \mapsto \mathbb{R}^{q \times m_d}$, $D_f(t) \equiv D_f(t,\omega) : [t_0, t_f] \times \Omega \mapsto \mathbb{R}^{q \times m_f}$ and $G(t) \equiv G(t,\omega) : [t_0, t_f] \times \Omega \mapsto \mathbb{R}^{n \times p}$ are the continuous-time matrix functions. In addition, the

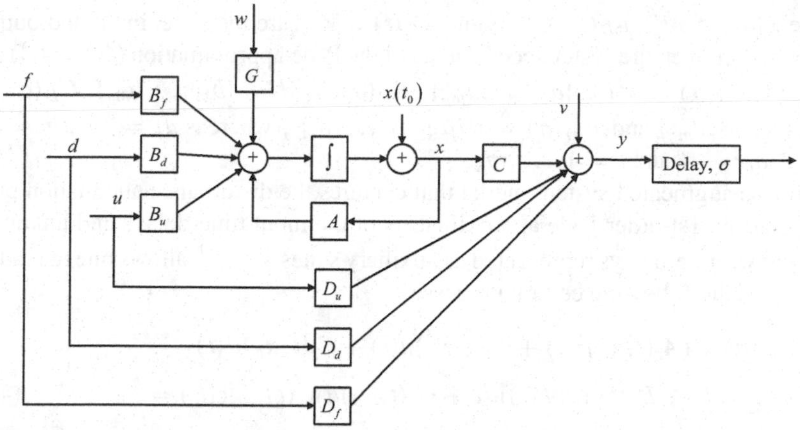

Fig. 4.1 Arrangement of disturbance attenuation studied

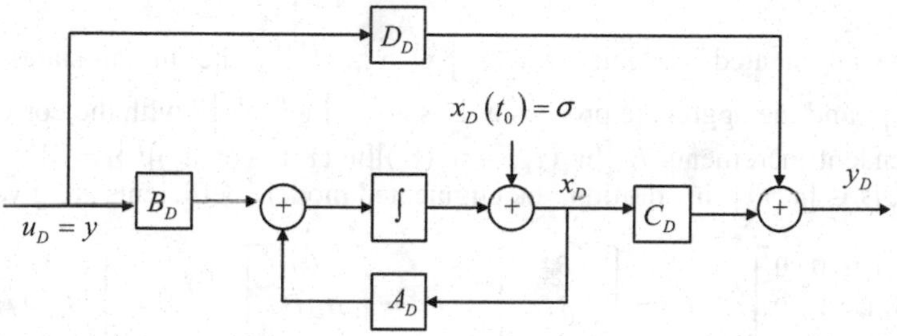

Fig. 4.2 Pade approximation to a pure time delay

random process and measurement noises $w(t) \equiv w(t,\omega)$ and $v(t) \equiv v(t,\omega)$ are the mutually uncorrelated stationary Wiener processes with the correlations of independent increments for all $\tau_1, \tau_2 \in [t_0, t_f]$

$$E\left\{[w(\tau_1) - w(\tau_2)][w(\tau_1) - w(\tau_2)]^T\right\} = W|\tau_1 - \tau_2|$$

$$E\left\{[v(\tau_1) - v(\tau_2)][v(\tau_1) - v(\tau_2)]^T\right\} = V|\tau_1 - \tau_2|$$

whose a-priori second-order statistics $W > 0$ and $V > 0$ are also assumed known.

Temporal features of the dynamical system (4.1)–(4.2) govern its dynamics in a wide variety of ways. For instance, measurement feedback delays help to account for the system's responsiveness to adaptive updating. Hereafter, a constant but unknown time-delay of σ seconds with the Laplace-domain representation of $e^{-s\sigma}$ is used to model the inherent network reaction time delay for feedback measurements. As shown in Fig. 4.2, the 1st-order Pade approximation to a pure time delay gives an account of

$$dx_D(t) = (A_D(\sigma)x_D(t) + B_D(\sigma)u_D(t))dt \tag{4.3}$$

$$dy_D(t) = (C_D(\sigma)x_D(t) + D_D(\sigma)u_D(t))dt \tag{4.4}$$

where $x_D(t) \in \mathbb{R}^{m_u}$, $u_D(t) \in \mathbb{R}^{m_u}$ and $y_D(t) \in \mathbb{R}^{m_u}$ are the state, input and output vectors. Moreover, the system coefficients of the Pade approximation (4.3)–(4.4) are given by $A_D(\sigma) \triangleq diag\{a_1,\ldots,a_{m_u}\}$, $B_D(\sigma) \triangleq diag\{b_1,\ldots,b_{m_u}\}$, $C_D(\sigma) \triangleq diag\{c_1,\ldots,c_{m_u}\}$ and $D_D(\sigma) \triangleq diag\{d_1,\ldots,d_{m_u}\}$ whereas $a_i = -\frac{2}{\sigma}$, $b_i = \frac{2}{\sigma}$, $c_i = 2$, $d_i = -1$ and $i = 1,\ldots,m_u$.

Next an augmented system model that captures the disturbance attenuation phenomenon, the 1st-order Pade approximation of the input time-delays and unknown yet constant time-delays represented as auxiliary states $x_\sigma \triangleq \frac{1}{\sigma}$ allows one to readily understand the following central processes

$$dz(t) =(A_z(t,x_\sigma)z(t) + B_z(t,x_\sigma)u(t) + C_z(t,x_\sigma)d(t) + D_z(t,x_\sigma)f(t))dt + G_z(t,x_\sigma)dw_z(t), \quad z(t_0) = z_0 \tag{4.5}$$

$$dy_D(t) = (H_z(t,x_\sigma)z(t) + I_z(t,x_\sigma)u(t) + J_z(t,x_\sigma)d(t) + K_z(t,x_\sigma)f(t))dt + D_D(t,x_\sigma)dv(t) \tag{4.6}$$

where the augmented variables $z \triangleq \left[x^T\ x_D^T\ x_\sigma^T\right]^T$, the initial states $z_0 \triangleq \left[x_0^T\ 0\ \frac{1}{\sigma}\right]^T$ and the aggregate process noises $w_z \triangleq \left[w^T\ v^T\right]^T$ with the correlations of independent increments $E\left\{[w_z(\tau_1) - w_z(\tau_2)][w_z(\tau_1) - w_z(\tau_2)]^T\right\} = W_z|\tau_1 - \tau_2|$ for all $\tau_1, \tau_2 \in [t_0, t_f]$. In addition, the augmented model coefficients are given by

$$A_z = \begin{bmatrix} A & 0 & 0 \\ B_D C & A_D & 0 \\ 0 & 0 & 0 \end{bmatrix}; \quad B_z = \begin{bmatrix} B_u \\ B_D D_u \\ 0 \end{bmatrix}; \quad C_z = \begin{bmatrix} B_d \\ B_D D_d \\ 0 \end{bmatrix}; \quad D_z = \begin{bmatrix} B_f \\ B_D D_f \\ 0 \end{bmatrix}$$

$$G_z = \begin{bmatrix} G & 0 \\ 0 & B_D \\ 0 & 0 \end{bmatrix}; \quad H_z = \begin{bmatrix} D_D C & C_D & 0 \end{bmatrix}; \quad I_z = D_D D_u; \quad J_z = D_D D_d; \quad K_z = D_D D_f.$$

And the implementation is given in Fig. 4.3.

Furthermore, the development hereafter continually probes beneath the surface of the following σ-algebras

$$\mathbb{F}_{t_0,t} \triangleq \sigma\{(w(\tau), v(\tau)) : t_0 \le \tau \le t\}$$

$$\mathcal{G}^y_{t_0,t} \triangleq \sigma\{y(\tau) : t_0 \le \tau \le t\}, \quad t \in [t_0, t_f]$$

to determine their underlying causes and to find a suitable state estimator, $\hat{z}(t) \triangleq E\{z(t)|\mathcal{G}^y_{t_f}\}$. As depicted in Fig. 4.4, a Kalman-like estimator that later forms part of the estimate-based controller preserves, even earlier, the inherent linear Gaussian structure of (4.5)–(4.6), but this, too, takes into account of the information available $\mathcal{G}^y_{t_f} \triangleq \{\mathcal{G}^y_{t_0,t} : t \in [t_0, t_f]\} \subset \{\mathbb{F}_{t_0,t} : t \in [t_0, t_f]\}$; e.g.,

$$d\hat{z}(t) = (A_z(t,x_\sigma)\hat{z}(t)+B_z(t,x_\sigma)u(t)+C_z(t,x_\sigma)d(t)+D_z(t,x_\sigma)f(t))dt + L_z(t)(dy_D(t)-(H_z(t,x_\sigma)\hat{z}(t)+I_z(t,x_\sigma)u(t)+J_z(t,x_\sigma)d(t)+K_z(t,x_\sigma)f(t))dt) \tag{4.7}$$

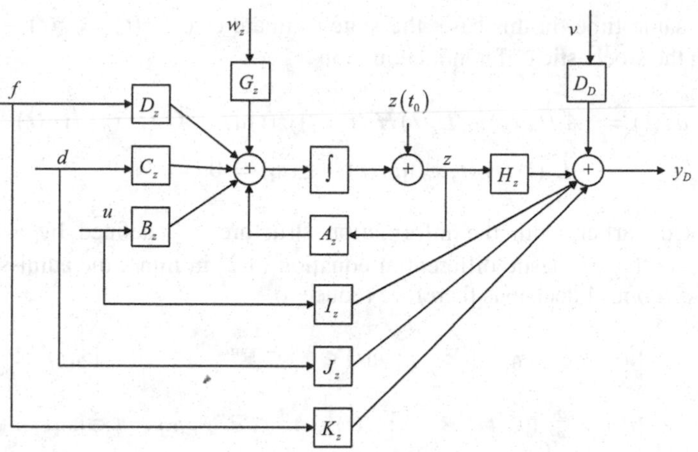

Fig. 4.3 Structure of stochastic disturbance attenuation system with Pade approximation for delayed feedback measurements

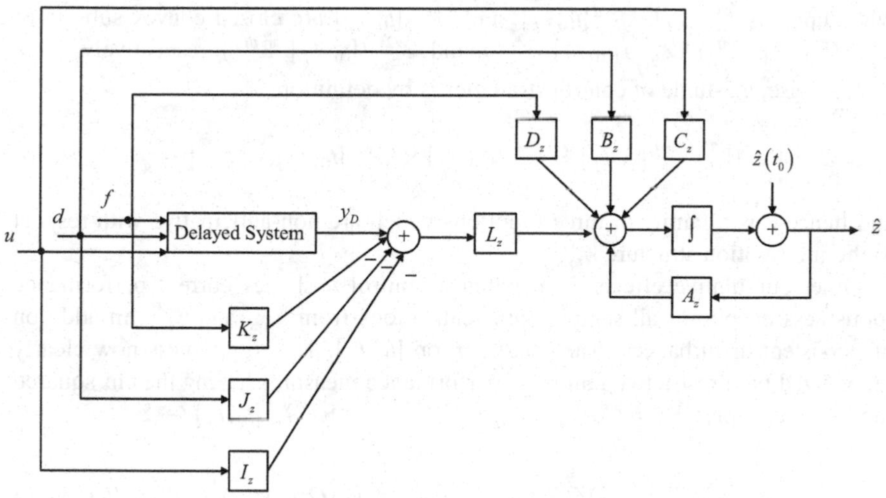

Fig. 4.4 Realization of Kalman-like estimator

where $\hat{z}(t_0) = z(t_0)$ and whereas the filtering gain $L_z(t) = \Sigma_z(t) H_z^T(t, x_\sigma) V^{-1}$ and state-estimate error covariance, $\Sigma_z(t) \triangleq E\{[z(t) - \hat{z}(t)][z(t) - \hat{z}(t)]^T | \mathcal{G}_{t_f}^y\}$ is given by

$$\frac{d}{dt}\Sigma_z(t) = A_z(t, x_\sigma)\Sigma_z(t) + \Sigma_z(t)A_z^T(t, x_\sigma) - \Sigma_z(t)H_z^T(t, x_\sigma)V^{-1}H_z(t, x_\sigma)\Sigma_z(t)$$
$$+ G_z(t, x_\sigma)W_z G_z^T(t, x_\sigma), \quad \Sigma_z(t_0) = 0. \tag{4.8}$$

At the same time, in this case, the state-estimate errors $\tilde{z}(t) \triangleq z(t) - \hat{z}(t)$ are satisfying the stochastic differential equation

$$d\tilde{z}(t) = (A_z(t, x_\sigma) - L_z(t)H_z(t, x_\sigma))\tilde{z}(t)dt + G_z(t, x_\sigma)dw_z(t)$$
$$-L_z(t)D_D(t, x_\sigma)dv(t), \quad \tilde{z}(t_0) = 0. \tag{4.9}$$

As noted earlier, with the information structure $\mathcal{G}^y_{t_f}$ defined by a memory feedback via the stochastic differential equation (4.2) in mind, the admissible sets of feedback control decisions therefore reduce to

$$\mathbb{U}^{y,u}[t_0, t_f] \triangleq \{u \in \mathcal{L}^2_{\mathcal{G}^y_{t_f}}([t_0, t_f], \mathbb{R}^{m_u}) : u(t) \in \mathbb{A}^u \subset \mathbb{R}^{m_u}, \ a.e. \quad t \in [t_0, t_f], \ \mathbb{P} - a.s.\}$$

$$\mathbb{U}^{y,d}[t_0, t_f] \triangleq \{d \in \mathcal{L}^2_{\mathcal{G}^y_{t_f}}([t_0, t_f], \mathbb{R}^{m_d}) : d(t) \in \mathbb{A}^d \subset \mathbb{R}^{m_d}, \ a.e. \quad t \in [t_0, t_f], \ \mathbb{P} - a.s.\}$$

$$\mathbb{U}^{y,f}[t_0, t_f] \triangleq \{f \in \mathcal{L}^2_{\mathcal{G}^y_{t_f}}([t_0, t_f], \mathbb{R}^{m_f}) : f(t) \in \mathbb{A}^f \subset \mathbb{R}^{m_f}, \ a.e. \quad t \in [t_0, t_f], \ \mathbb{P} - a.s.\}$$

whereupon $\mathbb{U}^{y,u}[t_0, t_f]$, $\mathbb{U}^{y,d}[t_0, t_f]$, and $\mathbb{U}^{y,f}[t_0, t_f]$ are closed convex subsets of $\mathcal{L}^2_{\mathbb{F}_{t_f}}([t_0, t_f], \mathbb{R}^{m_u})$, $\mathcal{L}^2_{\mathbb{F}_{t_f}}([t_0, t_f], \mathbb{R}^{m_d})$, and $\mathcal{L}^2_{\mathbb{F}_{t_f}}([t_0, t_f], \mathbb{R}^{m_f})$, respectively.

Likewise, a 3-tuple of control strategies is by definition

$$(u, d, f) \in \mathbb{U}^{(3), y^{u,d,f}}[t_0, t_f] \triangleq \mathbb{U}^{y,u}[t_0, t_f] \times \mathbb{U}^{y,d}[t_0, t_f] \times \mathbb{U}^{y,f}[t_0, t_f] \tag{4.10}$$

and hence it is a family of three functions which are non-anticipative with respect to the information structure $\mathcal{G}^y_{t_0,t}$.

Other cumulative effects arise when a control u shapes current performance robustness despite of all sample-path realizations from $\mathbb{F}_{t_0,t}$ and $\mathcal{G}^y_{t_0,t}$, in addition of persistent disturbances d and faults f on $[t_0, t_f]$. This situation is now clearly recognized by a restrictive family of performance measures having the chi-squared random behavior; e.g., $J : \mathbb{U}^{y,u}[t_0, t_f] \times \mathbb{U}^{y,d}[t_0, t_f] \times \mathbb{U}^{y,f}[t_0, t_f] \mapsto \mathbb{R}^+$

$$J(u(\cdot), d(\cdot), f(\cdot)) = z^T(t_f)Q_f z(t_f) + \int_{t_0}^{t_f} \Big\{ z^T(\tau)Q_z(\tau)z(\tau) + u^T(\tau)R_u(\tau)u(\tau)$$
$$+ \frac{1}{\delta}[d^T(\tau)S_d^{-1}(\tau)d(\tau) + f^T(\tau)S_f^{-1}(\tau)f(\tau)] \Big\} d\tau \tag{4.11}$$

where the terminal penalty weighting $Q_f \in \mathbb{R}^{n \times n}$, the state weighting $Q_z(t) \equiv Q_z(t, \omega) : [t_0, t_f] \times \Omega \mapsto \mathbb{R}^{n \times n}$, control weighting $R_u(t) \equiv R_u(t, \omega) : [t_0, t_f] \times \Omega \mapsto \mathbb{R}^{m_u \times m_u}$ and the disturbance sensitivity weightings $S_d(t) \equiv S_d(t, \omega) : [t_0, t_f] \times \Omega \mapsto \mathbb{R}^{m_d \times m_d}$ and $S_f(t) \equiv S_f(t, \omega) : [t_0, t_f] \times \Omega \mapsto \mathbb{R}^{m_f \times m_f}$ are the continuous-time matrix functions with the properties of symmetry and positive semi-definiteness. In addition, $R_u(t)$, $S_d(t)$ and $S_f(t)$ are invertible. And δ is a negative constant.

In the case of incomplete information, the admissible decision laws must be of the forms

$$u(t) = \eth_u(t, y_D(\tau)), \quad d(t) = \eth_d(t, y_D(\tau)), \quad f(t) = \eth_f(t, y_D(\tau)), \quad \forall t \in [t_0, t].$$

In general, the conditional probability density $p(z(t)|\mathcal{G}^y_{t_0,t})$ which is the probability density of $z(t)$ conditioned on $\mathcal{G}^y_{t_0,t}$ (i.e., induced by the observation $\{y_D(\tau) : \tau \in [t_0, t]\}$) represents the sufficient statistics for describing the conditional stochastic effects of future control action. Under the Gaussian assumption, the conditional probability density $p(z(t)|\mathcal{G}^y_{t_0,t})$ is parameterized by its conditional mean $\hat{z}(t) \triangleq E\{z(t)|\mathcal{G}^y_{t_0,t}\}$ and covariance $\Sigma_z(t) \triangleq E\{[z(t) - \hat{z}(t)][z(t) - \hat{z}(t)]^T|\mathcal{G}^y_{t_0,t}\}$. With respect to the linear-Gaussian conditions, the covariance is independent of controls and observations. Therefore, to look for optimal control laws of the aforementioned forms, it is only required that

$$u(t) = \gamma_u(t, \hat{z}(t)), \quad d(t) = \gamma_d(t, \hat{z}(t)), \quad f(t) = \gamma_f(t, \hat{z}(t)), \quad \forall t \in [t_0, t].$$

As shown in [4], the search for optimal control solutions to the disturbance attenuation problem with unknown time-delays is restricted to linear time-varying feedback laws generated from the accessible states $\hat{z}(t)$ by

$$u(t) = K_u(t)\hat{z}(t), \quad d(t) = K_d(t)\hat{z}(t), \quad f(t) = K_f(t)\hat{z}(t), \quad \forall t \in [t_0, t_f] \tag{4.12}$$

with the admissible gains $K_u \in \mathcal{C}([t_0, t_f]; \mathbb{R}^{m_u \times n_z})$, $K_d \in \mathcal{C}([t_0, t_f]; \mathbb{R}^{m_d \times n_z})$, $K_f \in \mathcal{C}([t_0, t_f]; \mathbb{R}^{m_f \times n_z})$ and $n_z \triangleq n + m_u + 1$ to which further defining properties will be defined subsequently.

For the admissible $K_u(\cdot)$, $K_d(\cdot)$, $K_f(\cdot)$ and the pair (t_0, z_0), it gives a sufficient condition for the existence of $\hat{z}(\cdot)$ and $\tilde{z}(\cdot)$ in (4.7) and (4.9), respectively. In view of the control policies (4.12), the controlled system (4.5)–(4.11) is rewritten as

$$dm(t) = F_m(t)m(t)dt + G_m(t)dw_m(t), \quad m(t_0) \tag{4.13}$$

where, for each $t \in [t_0, t_f]$, the augmented state variables $m^T \triangleq \begin{bmatrix} \hat{z}^T & \tilde{z}^T \end{bmatrix}$, the initial condition $m(t_0) \triangleq m_0 = \begin{bmatrix} z_0^T & 0 \end{bmatrix}$ and the process noises $w_m^T \triangleq \begin{bmatrix} w_z^T & v^T \end{bmatrix}$ with the correlation of independent increments $E\left\{[w_m(\tau_1)-w_m(\tau_2)][w_m(\tau_1)-w_m(\tau_2)]^T\right\} = W_m|\tau_1 - \tau_2|$ for all $\tau_1, \tau_2 \in [t_0, t_f]$ and $W_m > 0$. Moreover, the augmented system coefficients are given by

$$F_m = \begin{bmatrix} A_z + B_z K_u + C_z K_d + D_z K_f & L_z H_z \\ 0 & A_z - L_z H_z \end{bmatrix}, \quad G_m = \begin{bmatrix} 0 & L_z D_D \\ G_z & -L_z D_D \end{bmatrix}, \quad W_m = \begin{bmatrix} W & 0 & 0 \\ 0 & V & 0 \\ 0 & 0 & V \end{bmatrix}.$$

Likewise, the finite-horizon integral-quadratic form cost (4.11) follows

$$J(K_u, K_d, K_f) = m^T(t_f)N_f m(t_f) + \int_{t_0}^{t_f} m^T(\tau)N(\tau)m(\tau)d\tau \tag{4.14}$$

where the continuous-time weightings are given by

$$N_f = \begin{bmatrix} Q_f & Q_f \\ Q_f & Q_f \end{bmatrix}, \quad N = \begin{bmatrix} Q_z + K_u^T R_u K_u + K_d^T(\delta S_d)^{-1}K_d + K_f^T(\delta S_f)^{-1}K_f & Q_z \\ Q_z & Q_z \end{bmatrix}.$$

Interestingly, it is evident that the closed-loop performance (4.14) pertaining to the problem of disturbance attenuation here is a random variable of the generalized chi-squared type. This insight occupies a central role in selecting an appropriate criterion for the goals of the controller, persistent disturbances and faults. They all prefer to avoid the risks associated with closed-loop performance variations. And, for the reason of measure of effectiveness, much of the discussion that follows will concern the case involving performance values and risks.

Several revealing indicators of the degrees of higher-order characteristics influence the closed-loop system performance distributions have been their circumscribed roles in the performance-measure statistics. Realistically, these mathematical statistics can be generated by the use of a Maclaurin series expansion of the second-order characteristic function or equivalently, the cumulant-generating function of (4.14).

Theorem 4.2.1 (Cumulant-Generating Function). *Let $m(\cdot)$ be a state variable of the stochastic disturbance attenuation dynamics concerning unknown input time-delays (4.13) with initial values $m(\tau) \equiv m_\tau$ and $\tau \in [t_0, t_f]$. Further let the moment-generating function be denoted by*

$$\varphi(\tau, m_\tau, \theta) = \varrho(\tau, \theta) \exp\{m_\tau^T \Upsilon(\tau, \theta) m_\tau\} \tag{4.15}$$

$$\upsilon(\tau, \theta) = \ln\{\varrho(\tau, \theta)\}, \qquad \theta \in \mathbb{R}^+. \tag{4.16}$$

Then, the cumulant-generating function has the form of quadratic affine

$$\psi(\tau, m_\tau; \theta) = m_\tau^T \Upsilon(\tau, \theta) m_\tau + \upsilon(\tau, \theta) \tag{4.17}$$

where the scalar solution $\upsilon(\tau, \theta)$ solves the backward-in-time differential equation

$$\frac{d}{d\tau}\upsilon(\tau, \theta) = -\mathrm{Tr}\left\{\Upsilon(\tau, \theta) G_m(\tau) W_m G_m^T(\tau)\right\}, \quad \upsilon(t_f, \theta) = 0 \tag{4.18}$$

and the matrix solution $\Upsilon(\tau, \theta)$ satisfies the backward-in-time differential equation

$$\begin{aligned}\frac{d}{d\tau}\Upsilon(\tau, \theta) = &- F_m^T(\tau)\Upsilon(\tau, \theta) - \Upsilon(\tau, \theta)F_m(\tau) \\ &- 2\Upsilon(\tau, \theta)G_m(\tau)W_m G_m^T(\tau)\Upsilon(\tau, \theta) - \theta N(\tau), \quad \Upsilon(t_f, \theta) = \theta N_f.\end{aligned} \tag{4.19}$$

Also, the scalar solution $\varrho(\tau,\theta)$ satisfies the time-backward differential equation

$$\frac{d}{d\tau}\varrho(\tau,\theta) = -\varrho(\tau,\theta)\,\mathrm{Tr}\{\Upsilon(\tau,\theta)G_m(\tau)\,W_m G_m^T(\tau)\}, \quad \varrho(t_f,\theta) = 1. \tag{4.20}$$

Proof. In the course of the proof, it is convenient to have $\varpi(\tau, m_\tau;\theta) \triangleq e^{\{\theta J(\tau,m_\tau)\}}$ in which the performance measure (4.14) is rewritten as the cost-to-go function from an arbitrary state m_τ at a running time $\tau \in [t_0, t_f]$, that is,

$$J(\tau, m_\tau) = m^T(t_f)N_f m(t_f) + \int_\tau^{t_f} m^T(t)N(t)m(t)dt \tag{4.21}$$

subject to

$$dm(t) = F_m(t)m(t)dt + G_m(t)dw_m(t), \quad m(\tau) = m_\tau. \tag{4.22}$$

It is thus notably that the moment-generating function is defined by

$$\varphi(\tau, m_\tau;\theta) \triangleq E\{\varpi(\tau, m_\tau;\theta)\}.$$

Beyond this, the total time derivative of $\varphi(\tau, m_\tau;\theta)$ is further obtained as follows

$$\frac{d}{d\tau}\varphi(\tau, m_\tau, \theta) = -\varphi(\tau, m_\tau, \theta)\,\theta m_\tau^T N(\tau)m_\tau.$$

Particularly where the standard formula of Ito calculus is applied, one can easily show the following result

$$\begin{aligned} d\varphi(\tau, m_\tau;\theta) &= E\{d\varpi(\tau, m_\tau;\theta)\} \\ &= E\Big\{\varpi_\tau(\tau, m_\tau;\theta)\,d\tau + \varpi_{m_\tau}(\tau, m_\tau;\theta)\,dm_\tau \\ &\quad + \frac{1}{2}\mathrm{Tr}\left\{\varpi_{m_\tau m_\tau}(\tau, m_\tau;\theta)G_m(\tau)W_m G_m^T(\tau)\right\}d\tau\Big\} \\ &= \varphi_\tau(\tau, m_\tau;\theta)d\tau + \varphi_{m_\tau}(\tau, m_\tau;\theta)F_m(\tau)m_\tau d\tau \\ &\quad + \frac{1}{2}\mathrm{Tr}\left\{\varphi_{m_\tau m_\tau}(\tau, m_\tau;\theta)G_m(\tau)W_m G_m^T(\tau)\right\}d\tau \end{aligned}$$

which under the definition of

$$\varphi(\tau, m_\tau;\theta) \triangleq \varrho(\tau,\theta)\exp\{m_\tau^T \Upsilon(\tau,\theta)m_\tau\}$$

and its partial derivatives leads to the expression

$$-\theta m_\tau^T N(\tau) m_\tau \varphi(\tau, m_\tau; \theta) = \Big\{ \frac{\frac{d}{d\tau}\varrho(\tau,\theta)}{\varrho(\tau,\theta)} + \operatorname{Tr}\{\Upsilon(\tau,\theta) G_m(\tau) W_m G_m^T(\tau)\}$$
$$+ m_\tau^T [\frac{d}{d\tau}\Upsilon(\tau,\theta) + F_m^T(\tau)\Upsilon(\tau,\theta) + \Upsilon(\tau,\theta) F_m(\tau)] m_\tau$$
$$+ 2 m_\tau^T \Upsilon(\tau,\theta) G_m(\tau) W_m G_m^T(\tau) \Upsilon(\tau,\theta) m_\tau \Big\} \varphi(\tau, m_\tau; \theta).$$

It is now natural to wonder whether it would be possible to have the constant and quadratic terms being independent of arbitrary initial values m_τ. To this end one needs the following equations hold true

$$\frac{d}{d\tau}\Upsilon(\tau,\theta) = -F_m^T(\tau)\Upsilon(\tau,\theta) - \Upsilon(\tau,\theta) F_m(\tau)$$
$$- 2\Upsilon(\tau,\theta) G_m(\tau) W_m G_m^T(\tau) \Upsilon(\tau,\theta) - \theta N(\tau), \quad \Upsilon(t_f,\theta) = \theta N_f$$

$$\frac{d}{d\tau}\varrho(\tau,\theta) = -\varrho(\tau,\theta) \operatorname{Tr}\{\Upsilon(\tau,\theta) G_m(\tau) W_m G_m^T(\tau)\}, \quad \varrho(t_f,\theta) = 1.$$

In addition, the differential equation satisfied by the solution $\upsilon(\tau,\theta)$ is obtained

$$\frac{d}{d\tau}\upsilon(\tau,\theta) = -\operatorname{Tr}\{\Upsilon(\tau,\theta) G_m(\tau) W_m G_m^T(\tau)\}, \quad \upsilon(t_f,\theta) = 0$$

which completes the proof.

At this stage it is also plausible to investigate some properties associated with the solution to the nonlinear matrix-valued differential equation (4.19) to be a matter that is appropriate for further understanding of the mathematical statistics of (4.14).

Proposition 4.2.1 (Solution Representation). *Let $(A, [B_u, B_d, B_f])$ be uniformly stabilizable and (A, C) be uniformly detectable. Then the solution to the matrix-valued differential equation*

$$\frac{d}{d\tau}\Upsilon(\tau,\theta) = -F_m^T(\tau)\Upsilon(\tau,\theta) - \Upsilon(\tau,\theta) F_m(\tau)$$
$$- 2\Upsilon(\tau,\theta) G_m(\tau) W_m G_m^T(\tau) \Upsilon(\tau,\theta) - \theta N(\tau), \quad \Upsilon(t_f,\theta) = \theta N_f \tag{4.23}$$

can be rewritten in the form of

$$\Upsilon(\tau,\theta) = \Phi^T(t_f,\tau) X(\tau,\theta) \Phi(t_f,\tau), \qquad \tau \in [t_0, t_f] \tag{4.24}$$

where $\Phi(\cdot,\cdot)$ is the state transition matrix of the closed-loop stable $F_m(\tau)$ and whereas the supporting solution $X(\tau,\theta)$ is satisfying its own backward-in-time matrix-valued differential equation

$$\frac{d}{d\tau}X(\tau,\theta) = -2X(\tau,\theta)\,\Phi(t_f,\tau)\,G_m(\tau)\,W_m G_m^T(\tau)\,\Phi^T(t_f,\tau)\,X(\tau,\theta) - \Phi^T(\tau,t_f)\,\theta N(\tau)\Phi(\tau,t_f), \quad X(t_f,\theta) = \theta N_f. \tag{4.25}$$

Proof. In the face of such a matrix transformation (4.24), the time derivative of $X(\tau,\theta)$ follows

$$\frac{d}{d\tau}X(\tau,\theta) = \Phi^T(\tau,t_f)\,F_m^T(\tau)\,\Upsilon(\tau,\theta)\Phi(\tau,t_f) + \Phi^T(\tau,t_f)\frac{d}{d\tau}\Upsilon(\tau,\theta)\Phi(\tau,t_f) + \Phi^T(\tau,t_f)\,\Upsilon(\tau,\theta)F_m(\tau)\,\Phi(\tau,t_f).$$

In view of (4.23), it is clear to see that

$$\frac{d}{d\tau}X(\tau,\theta) = -\Phi^T(\tau,t_f)\,[2\Upsilon(\tau,\theta)G_m(\tau)\,W_m G_m^T(\tau)\,\Upsilon(\tau,\theta) + \theta N(\tau)]\Phi(\tau,t_f) \tag{4.26}$$

where the terminal-value condition $X(t_f,\theta) = \theta N_f$. Therefore, the replacement of the expression (4.24) for $\Upsilon(\tau,\theta)$ into the preceding equation (4.26) yields the matrix differential equation (4.25).

In effect, the monotone behavior of the matrix solution $\Upsilon(\tau,\theta)$ for $\tau \in [t_0,t_f]$ and $\theta > 0$ can be shown qualitatively.

Proposition 4.2.2 (Monotonicity of Solution). *Let $(A,[B_u,B_d,B_f])$ be uniformly stabilizable and (A,C) be uniformly detectable. Then, the backward-in-time matrix-valued differential equation (4.23) tends to have a unique steady-state positive semidefinite solution as $t_f \to \infty$.*

Proof. What seems evident is that if the solution $X(\tau,\theta)$ for all $\tau \in [t_0,t_f]$ satisfying Eq. (4.25) was positive semidefinite and monotone increasing from t_f to t_0. Then the assertion of the proposition would be justified due to the relationship (4.24) between $X(\tau,\theta)$ and $\Upsilon(\tau,\theta)$.

Indeed, the definiteness and monotone behavior of $X(\tau,\theta)$ can be easily seen by putting Eq. (4.25) in a quadratic form, for any nonzero vector z

$$\frac{d}{d\tau}\{z^T X(\tau,\theta)\,z\} = -z^T X(\tau,\theta)\,\Phi(t_f,t)\,G_m(\tau)\,W_m G_m^T(\tau)\,\Phi^T(t_f,\tau)\,X(\tau,\theta)\,z - z^T\Phi^T(\tau,t_f)\,\theta N(\tau)\Phi(\tau,t_f)\,z$$

together with the terminal-value condition $z^T X(t_f,\theta)\,z = z^T\theta N_f z \geq 0$.

Given the interest of $\theta > 0$ and the fact of $G_m(\tau) W_m G_m^T(\tau)$ and $N(\tau)$ positive semidefinite, it follows that

$$\frac{d}{d\tau}\{z^T X(\tau,\theta) z\} \leq 0, \qquad z^T X(t_f,\theta) z = z^T \theta N_f z \geq 0.$$

Furthermore, the scalar-valued function $z^T X(\tau,\theta) z$ is also absolutely continuous on $[t_0, t_f]$. Thus, it suffices to conclude that $z^T X(\tau,\theta) z$ is monotonically increasing when going to the left from t_f to t_0. In other words, both $X(\tau,\theta) \geq 0$ and $\Upsilon(\tau,\theta) \geq 0$ are unique and monotone increasing from t_f to t_0.

Under the assumptions of $(A, [B_u, B_d, B_f])$ is uniformly stabilizable and (A, C) is uniformly detectable, there always exist some feedback control and filter gains such that the composite state matrix $F_m(\tau)$ is exponentially stable. According to the results in [7, pp. 190], it means that the associated state transition matrix

$$\lim_{t_f \to \infty} \|\Phi(t_f,\sigma)\| = 0$$

$$\lim_{t_f \to \infty} \int_{t_0}^{t_f} \|\Phi(t_f,\sigma)\|^2 d\sigma < \infty, \quad \forall \sigma \in [t_0, t_f].$$

Moreover, the solution to the matrix differential equation (4.25) can be expressed as

$$X(\tau,\theta) = \int_{\tau}^{t_f} [2X(\sigma,\theta)\Phi(t_f,\sigma)G_m(\sigma)W_m G_m^T(\sigma)\Phi^T(t_f,\sigma)X(\sigma,\theta) + \Phi^T(\sigma,t_f)\theta N(\sigma)\Phi(\sigma,t_f)]d\sigma.$$

By the transformation stated in (4.24), it follows that

$$\Upsilon(\tau,\theta) = \Phi^T(t_f,\tau) \int_{\tau}^{t_f} \big[2X(\sigma,\theta)\Phi(t_f,\sigma)G_m(\sigma)W_m G_m^T(\sigma)\Phi^T(t_f,\sigma)X(\sigma,\theta) + \Phi^T(\sigma,t_f)\theta N(\sigma)\Phi(\sigma,t_f)\big]d\sigma \, \Phi(t_f,\tau).$$

As long as the growth rate of the integral is not faster than the exponentially decreasing rate of two factors $\Phi(t_f,\tau)$, it is concluded that there exists an upper bound on the solution $\Upsilon(\tau,\theta)$ for any time interval $[t_0, t_f]$. Therefore, there have been shown that the positive semidefinite solution $\Upsilon(\tau,\theta)$ is monotone increasing to the left on the interval $[t_0, t_f]$ and bounded above. By the monotone convergence theorem, Eq. (4.23) has its unique steady-state solution when t_f gets arbitrarily large.

From the statistical optimal control point of view, the principled approach is to translate the goal of decision making for performance reliability into all the mathematical statistics associated with (4.14). Uncertainties in performance distributions contained in the probability density function of (4.14) are now transferred as efficiently as possible to these performance-measure statistics of (4.14). Therefore, one expects to extract as much performance information as possible from the mathematical statistics of (4.14) by means of a Maclaurin series expansion of (4.17)

$$\psi(\tau, m_\tau; \theta) = \sum_{r=1}^{\infty} \frac{\partial^{(r)}}{\partial\theta^{(r)}} \psi(\tau, m_\tau; \theta)\bigg|_{\theta=0} \frac{\theta^r}{r!} \tag{4.27}$$

in which all $\kappa_r \triangleq \frac{\partial^{(r)}}{\partial\theta^{(r)}} \psi(\tau, m_\tau, \theta)\big|_{\theta=0}$ are called rth-order performance-measure statistics. Moreover, the series expansion coefficients are computed by using the cumulant-generating function (4.17)

$$\frac{\partial^{(r)}}{\partial\theta^{(r)}} \psi(\tau, m_\tau; \theta)\bigg|_{\theta=0} = m_\tau^T \frac{\partial^{(r)}}{\partial\theta^{(r)}} \Upsilon(\tau, \theta)\bigg|_{\theta=0} m_\tau + \frac{\partial^{(r)}}{\partial\theta^{(r)}} \upsilon(\tau, \theta)\bigg|_{\theta=0} . \tag{4.28}$$

In view of the definition (4.27), the rth performance-measure statistic follows

$$\kappa_r = m_\tau^T \frac{\partial^{(r)}}{\partial\theta^{(r)}} \Upsilon(\tau, \theta)\bigg|_{\theta=0} m_\tau + \frac{\partial^{(r)}}{\partial\theta^{(r)}} \upsilon(\tau, \theta)\bigg|_{\theta=0} \tag{4.29}$$

for any finite $1 \leq r < \infty$. For notational convenience, there is, in turn, a need of changing notations as illustrated as below

$$H_r(\tau) \triangleq \frac{\partial^{(r)} \Upsilon(\tau, \theta)}{\partial\theta^{(r)}}\bigg|_{\theta=0}, \qquad D_r(\tau) \triangleq \frac{\partial^{(r)} \upsilon(\tau, \theta)}{\partial\theta^{(r)}}\bigg|_{\theta=0} \tag{4.30}$$

so that the next result provides an effective and accurate capability for forecasting all the higher-order characteristics associated with performance uncertainty. Therefore, via these performance-measure statistics and adaptive decision making, it is anticipated that future performance variations will lose the element of surprise due to the inherent property of decision and feedback control solutions that are readily capable of reshaping the cumulative probability distribution of closed-loop performance.

Theorem 4.2.2 (Performance-Measure Statistics). *Let the disturbance attenuation system with unknown input time-delays be governed by (4.13)–(4.14) wherein the pairs* $(A, [B_u, B_d, B_f])$ *and* (A, C) *are uniformly stabilizable and detectable, respectively. For* $k \in \mathbb{N}$ *fixed, the* k*th statistics of performance measure (4.14) of the chi-squared type is*

$$\kappa_k = m_0^T H_k(t_0) m_0 + D_k(t_0) \tag{4.31}$$

where the supporting variables $\{H_r(\tau)\}_{r=1}^k$ *and* $\{D_r(\tau)\}_{r=1}^k$ *evaluated at* $\tau = t_0$ *satisfy the backward-in-time matrix and scalar-valued differential equations (with the dependence of* $H_r(\tau)$ *and* $D_r(\tau)$ *upon* $K_u(\tau)$, $K_d(\tau)$ *and* $K_f(\tau)$ *suppressed)*

$$\frac{d}{d\tau}H_1(\tau) = -F_m^T(\tau)H_1(\tau) - H_1(\tau)F_m(\tau) - N(\tau) \tag{4.32}$$

$$\frac{d}{d\tau}H_r(\tau) = -F_m^T(\tau)H_r(\tau) - H_r(\tau)F_m(\tau) - \sum_{s=1}^{r-1}\frac{2r!}{s!(r-s)!}H_s(\tau)G_m(\tau)W_m G_m^T(\tau)H_{r-s}(\tau), \quad 2 \le r \le k \tag{4.33}$$

$$\frac{d}{d\tau}D_r(\tau) = -\operatorname{Tr}\left\{H_r(\tau)G_m(\tau)W_m G_m^T(\tau)\right\}, \quad 1 \le r \le k \tag{4.34}$$

where the terminal-value conditions $H_1(t_f) = N_f$, $H_r(t_f) = 0$ for $2 \le r \le k$ and $D_r(t_f) = 0$ for $1 \le r \le k$.

Proof. The expression of performance-measure statistics described in (4.31) is readily justified by using result (4.29) and definition (4.30). What remains is to show that the solutions $H_r(\tau)$ and $D_r(\tau)$ for $1 \le r \le k$ indeed satisfy the backward-in-time dynamical equations (4.32)–(4.34). Notice that the dynamical equations (4.32)–(4.34) satisfied by the solutions $H_r(\tau)$ and $D_r(\tau)$ are therefore obtained by successively taking derivatives with respect to θ of the backward-in-time matrix-valued differential equations (4.18)–(4.19) under the assumptions of $(A, [B_u, B_d, B_f])$ and (A, C) being uniformly stabilizable and detectable on $[t_0, t_f]$.

Making use of the result stated above, the theorem that follows will show some effects of different terminal values on the existence of solutions to the cumulant-generating equations.

Theorem 4.2.3 (Effect of Terminal Values on Global Existence). *Suppose that $(A, [B_u, B_d, B_f])$ is uniformly stabilizable and (A, C) is uniformly detectable. The solutions $\{H_r(\tau)\}_{r=1}^k$ to the cumulant-generating equations (4.32)–(4.33) assume two distinct terminal values N_1 and N_2 at the terminal time t_f. If $N_1 - N_2$ is nonnegative definite. Then for all $\tau < t_f$, the difference between solutions $H_r(\tau, t_f, N_1) - H_r(\tau, t_f, N_2)$ is also nonnegative definite as long as both solutions $H_r(\tau, t_f, N_1)$ and $H_r(\tau, t_f, N_2)$ exist. In other words, if the solution $H_r(\tau, t_f, N_2)$ passing through N_2 exists on $t_0 \le \tau \le t_f$ then the solution $H_r(\tau, t_f, N_1)$ passing through N_1 exists on the interval extending back at least as far as t_0.*

Proof. Let $\delta H_r(\tau) \triangleq H_r(\tau, t_f, N_1) - H_r(\tau, t_f, N_2)$ for $1 \le r \le k$. Then it is easy to verify that for all $\tau \in [t_0, t_f]$, the difference $\delta H_r(\tau)$ satisfies the following linear matrix differential equation

$$\frac{d}{d\tau}\delta H_r(\tau) = -F_m^T(\tau)\delta H_r(\tau) - \delta H_r(\tau)F_m(\tau), \quad \delta H_r(t_f) = N_1 - N_2. \tag{4.35}$$

In addition, it is fruitful to introduce the following matrix transformation

$$\delta V_r(\tau) = \Phi^T(\tau, t_f)\delta H_r(\tau)\Phi_r(\tau, t_f), \quad \forall \tau \in [t_0, t_f] \tag{4.36}$$

and its time derivative becomes

$$\frac{d}{d\tau}\delta V_r(\tau) = \Phi(\tau, t_f)\{\frac{d}{d\tau}\delta H_r(\tau) + F_m^T(\tau)\delta H_r(\tau) + \delta H_r(\tau)F_m(\tau)\}\Phi(\tau, t_f). \quad (4.37)$$

Placing the expression for $\frac{d}{d\alpha}\delta H_r(\tau)$ as shown in (4.35) inside the bracket of (4.37) yields the following result

$$\frac{d}{d\tau}\delta V_r(\tau) = 0, \quad \delta V_r(t_f) = N_1 - N_2. \quad (4.38)$$

It follows immediately that the solution of Eq. (4.38) has the form of

$$\delta V_r(\tau) = N_1 - N_2, \forall \tau \in [t_0, t_f].$$

Therefore, for $1 \le r \le k$ it follows that

$$H_r(\tau, t_f, N_1) - H_r(\tau, t_f, N_2) = \Phi^T(t_f, \tau)(N_1 - N_2)\Phi(t_f, \tau) \ge 0,$$

if $N_1 - N_2$ is non-negative definite. The proof is finally complete.

4.3 Problem Statements of Lower and Upper Values

Paradoxically, in much of the optimization, estimation and control work relevant to disturbance attenuation applications, robust decision making and control strategies under performance uncertainty and risk is seldom addressed beyond the widely used measure of statistical average or mean to summarize the underlying performance variations. The emergent decision optimization with performance risk aversion here is distinguished by the fact that the cognitive controller, persistent disturbances and faults are no longer content with the expected performance and thus decide to mitigate performance riskiness by the use of a new performance index. The so-called risk-value aware performance index will ensure how much of the inherent or design-in reliability actually ends up in the developmental and operational phases.

Looking back from the vantage point of the statements related to the statistical optimal control, the significance of compact notations and thereby aiding the subsequent mathematical manipulations lies so much in the k-tuple state variables

$$\mathscr{H}(\cdot) \triangleq (\mathscr{H}_1(\cdot), \ldots, \mathscr{H}_k(\cdot)), \qquad \mathscr{D}(\cdot) \triangleq (\mathscr{D}_1(\cdot), \ldots, \mathscr{D}_k(\cdot))$$

whose components $\mathscr{H}_r \in \mathscr{C}^1([t_0, t_f]; \mathbb{R}^{2n_z \times 2n_z})$ and $\mathscr{D}_r \in \mathscr{C}^1([t_0, t_f]; \mathbb{R})$ are continuously differentiable and have the representations $\mathscr{H}_r(\cdot) \triangleq H_r(\cdot)$ and $\mathscr{D}_r(\cdot) \triangleq D_r(\cdot)$ with the right members satisfying the dynamics (4.32)–(4.34) on $[t_0, t_f]$.

In the remainder of the development, the bounded and Lipschitz continuous mappings are introduced

$$\mathscr{F}_r : [t_0, t_f] \times (\mathbb{R}^{2n_z \times 2n_z})^k \mapsto \mathbb{R}^{2n_z \times 2n_z}$$
$$\mathscr{G}_r : [t_0, t_f] \times (\mathbb{R}^{2n_z \times 2n_z})^k \mapsto \mathbb{R}$$

where the rules of action are given by

$$\mathscr{F}_1(\tau, \mathscr{H}) \triangleq -F_m^T(\tau)\mathscr{H}_1(\tau) - \mathscr{H}_1(\tau)F_m(\tau) - N(\tau)$$
$$\mathscr{F}_r(\tau, \mathscr{H}) \triangleq -F_m^T(\tau)\mathscr{H}_r(\tau) - \mathscr{H}_r(\tau)F_m(\tau)$$
$$- \sum_{s=1}^{r-1} \frac{2r!}{s!(r-s)!} \mathscr{H}_s(\tau)G_m(\tau)W_m G_m^T(\tau)\mathscr{H}_{r-s}(\tau), \quad 2 \le r \le k$$
$$\mathscr{G}_r(\tau, \mathscr{H}) \triangleq -\mathrm{Tr}\left\{\mathscr{H}_r(\tau)G_m(\tau)W_m G_m^T(\tau)\right\}, \quad 1 \le r \le k.$$

The product mappings that necessarily follow are for a compact formulation

$$\mathscr{F}_1 \times \cdots \times \mathscr{F}_k : [t_0, t_f] \times (\mathbb{R}^{2n_z \times 2n_z})^k \mapsto (\mathbb{R}^{2n_z \times 2n_z})^k$$
$$\mathscr{G}_1 \times \cdots \times \mathscr{G}_k : [t_0, t_f] \times (\mathbb{R}^{2n_z \times 2n_z})^k \mapsto \mathbb{R}^k$$

where the corresponding notations

$$\mathscr{F} \triangleq \mathscr{F}_1 \times \cdots \times \mathscr{F}_k,$$
$$\mathscr{G} \triangleq \mathscr{G}_1 \times \cdots \times \mathscr{G}_k$$

are used. Thus, the dynamic equations of motion (4.32)–(4.34) are rewritten as follows

$$\frac{d}{d\tau}\mathscr{H}(\tau) = \mathscr{F}(\tau, \mathscr{H}(\tau)), \qquad \mathscr{H}(t_f) \equiv \mathscr{H}_f \tag{4.39}$$

$$\frac{d}{d\tau}\mathscr{D}(\tau) = \mathscr{G}(\tau, \mathscr{H}(\tau)), \qquad \mathscr{D}(t_f) \equiv \mathscr{D}_f \tag{4.40}$$

wherein the terminal-value conditions

$$\mathscr{H}_f \triangleq \mathscr{H}(t_f) = (N_f, 0, \ldots, 0),$$
$$\mathscr{D}_f \triangleq \mathscr{D}(t_f) = (0, \ldots, 0).$$

Theoretically, the product system (4.39)–(4.40) uniquely determines the state matrices $\mathscr{H}$ and $\mathscr{D}$ once the admissible feedback strategy gains K_u, K_d and K_f

being specified. Henceforth, emphasis should be given to these state variables being considered as $\mathscr{H}(\cdot) \equiv \mathscr{H}(\cdot, K_u, K_d, K_f)$ and $\mathscr{D}(\cdot) \equiv \mathscr{D}(\cdot, K_u, K_d, K_f)$.

For historical reasons, in conformity with the use of the principle of optimality, the developments that follow are important. For the given terminal data $(\varepsilon, \mathscr{Y}, \mathscr{Z})$, the classes of admissible feedback strategy gains are next defined.

Definition 4.3.1 (Non-anticipative Feedback Strategy Gains). For the given $k \in \mathbb{Z}^+$ and sequence $\mu = \{\mu_r \geq 0\}_{r=1}^k$ with $\mu_1 > 0$, a non-anticipative feedback strategy gain $K_u \in \mathscr{K}^u_{\varepsilon,\mathscr{Y},\mathscr{Z};\mu} \triangleq \{\mathscr{C}([t_0,\varepsilon];\mathbb{R}^{m_d \times n_z}) \times \mathscr{C}([t_0,\varepsilon];\mathbb{R}^{m_f \times n_z})$ such that for any $K_d, \tilde{K}_d \in \mathscr{C}([t_0,\varepsilon];\mathbb{R}^{m_d \times n_z})$ and $K_f, \tilde{K}_f \in \mathscr{C}([t_0,\varepsilon];\mathbb{R}^{m_f \times n_z})$ and $K_d(\tau) \approx \tilde{K}_d(\tau)$ and $K_f(\tau) \approx \tilde{K}_f(\tau)$ for $t_0 \leq \epsilon \leq \tau \leq \varepsilon \leq t_f$ imply $K_u[K_d, K_f](\tau) \approx K_u[\tilde{K}_d, \tilde{K}_f](\tau)$ for $t_0 \leq \epsilon \leq \tau \leq \varepsilon \leq t_f\}$, whereby $K_d(\tau) \approx \tilde{K}_d(\tau)$ on $[\epsilon, \varepsilon]$ if the probability $P(K_d(\tau) = \tilde{K}_d(\tau)$ almost everywhere in $[\epsilon,\varepsilon]) = 1$. Similarly, non-anticipative feedback strategy gains for persistent disturbances and faults are defined as $K_d \in \mathscr{K}^d_{\varepsilon,\mathscr{Y},\mathscr{Z};\mu}$ and $K_f \in \mathscr{K}^f_{\varepsilon,\mathscr{Y},\mathscr{Z};\mu}$. In addition, $\mathscr{K}^u_{\varepsilon,\mathscr{Y},\mathscr{Z};\mu}$, $\mathscr{K}^d_{\varepsilon,\mathscr{Y},\mathscr{Z};\mu}$ and $\mathscr{K}^f_{\varepsilon,\mathscr{Y},\mathscr{Z};\mu}$ assume the respective values in $\overline{K}_u \in \mathbb{R}^{m_u \times n_z}$, $\overline{K}_d \in \mathbb{R}^{m_d \times n_z}$ and $\overline{K}_f \in \mathbb{R}^{m_f \times n_z}$ for which the solutions to Eqs. (4.39)–(4.40) with the terminal-value conditions $\mathscr{H}(\varepsilon) = \mathscr{Y}$ and $\mathscr{D}(\varepsilon) = \mathscr{Z}$ exist on $[t_0, \varepsilon]$.

On $\mathscr{K}^u_{t_f,\mathscr{H}_f,\mathscr{D}_f;\mu}$, $\mathscr{K}^d_{t_f,\mathscr{H}_f,\mathscr{D}_f;\mu}$ and $\mathscr{K}^f_{t_f,\mathscr{H}_f,\mathscr{D}_f;\mu}$ the aspects of asymmetry or skewness of the probabilistic performance distributions are reviewed more closely as the focal concern of the performance index with mean-risk awareness in the disturbance attenuation problem with unknown time-delays is defined next.

Definition 4.3.2 (Mean-Risk Aware Performance Index). Fix $k \in \mathbb{N}$ and the sequence of scalar coefficients $\mu = \{\mu_r \geq 0\}_{r=1}^k$ with $\mu_1 > 0$. Then for the given m_0, the mean-risk aware performance index $\phi_0 : \{t_0\} \times (\mathbb{R}^{2n_z \times 2n_z})^k \times \mathbb{R}^k \mapsto \mathbb{R}^+$ pertaining to risk-averse decision problem on $[t_0, t_f]$ is

$$\phi_0(t_0, \mathscr{H}(t_0), \mathscr{D}(t_0)) \triangleq \underbrace{\mu_1 \kappa_1}_{\text{Mean Measure}} + \underbrace{\mu_2 \kappa_2 + \cdots + \mu_k \kappa_k}_{\text{Risk Measures}}$$

$$= \sum_{r=1}^{k} \mu_r \left[m_0^T \mathscr{H}_r(t_0) m_0 + \mathscr{D}_r(t_0) \right] \tag{4.41}$$

where additional design of freedom by means of μ_r's are sufficient to meet different levels of performance reliability; for instance, mean (i.e., the average of performance measure), variance (i.e., the dispersion of values of performance measure around its mean), skewness (i.e., the anti-symmetry of the probability density of performance measure), kurtosis (i.e., the heaviness in the probability density tails of performance measure), etc.; while the component solutions $\{\mathscr{H}_r(\tau)\}_{r=1}^k$ and $\{\mathscr{D}_r(\tau)\}_{r=1}^k$ evaluated at $\tau = t_0$ satisfy the dynamical equations (4.39)–(4.40).

When necessary to indicate the dependence of the mean-risk aware performance index (4.41) expressed in Mayer form on $K_u \in \mathscr{K}^u_{t_f,\mathscr{H}_f,\mathscr{D}_f;\mu}$, $K_d \in \mathscr{K}^d_{t_f,\mathscr{H}_f,\mathscr{D}_f;\mu}$ and $K_f \in \mathscr{K}^f_{t_f,\mathscr{H}_f,\mathscr{D}_f;\mu}$, it is then rewritten explicitly as $\phi_0(K_u, K_d, K_f)$. Notice that the research investigation herein is concerned with the application of multi-person zero-sum differential game wherein player 1 (u or K_u) is trying to minimize (4.41) and players 2 and 3 (d and f or K_d and K_f) are attempting to maximize (4.41). If the game admits a saddle-point equilibrium, there exist best responses K_u^*, K_d^* and K_f^* which are satisfying the following inequalities

$$\phi_0(K_u^*, K_d, K_f) \leq \phi_0(K_u^*, K_d^*, K_f^*) \leq \phi_0(K_u, K_d^*, K_f^*)$$

for all non-anticipative feedback strategy gains $K_u \in \mathscr{K}^u_{t_f,\mathscr{H}_f,\mathscr{D}_f;\mu}$, $K_d \in \mathscr{K}^d_{t_f,\mathscr{H}_f,\mathscr{D}_f;\mu}$ and $K_f \in \mathscr{K}^f_{t_f,\mathscr{H}_f,\mathscr{D}_f;\mu}$.

Theorem 4.3.1 (Existence of a Saddle-Point Equilibrium). *Assume $\mathscr{K}^u_{t_f,\mathscr{H}_f,\mathscr{D}_f;\mu}$, $\mathscr{K}^d_{t_f,\mathscr{H}_f,\mathscr{D}_f;\mu}$ and $\mathscr{K}^f_{t_f,\mathscr{H}_f,\mathscr{D}_f;\mu}$ are nonempty, compact and convex. And the continuous performance index (4.41) is strictly convex in K_u and strictly concave in K_d and K_f. Then, the zero-sum differential game associated with the disturbance attenuation problem with unknown input time-delays admits a saddle-point equilibrium in pure strategies.*

Proof. The sets of non-anticipative feedback strategies $\mathscr{K}^u_{t_f,\mathscr{H}_f,\mathscr{D}_f;\mu}$, $\mathscr{K}^d_{t_f,\mathscr{H}_f,\mathscr{D}_f;\mu}$ and $\mathscr{K}^f_{t_f,\mathscr{H}_f,\mathscr{D}_f;\mu}$ are nonempty, compact and convex. What remains is to show the continuous function $\phi_0(K_u, K_d, K_f)$ is strictly convex in K_u and strictly concave in K_d and K_f. Such a case is illustrated by aggregating Eqs. (4.32)–(4.33)

$$\frac{d}{d\tau}\Lambda(\tau) = -F_m^T(\tau)\Lambda(\tau) - \Lambda(\tau)F_m(\tau) - \mu_1 N(\tau)$$
$$-\sum_{r=2}^{k}\mu_r\sum_{s=1}^{r-1}\frac{2r!}{s!(r-s)!}\mathscr{H}_s(\tau)G_m(\tau)W_m G_m^T(\tau)\mathscr{H}_{r-s}(\tau), \quad \forall\,\tau \in [t_0, t_f] \tag{4.42}$$

where $\Lambda(\tau) \triangleq \sum_{r=1}^{k}\mu_r\mathscr{H}_r(\tau)$ and $\Lambda(t_f) = \mu_1 N_f$. The fundamental theorem of calculus and stochastic differential rule applied to $m^T(\tau)\Lambda(\tau)m(\tau)$ yield the result

$$E\left\{m^T(t_f)\mu_1 N_f m(t_f)\right\} - m_0^T\Lambda(t_0)m_0 = E\left\{\int_{t_0}^{t_f} d\left[m^T(\tau)\Lambda(\tau)m(\tau)\right]\right\}$$
$$= E\left\{\int_{t_0}^{t_f}[dm^T(\tau)\Lambda(\tau)m(\tau) + m^T(\tau)\Lambda(\tau)dm(\tau)\right.$$
$$\left. m^T(\tau)\frac{d}{d\tau}\Lambda(\tau)m(\tau)d\tau + dm^T(\tau)\Lambda(\tau)dm(\tau)]\right\}.$$

After some manipulations, it follows that

$$E\left\{m^T(t_f)\mu_1 N_f m(t_f)\right\} - m_0^T \Lambda(t_0) m_0 = \int_{t_0}^{t_f} \mathrm{Tr}\left\{\Lambda(\tau) G_m(\tau) W_m G_m^T(\tau)\right\} d\tau$$

$$+ E\left\{\int_{t_0}^{t_f} m^T(\tau)[F_m(\tau)^T \Lambda(\tau) + \Lambda(\tau) F_m(\tau) + \frac{d}{d\tau}\Lambda(\tau)] m(\tau) d\tau\right\} \quad (4.43)$$

Notice that the solution of (4.34) is written by an integral form

$$\mathscr{D}_r(t_0) = \int_{t_0}^{t_f} \mathrm{Tr}\left\{\mathscr{H}_r(\tau) G_m(\tau) W_m G_m^T(\tau)\right\} d\tau, \quad 1 \le r \le k$$

In view of the definition of $\Lambda(\cdot)$, it is then easy to see that

$$\sum_{r=1}^{k} \mu_r \mathscr{D}_r(t_0) = \int_{t_0}^{t_f} \mathrm{Tr}\left\{\Lambda(\tau) G_m(\tau) W_m G_m^T(\tau)\right\} d\tau.$$

As a contribution to this development, the performance index (4.41) receives further consideration and is now rewritten as follows

$$\phi_0(K_u, K_d, K_f) = m_0^T \Lambda(t_0) m_0 + \int_{t_0}^{t_f} \mathrm{Tr}\left\{\Lambda(\tau) G_m(\tau) W_m G_m^T(\tau)\right\} d\tau. \quad (4.44)$$

Following the adoption of the results (4.42) and (4.44) oriented towards the mean-risk aware performance index (4.43), it is evident that

$$\phi_0(K_u, K_d, K_f) = E\left\{m^T(t_f)\mu_1 N_f m(t_f)\right\} + E\left\{\int_{t_0}^{t_f} m^T(\tau)[\mu_1 N(\tau) + \sum_{r=2}^{k} \mu_r \sum_{s=1}^{r-1} \frac{2r!}{s!(r-s)!} \mathscr{H}_s(\tau) G_m(\tau) W_m G_m^T(\tau) \mathscr{H}_{r-s}(\tau)] m(\tau) d\tau\right\} \quad (4.45)$$

which further leads to

$$\phi_0(K_u, K_d, K_f) = \mathrm{Tr}\left\{\mu_1 N_f P(t_f)\right\} + \mathrm{Tr}\left\{\int_{t_0}^{t_f} [\mu_1 N(\tau) + \sum_{r=2}^{k} \mu_r \sum_{s=1}^{r-1} \frac{2r!}{s!(r-s)!} \mathscr{H}_s(\tau) G_m(\tau) W_m G_m^T(\tau) \mathscr{H}_{r-s}(\tau)] P(\tau) d\tau\right\} \quad (4.46)$$

where the positive-definite $P(\cdot) \triangleq E\{m(\cdot)m^T(\cdot)\}$ is satisfying the forward-in-time matrix-valued differential equation

$$\frac{d}{d\tau}P(\tau) = P(\tau)F_m^T(\tau) + F_m(\tau)P(\tau) + G_m(\tau)W_m G_m^T(\tau), \quad P(t_0) = m_0 m_0^T. \tag{4.47}$$

Notice that within the integrand of (4.46), $N(\cdot)$ is strictly convex in K_u and strictly concave in K_d and K_f while other factors are positive semi-definite. Henceforth, the performance index (4.46), or equivalently (4.41) is strictly convex in K_u and strictly concave in K_d and K_f. Subsequently, the multi-person zero-sum differential game considered here admits a saddle-point equilibrium in accordance with Proposition 3.3 from [8].

As a tenet of transition from the principle of optimality, a family of games based on different starting points is now of concerned. With an intermission of time, ε in mid-play considered here, the path has reached some definitive point at its commencement. There exist some potential trajectories $(\mathscr{Y}, \mathscr{Z})$ which may be reached at the end of the intermission for all possible choices of (K_u, K_d, K_f). Hence, the concept of *playable set* and *value function* is defined as follows.

Definition 4.3.3 (Playable Set). Let $\mathscr{Q} \triangleq \Big\{(\varepsilon, \mathscr{Y}, \mathscr{Z}) \in [t_0, t_f] \times (\mathbb{R}^{2n_z \times 2n_z})^k \times \mathbb{R}^k$ such that $\mathscr{K}^u_{\varepsilon,\mathscr{Y},\mathscr{Z};\mu} \times \mathscr{K}^d_{\varepsilon,\mathscr{Y},\mathscr{Z};\mu} \times \mathscr{K}^f_{\varepsilon,\mathscr{Y},\mathscr{Z};\mu} \neq \emptyset\Big\}$.

Definition 4.3.4 (Value Function). Suppose that $(\varepsilon, \mathscr{Y}, \mathscr{Z}) \in [t_0, t_f] \times (\mathbb{R}^{n_m \times n_m})^k \times \mathbb{R}^k$ is given and fixed. A saddle-point equilibrium is obtained by solving two optimal decision problems for the lower and upper values of the game

$$\underline{\mathscr{V}}(\varepsilon, \mathscr{Y}, \mathscr{Z}) \triangleq \inf_{K_u \in \mathscr{K}^u_{\varepsilon,\mathscr{Y},\mathscr{Z};\mu}} \sup_{K_d(\cdot) \in \overline{K}_d} \sup_{K_f(\cdot) \in \overline{K}_f}$$

$$\phi_0(t_0, \mathscr{H}(t_0, K_u[K_d, K_f], K_d, K_f), \mathscr{D}(t_0, K_u[K_d, K_f], K_d, K_f))$$

$$\overline{\mathscr{V}}(\varepsilon, \mathscr{Y}, \mathscr{Z}) \triangleq \sup_{K_d \in \mathscr{K}^d_{\varepsilon,\mathscr{Y},\mathscr{Z};\mu}} \sup_{K_f \in \mathscr{K}^f_{\varepsilon,\mathscr{Y},\mathscr{Z};\mu}} \inf_{K_u(\cdot) \in \overline{K}_u}$$

$$\phi_0(t_0, \mathscr{H}(t_0, K_u, K_d[K_u], K_f[K_u]), \mathscr{D}(t_0, K_u, K_d[K_u], K_f[K_u])).$$

Then, a saddle-point equilibrium exists when the following Issacs condition holds

$$\mathscr{V}(\varepsilon, \mathscr{Y}, \mathscr{Z}) \triangleq \overline{\mathscr{V}}(\varepsilon, \mathscr{Y}, \mathscr{Z}) = \underline{\mathscr{V}}(\varepsilon, \mathscr{Y}, \mathscr{Z})$$

where $\mathscr{V}(\varepsilon, \mathscr{Y}, \mathscr{Z})$ is also called the value of the game. In general, $\underline{\mathscr{V}}(\varepsilon, \mathscr{Y}, \mathscr{Z}) \leq \overline{\mathscr{V}}(\varepsilon, \mathscr{Y}, \mathscr{Z})$.

For each endpoint in $\mathcal{Q}$, the game beginning there has already been solved. Then, the lower and upper values $\underline{\mathcal{V}}(\varepsilon, \mathcal{Y}, \mathcal{Z})$ and $\overline{\mathcal{V}}(\varepsilon, \mathcal{Y}, \mathcal{Z})$ resulted from each choice of (K_u, K_d, K_f) are focused without the assumption of the Issacs condition.

Theorem 4.3.2 (Hamilton-Jacobi-Issacs (HJI) Equation for Mayer Problem). *Let $(\varepsilon, \mathcal{Y}, \mathcal{Z})$ be any interior point of the playable set $\mathcal{Q}$ at which the lower and upper value functions $\underline{\mathcal{V}}(\varepsilon, \mathcal{Y}, \mathcal{Z})$ and $\overline{\mathcal{V}}(\varepsilon, \mathcal{Y}, \mathcal{Z})$ are differentiable. If there exists a saddle point $(K_u^*, K_d^*, K_f^*) \in \mathcal{K}^u_{\varepsilon,\mathcal{Y},\mathcal{Z};\mu} \times \mathcal{K}^d_{\varepsilon,\mathcal{Y},\mathcal{Z};\mu} \times \mathcal{K}^f_{\varepsilon,\mathcal{Y},\mathcal{Z};\mu}$, then the partial differential equations of the zero-sum game*

$$0 = \min_{K_u \in \overline{K}_u} \max_{K_d \in \overline{K}_d} \max_{K_f \in \overline{K}_f} \Big\{ \frac{\partial}{\partial \varepsilon} \underline{\mathcal{V}}(\varepsilon, \mathcal{Y}, \mathcal{Z}) + \frac{\partial}{\partial\, \mathrm{vec}(\mathcal{Z})} \underline{\mathcal{V}}(\varepsilon, \mathcal{Y}, \mathcal{Z}) \cdot \mathrm{vec}(\mathcal{G}(\varepsilon, \mathcal{Y})) + \frac{\partial}{\partial\, \mathrm{vec}(\mathcal{Y})} \underline{\mathcal{V}}(\varepsilon, \mathcal{Y}, \mathcal{Z}) \cdot \mathrm{vec}(\mathcal{F}(\varepsilon, \mathcal{Y}, K_u, K_d, K_f)) \Big\} \tag{4.48}$$

and

$$0 = \max_{K_d \in \overline{K}_d} \max_{K_f \in \overline{K}_f} \min_{K_u \in \overline{K}_u} \Big\{ \frac{\partial}{\partial \varepsilon} \overline{\mathcal{V}}(\varepsilon, \mathcal{Y}, \mathcal{Z}) + \frac{\partial}{\partial\, \mathrm{vec}(\mathcal{Z})} \overline{\mathcal{V}}(\varepsilon, \mathcal{Y}, \mathcal{Z}) \cdot \mathrm{vec}(\mathcal{G}(\varepsilon, \mathcal{Y})) + \frac{\partial}{\partial\, \mathrm{vec}(\mathcal{Y})} \overline{\mathcal{V}}(\varepsilon, \mathcal{Y}, \mathcal{Z}) \cdot \mathrm{vec}(\mathcal{F}(\varepsilon, \mathcal{Y}, K_u, K_d, K_f)) \Big\} \tag{4.49}$$

are satisfied together with the boundary conditions

$$\underline{\mathcal{V}}(t_0, \mathcal{H}(t_0), \mathcal{D}(t_0)) = \overline{\mathcal{V}}(t_0, \mathcal{H}(t_0), \mathcal{D}(t_0)) \equiv \phi_0(t_0, \mathcal{H}(t_0), \mathcal{D}(t_0))$$

and whereas vec$(\cdot)$ is the vectorizing operator of enclosed entities.

Proof. The proof can be shown by extending Theorem 4.1 on page 159 in [9].

4.4 Saddle-Point Strategies with Risk Aversion

The existence of an explicit, closed-form solution for the problem of disturbance attenuation with unknown input time-delays is now investigated. The terminal time and states $(t_f, \mathcal{H}_f, \mathcal{D}_f)$ are subsequently parameterized as $(\varepsilon, \mathcal{Y}, \mathcal{Z})$ for a family of optimization problems. For instance, the states (4.39)–(4.40) defined on the interval $[t_0, \varepsilon]$ now have terminal values denoted by $\mathcal{H}(\varepsilon) \equiv \mathcal{Y}$ and $\mathcal{D}(\varepsilon) \equiv \mathcal{Z}$, where $\varepsilon \in [t_0, t_f]$. Furthermore, with $k \in \mathbb{Z}^+$ and $(\varepsilon, \mathcal{Y}, \mathcal{Z})$ in $\mathcal{Q}$, the mathematical convenience of (4.41) provides the conceptual necessity concerning a real-valued function candidate, e.g.,

$$\mathcal{W}(\varepsilon, \mathcal{Y}, \mathcal{Z}) = m_0^T \sum_{r=1}^{k} \mu_r (\mathcal{Y}_r + \mathcal{E}_r(\varepsilon)) m_0 + \sum_{r=1}^{k} \mu_r (\mathcal{Z}_r + \mathcal{T}_r(\varepsilon)) \tag{4.50}$$

for the (lower/upper) value function. This, in particular, includes the time derivative of $\mathscr{W}(\varepsilon, \mathscr{Y}, \mathscr{Z})$ which can also be shown of the form

$$\frac{d}{d\varepsilon}\mathscr{W}(\varepsilon, \mathscr{Y}, \mathscr{Z}) = m_0^T \sum_{r=1}^{k} \mu_r (\mathscr{F}_r(\varepsilon, \mathscr{Y}, K_u, K_d, K_f) + \frac{d}{d\varepsilon}\mathscr{E}_r(\varepsilon)) m_0$$
$$+ \sum_{r=1}^{k} \mu_r (\mathscr{G}_r(\varepsilon, \mathscr{Y}) + \frac{d}{d\varepsilon}\mathscr{T}_r(\varepsilon))$$

where the parametric functions $\mathscr{E}_r \in \mathscr{C}^1([t_0, t_f]; \mathbb{R}^{2n_z \times 2n_z})$ and $\mathscr{T}_r \in \mathscr{C}^1([t_0, t_f]; \mathbb{R})$ are yet to be determined.

In order to realize the principle of learning by doing, the boundary condition $\mathscr{W}(t_0, \mathscr{H}(t_0), \mathscr{D}(t_0)) = \phi_0(t_0, \mathscr{H}(t_0), \mathscr{D}(t_0))$ results in the following result

$$m_0^T \sum_{r=1}^{k} \mu_r (\mathscr{H}_r(t_0) + \mathscr{E}_r(t_0)) m_0 + \sum_{r=1}^{k} \mu_r (\mathscr{D}_r(t_0) + \mathscr{T}_r(t_0))$$
$$= m_0^T \sum_{r=1}^{k} \mu_r \mathscr{H}_r(t_0) m_0 + \sum_{r=1}^{k} \mu_r \mathscr{D}_r(t_0). \tag{4.51}$$

By matching the boundary condition (4.51), it yields that the time parameter functions $\mathscr{E}_r(t_0) = 0$ and $\mathscr{T}_r(t_0) = 0$ for $1 \leq r \leq k$. Next it is necessary to verify that this candidate value function satisfies (4.48) along the corresponding trajectories produced by the feedback strategy gains K_u, K_d and K_f

$$0 = \min_{K_u \in \overline{K}_u} \max_{K_d \in \overline{K}_d} \max_{K_f \in \overline{K}_f} \Big\{ m_0^T \sum_{r=1}^{k} \mu_r \frac{d}{d\varepsilon}\mathscr{E}_r(\varepsilon)\, m_0$$
$$+ m_0^T \sum_{r=1}^{k} \mu_r \mathscr{F}_r(\varepsilon, \mathscr{Y}, K_u, K_d, K_f)\, m_0 + \sum_{r=1}^{k} \mu_r \mathscr{G}_r(\varepsilon, \mathscr{Y}) + \sum_{r=1}^{k} \mu_r \frac{d}{d\varepsilon}\mathscr{T}_r(\varepsilon) \Big\}. \tag{4.52}$$

Now the aggregate matrix coefficients $F_m(t)$ and $N(t)$ for $t \in [t_0, t_f]$ of the disturbance attenuation problem (4.13)–(4.14) with unknown input time-delays are next partitioned to conform with the n_z-dimensional structure of (4.5) as follows: $I_0^T \triangleq \begin{bmatrix} I & 0 \end{bmatrix}$ and $I_1^T \triangleq \begin{bmatrix} 0 & I \end{bmatrix}$, whereby I is an $n_z \times n_z$ identity matrix and

$$F_m = I_0(A_z + B_z K_u + C_z K_d + D_z K_f) I_0^T + I_0 L_z H_z I_1^T + I_1 (A_z - L_z H_z) I_1^T \tag{4.53}$$

$$N = I_0 (K_u^T R_u K_u + K_d^T (\delta S_d)^{-1} K_d + K_f^T (\delta S_f)^{-1} K_f + Q_z) I_0^T$$
$$+ I_0 Q_z I_1^T + I_1 Q_z I_0^T + I_1 Q_z I_1^T. \tag{4.54}$$

Subsequently, the derivatives of the expression in (4.52) with respect to the admissible feedback strategy gains K_u, K_d and K_f yield the necessary conditions for a minimax extremum of (4.52) on $[t_0, \varepsilon]$

$$K_u = -R_u^{-1}(\varepsilon)B_z^T(\varepsilon)I_0^T \sum_{s=1}^{k} \hat{\mu}_s \mathscr{Y}_s I_0((I_0^T I_0)^{-1})^T \tag{4.55}$$

$$K_d = -\delta S_d(\varepsilon)C_z^T(\varepsilon)I_0^T \sum_{s=1}^{k} \hat{\mu}_s \mathscr{Y}_s I_0((I_0^T I_0)^{-1})^T \tag{4.56}$$

$$K_f = -\delta S_f(\varepsilon)D_z^T(\varepsilon)I_0^T \sum_{s=1}^{k} \hat{\mu}_s \mathscr{Y}_s I_0((I_0^T I_0)^{-1})^T, \quad \hat{\mu}_s = \frac{\mu_s}{\mu_1}. \tag{4.57}$$

With the feedback strategy gains (4.55)–(4.57) replaced in the expression of the bracket (4.52) and having $\{\mathscr{Y}_s\}_{s=1}^k$ evaluated on the solution trajectories (4.39)–(4.40), the time dependent functions $\mathscr{E}_r(\varepsilon)$ and $\mathscr{T}_r(\varepsilon)$ are therefore chosen such that the HJI equation (4.48) is satisfied in the presence of the arbitrary value of m_0

$$\frac{d}{d\varepsilon}\mathscr{E}_1(\varepsilon) = F_m^T(\varepsilon)\mathscr{H}_1(\varepsilon) + \mathscr{H}_1(\varepsilon)F_m(\varepsilon) + N(\varepsilon)$$

$$\frac{d}{d\varepsilon}\mathscr{E}_r(\varepsilon) = F_m^T(\varepsilon)\mathscr{H}_r(\varepsilon) + \mathscr{H}_r(\varepsilon)F_m(\varepsilon)$$
$$+ \sum_{s=1}^{r-1} \frac{2r!}{s!(r-s)!}\mathscr{H}_s(\varepsilon)G_m(\varepsilon)W_m G_a^T(\varepsilon)\mathscr{H}_{r-s}(\varepsilon), \quad 2 \le r \le k$$

$$\frac{d}{d\varepsilon}\mathscr{T}_r(\varepsilon) = \mathrm{Tr}\left\{\mathscr{H}_r(\varepsilon)G_m(\varepsilon)W_m G_m^T(\varepsilon)\right\}, \quad 1 \le r \le k$$

with the initial-value conditions $\mathscr{E}_r(t_0) = 0$ and $\mathscr{T}_r(t_0) = 0$ for $1 \le r \le k$. Therefore, the HJI equation (4.48) is satisfied so that the feedback strategy gains (4.55)–(4.57) become optimal.

At present, the problem of disturbance attenuation with delayed feedback measurements is thus solved by a saddle-point solution with dynamical output-feedback. Such an understanding is in dispensable if a brief summary together with the system realization is in order for future practitioners of resilient controls towards the control problem class here and illustrated in Fig. 4.5.

Theorem 4.4.1 (Risk-Averse and Output Feedback Saddle-Point Equilibrium). *Let $(A, [B_u, B_d, B_f])$ and (A, C) be uniformly stablizable and detectable. Consider saddle-point strategies comprised by the controller $u^*(t) = K_u^*(t)\hat{z}^*(t)$ and persistent disturbances $d^*(t) = K_d^*(t)\hat{z}^*(t)$ and faults $f^*(t) = K_f^*(t)\hat{z}^*(t)$ where the time change of variable $t \triangleq t_0 + t_f - \tau$ and $\hat{z}^*(t)$ are the optimal state estimates of the augmented system model that captures the disturbance attenuation phenomenon, the 1st-order Pade approximation of input time-delays $e^{-s\sigma}$ and estimation for unknown input time-delays. Further let $k \in \mathbb{N}$ and the sequence* *of nonnegative coefficients* $\mu = \{\mu_r \geq 0\}_{r=1}^k$ *with* $\mu_1 > 0$. *Then, there exists a saddle-point equilibrium which strives to best respond to the risk-value awareness performance index (4.41)*

$$K_u^*(\tau) = -R_u^{-1}(\tau)B_z^T(\tau)I_0^T \sum_{r=1}^k \hat{\mu}_r \mathcal{H}_r^*(\tau)I_0((I_0^T I_0)^{-1})^T$$

$$K_d^*(\tau) = -\delta S_d(\tau)C_z^T(\tau)I_0^T \sum_{r=1}^k \hat{\mu}_r \mathcal{H}_r^*(\tau)I_0((I_0^T I_0)^{-1})^T$$

$$K_f^*(\tau) = -\delta S_f(\tau)D_z^T(\tau)I_0^T \sum_{r=1}^k \hat{\mu}_r \mathcal{H}_r^*(\tau)I_0((I_0^T I_0)^{-1})^T$$

where the normalized parametric design of freedom $\hat{\mu}_r \triangleq \frac{\mu_r}{\mu_1}$ *and the optimal solutions* $\{\mathcal{H}_r^*(\cdot)\}_{r=1}^k$ *supporting the risk-averse decisions satisfy the backward-in-time matrix valued differential equations*

$$\frac{d}{d\tau}\mathcal{H}_1^*(\tau) = -F_m^{*T}(\tau)\mathcal{H}_1^*(\tau) - \mathcal{H}_1^*(\tau)F_m^*(\tau) - N^*(\tau), \quad \mathcal{H}_1^*(t_f) = N_f \tag{4.58}$$

$$\begin{aligned}\frac{d}{d\tau}\mathcal{H}_r^*(\tau) = &- F_m^{*T}(\tau)\mathcal{H}_r^*(\tau) - \mathcal{H}_r^*(\tau)F_m^*(\tau)\\ &- \sum_{s=1}^{r-1} \frac{2r!}{s!(r-s)!}\mathcal{H}_s^*(\tau)G_m(\tau)W_m G_m^T(\tau)\mathcal{H}_{r-s}^*(\tau), \quad \mathcal{H}_r^*(t_f) = 0\end{aligned} \tag{4.59}$$

In addition, the optimal state estimates $\hat{z}^*(t)$ *under the saddle-point equilibrium are satisfying the forward-in-time vector-valued differential equation*

$$\begin{aligned}d\hat{z}^*(t) &= (A_z(t,\sigma)\hat{z}^*(t) + B_z(t,\sigma)u^*(t) + C_z(t,\sigma)d^*(t) + D_z(t,\sigma)f^*(t))dt\\ &\quad + L_z(t)(dy_D^*(t) - (H_z(t,\sigma)\hat{z}^*(t) + I_z(t,\sigma)u^*(t)\\ &\quad + J_z(t,\sigma)d^*(t) + K_z(t,\sigma)f^*(t))dt), \quad \hat{z}^*(t_0) = z_0\end{aligned} \tag{4.60}$$

where the filter gain $L_z(t) = \Sigma_z(t)H_z^T(t,\sigma)V^{-1}$ *and the error covariance* $\Sigma_z(t)$ *is obtained by the forward-in-time matrix-valued differential equation*

$$\begin{aligned}\frac{d}{dt}\Sigma_z(t) &= A_z(t,\sigma)\Sigma_z(t) + \Sigma_z(t)A_z^T(t,\sigma) + G_z(t,\sigma)W_z G_z^T(t,\sigma)\\ &\quad - \Sigma_z(t)H_z^T(t,\sigma)V^{-1}H_z(t,\sigma)\Sigma_z(t), \quad \Sigma_z(t_0) = 0.\end{aligned} \tag{4.61}$$

Even the aforementioned successes, however, the prospects for integrating the coupling backward-in-time matrix-valued differential equations (4.58)–(4.59) are

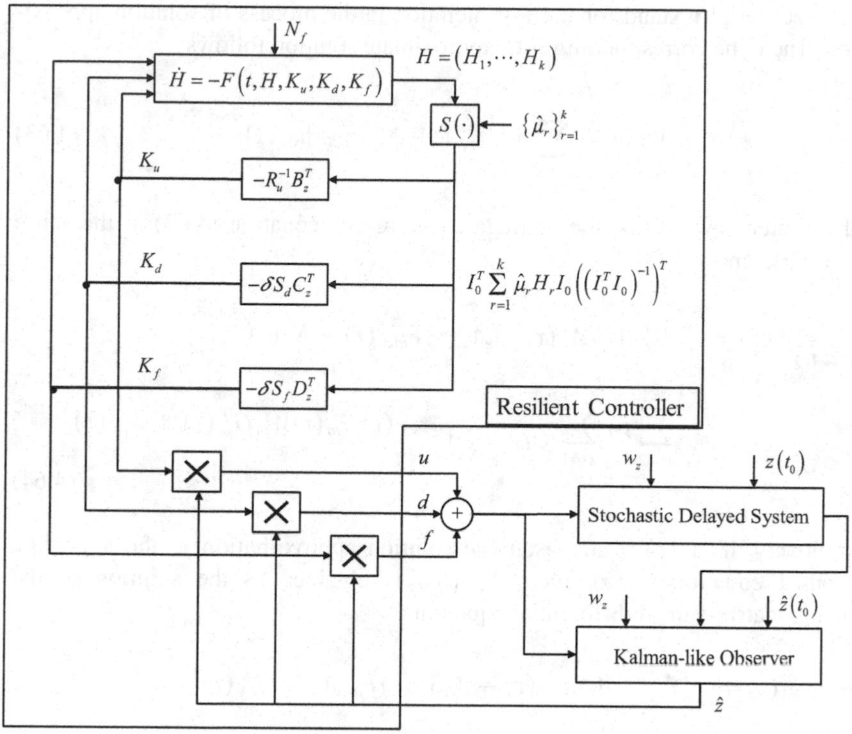

Fig. 4.5 Disturbance attenuation system with delayed feedback measurements integrated with resilient control towards cognitive disturbances

challenging. For one thing, the differential equation (4.58) is first solved and then its solution is used to find the next solutions in the higher-order differential equations (4.59) for $2 \leq r \leq k$. For another, in solving the risk-averse and output feedback saddle-point equilibrium the matrix-valued differential equations (4.58)–(4.59) must be simultaneously integrated over a finite time interval.

Despite these obstacles, certain favorable conditions exist with respect to the prospect for a solution approximation: the established reasonableness of successive substitutions and the iterative algorithm set by monotonic convergence. Given this background, then, it is necessary to add the i-th differential equation (4.59) multiplied by the associated $\hat{\mu}_r = \mu_r/\mu_1$ to Eq. (4.58) with i starting from 2

$$\frac{d}{d\tau}\sum_{r=1}^{k}\hat{\mu}_r\mathcal{H}_r^*(\tau) = -F_m^{*T}(\tau)\sum_{r=1}^{k}\hat{\mu}_r\mathcal{H}_r^*(\tau) - \sum_{r=1}^{k}\hat{\mu}_r\mathcal{H}_r^*(\tau)F_m^*(\tau)$$
$$- N^*(\tau) - \sum_{r=2}^{k}\hat{\mu}_r\sum_{s=1}^{r-1}\frac{2r!}{s!(r-s)!}\mathcal{H}_s^*(\tau)G_m(\tau)W_mG_m^T(\tau)\mathcal{H}_{r-s}^*(\tau) \tag{4.62}$$

with the terminal-value condition $\sum_{r=1}^{k}\hat{\mu}_r\mathcal{H}_r^*(t_f) = N_f$.

Let a subscript ν stand for the ν-th iteration in the process of solution approximation. Then, the corresponding ν-th approximate solution follows

$$\Lambda_\nu(\tau) \triangleq \sum_{r=1}^{k} \hat{\mu}_r \mathscr{H}_{r,\nu}(\tau), \quad \forall \tau \in [t_0, t_f]. \tag{4.63}$$

As the matter now stands, the aggregate differential equation (4.62) in the sense of (4.63) becomes

$$\begin{aligned}\frac{d}{d\tau}\Lambda_\nu(\tau) = &- F_{m,\nu}^T(\tau)\Lambda_\nu(\tau) - \Lambda_\nu(\tau)F_{m,\nu}(\tau) - N_\nu(\tau) \\ &- \sum_{r=2}^{k} \hat{\mu}_r \sum_{s=1}^{r-1} \frac{2r!}{s!(r-s)!} \mathscr{H}_{s,\nu}(\tau)G_m(\tau)W_m G_m^T(\tau)\mathscr{H}_{r-s,\nu}(\tau).\end{aligned} \tag{4.64}$$

Intuitively, if $\Lambda_\nu(\tau)$ is a reasonable solution approximation to the aggregate differential equation (4.65), then $\Lambda_{\nu+1}(\tau)$ is obtained as the solution of the following matrix-valued differential equation

$$\begin{aligned}\frac{d}{d\tau}\Lambda_{\nu+1}(\tau) = &- F_{m,\nu}^T(\tau)\Lambda_{\nu+1}(\tau) - \Lambda_{\nu+1}(\tau)F_{m,\nu}(\tau) - N_\nu(\tau) \\ &- \sum_{r=2}^{k} \hat{\mu}_r \sum_{s=1}^{r-1} \frac{2r!}{s!(r-s)!} \mathscr{H}_{s,\nu+1}(\tau)G_m(\tau)W_m G_m^T(\tau)\mathscr{H}_{r-s,\nu+1}(\tau)\end{aligned}$$

will be even a better approximation. More precisely, we state the following theorem.

Theorem 4.4.2 (Iterate Solutions for Saddle-point Equilibrium). *Let $k \in \mathbb{Z}^+$, $\nu \in \mathbb{N}$, and $\mu = \{\mu_i \geq 0\}_{i=1}^k$ with $\mu_1 > 0$. Assume that $(A, [B_u, C_d, D_f])$ and (A, C) are uniformly stabilizable and detectable, then the stabilizing feedback gains at ν-th iteration are denoted by*

$$K_{u,\nu}(\tau) \triangleq -R_u^{-1}(\tau)B_z^T(\tau)I_0^T \sum_{r=1}^{k} \hat{\mu}_r \mathscr{H}_{r,\nu}(\tau)I_0((I_0^T I_0)^{-1})^T \tag{4.65}$$

$$K_{d,\nu}(\tau) \triangleq -\delta S_d(\tau)C_z^T(\tau)I_0^T \sum_{r=1}^{k} \hat{\mu}_r \mathscr{H}_{r,\nu}(\tau)I_0((I_0^T I_0)^{-1})^T \tag{4.66}$$

$$K_{f,\nu}(\tau) \triangleq -\delta S_f(\tau)D_z^T(\tau)I_0^T \sum_{r=1}^{k} \hat{\mu}_r \mathscr{H}_{r,\nu}(\tau)I_0((I_0^T I_0)^{-1})^T, \quad \forall \tau \in [t_0, t_f]. \tag{4.67}$$

Further let

$$F_{m,\nu} \triangleq \begin{bmatrix} A_z + B_z K_{u,\nu} + C_z K_{d,\nu} + D_z K_{f,\nu} & L_z H_z \\ 0 & A_z - L_z H_z \end{bmatrix} \tag{4.68}$$

$$N_\nu \triangleq \begin{bmatrix} Q_z + K_{u,\nu}^T R_u K_{u,\nu} + K_{d,\nu}^T (\delta S_d)^{-1} K_{d,\nu} + K_{f,\nu}^T (\delta S_f)^{-1} K_{f,\nu} & Q_z \\ Q_z & Q_z \end{bmatrix} \tag{4.69}$$

$$M_{\nu+1}(\tau) \triangleq \sum_{r=2}^{k} \hat{\mu}_r \sum_{s=1}^{r-1} \frac{2r!}{s!(r-s)!} \mathcal{H}_{s,\nu+1}(\tau) G_m(\tau) W_m G_m^T(\tau) \mathcal{H}_{r-s,\nu+1}(\tau) \tag{4.70}$$

and $\Lambda_{\nu+1}(\tau)$ be the solution to the backward-in-time matrix-valued differential equation with the terminal-value condition $\Lambda_{\nu+1}(t_f) = N_f$

$$\frac{d}{d\tau}\Lambda_{\nu+1}(\tau) = -F_{m,\nu}^T(\tau)\Lambda_{\nu+1}(\tau) - \Lambda_{\nu+1}(\tau)F_{m,\nu}(\tau) - N_\nu(\tau) - M_{\nu+1}(\tau) \tag{4.71}$$

Whenever the inequality

$$M_{\nu+1}(\tau) \geq M_{\nu+2}(\tau) \tag{4.72}$$

holds. Then

(*a*) $\Lambda_{\nu+2}(\tau) \leq \Lambda_{\nu+1}(\tau)$,
(*b*) $\lim_{\nu\to\infty} \Lambda_\nu(\tau) = \underline{\Lambda}(\tau)$ *exists.*

Moreover, as $\nu \to \infty$ together with $\underline{F}_m(\tau) \triangleq \lim_{\nu\to\infty} F_{m,\nu}(\tau)$ and $\underline{N}(\tau) \triangleq \lim_{\nu\to\infty} N_\nu(\tau)$, the solutions $\underline{\mathcal{H}}_r(\tau)$ satisfy the limiting differential equations

$$\frac{d}{d\tau}\underline{\mathcal{H}}_1(\tau) = -\underline{F}_m^T(\tau)\underline{\mathcal{H}}_1(\tau) - \underline{\mathcal{H}}_1(\tau)\underline{F}_m(\tau) - \underline{N}(\tau), \tag{4.73}$$

$$\frac{d}{d\tau}\underline{\mathcal{H}}_r(\tau) = -\underline{F}_m^T(\tau)\underline{\mathcal{H}}_r(\tau) - \underline{\mathcal{H}}_r(\tau)\underline{F}_m(\tau) - \sum_{s=1}^{r-1} \frac{2r!}{s!(r-s)!} \underline{\mathcal{H}}_s(\tau) G_m(\tau) W_m G_m^T(\tau) \underline{\mathcal{H}}_{r-s}(\tau) \tag{4.74}$$

where the terminal-value conditions $\underline{\mathcal{H}}_1(t_f) = N_f$ and $\underline{\mathcal{H}}_r(t_f) = 0$ for $2 \leq r \leq k$.

Proof. In the background remains the first task of demonstrating the inequality of $\Lambda_{\nu+2}(\tau) \leq \Lambda_{\nu+1}(\tau)$ for all $\tau \in [t_0, t_f]$. Clearly, it is no difficult to see that

$$[K_{u,\nu}(\tau) - K_{u,\nu+1}(\tau)]^T R_u(\tau)[K_{u,\nu}(\tau) - K_{u,\nu+1}(\tau)] \geq 0$$
$$[K_{d,\nu}(\tau) - K_{d,\nu+1}(\tau)]^T (\delta S_d)^{-1}(\tau)[K_{d,\nu}(\tau) - K_{d,\nu+1}(\tau)] \geq 0$$
$$[K_{f,\nu}(\tau) - K_{f,\nu+1}(\tau)]^T (\delta S_f)^{-1}(\tau)[K_{f,\nu}(\tau) - K_{f,\nu+1}(\tau)] \geq 0$$

Differently expressed, these inequalities are rewritten as follows

$$\begin{aligned}
&K^T_{u,\nu}(\tau)R_u(\tau)K_{u,\nu}(\tau) \geq K^T_{u,\nu}(\tau)R_u(\tau)K_{u,\nu+1}(\tau) + K^T_{u,\nu+1}(\tau)R_u(\tau)K_{u,\nu}(\tau) \\
&\quad - K^T_{u,\nu+1}(\tau)R_u(\tau)K_{u,\nu+1}(\tau), \\
&K^T_{d,\nu}(\tau)(\delta S_d)^{-1}(\tau)K_{d,\nu}(\tau) \geq K^T_{d,\nu}(\tau)(\delta S_d)^{-1}(\tau)K_{d,\nu+1}(\tau) \\
&\quad + K^T_{d,\nu+1}(\tau)(\delta S_d)^{-1}(\tau)K_{d,\nu}(\tau) - K^T_{d,\nu+1}(\tau)(\delta S_d)^{-1}(\tau)K_{d,\nu+1}(\tau), \\
&K^T_{f,\nu}(\tau)(\delta S_f)^{-1}(\tau)K_{f,\nu}(\tau) \geq K^T_{f,\nu}(\tau)(\delta S_f)^{-1}(\tau)K_{f,\nu+1}(\tau) \\
&\quad + K^T_{f,\nu+1}(\tau)(\delta S_f)^{-1}(\tau)K_{f,\nu}(\tau) - K^T_{f,\nu+1}(\tau)(\delta S_f)^{-1}(\tau)K_{f,\nu+1}(\tau).
\end{aligned}$$

Furthermore, in such a circumstance, these inequalities can be rewritten as the corresponding equalities

$$\begin{aligned}
K^T_{u,\nu}(\tau)R_u(\tau)K_{u,\nu}(\tau) = &K^T_{u,\nu}(\tau)R_u(\tau)K_{u,\nu+1}(\tau) + K^T_{u,\nu+1}(\tau)R_u(\tau)K_{u,\nu}(\tau) \\
&- K^T_{u,\nu+1}(\tau)R_u(\tau)K_{u,\nu+1}(\tau) + R_{u,\nu}(\tau; K_{u,\nu}, K_{u,\nu+1})
\end{aligned} \tag{4.75}$$

$$\begin{aligned}
K^T_{d,\nu}(\tau)(\delta S_d)^{-1}(\tau)K_{d,\nu}(\tau) = &K^T_{d,\nu}(\tau)(\delta S_d)^{-1}(\tau)K_{d,\nu+1}(\tau) \\
&+ K^T_{d,\nu+1}(\tau)(\delta S_d)^{-1}(\tau)K_{d,\nu}(\tau) \\
&- K^T_{d,\nu+1}(\tau)(\delta S_d)^{-1}(\tau)K_{d,\nu+1}(\tau) \\
&+ R_{d,\nu}(\tau; K_{d,\nu}, K_{d,\nu+1})
\end{aligned} \tag{4.76}$$

$$\begin{aligned}
K^T_{f,\nu}(\tau)(\delta S_f)^{-1}(\tau)K_{f,\nu}(\tau) = &K^T_{f,\nu}(\tau)(\delta S_f)^{-1}(\tau)K_{f,\nu+1}(\tau) \\
&+ K^T_{f,\nu+1}(\tau)(\delta S_f)^{-1}(\tau)K_{f,\nu}(\tau) \\
&- K^T_{f,\nu+1}(\tau)(\delta S_f)^{-1}(\tau)K_{f,\nu+1}(\tau) \\
&+ R_{f,\nu}(\tau; K_{f,\nu}, K_{f,\nu+1}).
\end{aligned} \tag{4.77}$$

where the additional term $R_{u,\nu}(\tau; K_{u,\nu}, K_{u,\nu+1}) \geq 0$ is dependent on both $K_{u,\nu}(\tau)$ and $K_{u,\nu+1}(\tau)$. And, $R_{d,\nu}(\tau; K_{d,\nu}, K_{d,\nu+1})$ and $R_{f,\nu}(\tau; K_{f,\nu}, K_{f,\nu+1})$ are similarly defined.

Additionally, as the development continues, to the extent that the iterate differential equation (4.71) in response to the definition (4.69) and the equalities (4.75)–(4.77), one can show that

$$\frac{d}{d\tau}\Lambda_{\nu+1}(\tau) = -F_{m,\nu}^T(\tau)\Lambda_{\nu+1}(\tau) - \Lambda_{\nu+1}(\tau)F_{m,\nu}(\tau) - \Delta_{\nu,\nu+1}(\tau) - \Delta_{\nu+1,\nu}(\tau)$$
$$+ \Delta_{\nu+1,\nu+1}(\tau) - R_\nu(\tau) - O(\tau) - M_{\nu+1}(\tau). \quad (4.78)$$

where

$$\Delta_{\nu,\nu+1}(\tau) \triangleq \begin{bmatrix} d_{\nu,\nu+1}(\tau) & 0 \\ 0 & 0 \end{bmatrix}$$
$$d_{\nu,\nu+1}(\tau) \triangleq K_{u,\nu}^T(\tau)R_u(\tau)K_{u,\nu+1}(\tau) + K_{d,\nu}^T(\tau)(\delta S_d)^{-1}(\tau)K_{d,\nu+1}(\tau)$$
$$+ K_{f,\nu}^T(\tau)(\delta S_f)^{-1}(\tau)K_{f,\nu+1}(\tau)$$

and

$$\Delta_{\nu+1,\nu}(\tau) \triangleq \begin{bmatrix} d_{\nu+1,\nu}(\tau) & 0 \\ 0 & 0 \end{bmatrix}$$
$$d_{\nu+1,\nu}(\tau) \triangleq K_{u,\nu+1}^T(\tau)R_u(\tau)K_{u,\nu}(\tau) + K_{d,\nu+1}^T(\tau)(\delta S_d)^{-1}(\tau)K_{d,\nu}(\tau)$$
$$+ K_{f,\nu+1}^T(\tau)(\delta S_f)^{-1}(\tau)K_{f,\nu}(\tau)$$

and

$$\Delta_{\nu+1,\nu+1}(\tau) \triangleq \begin{bmatrix} d_{\nu+1,\nu+1}(\tau) & 0 \\ 0 & 0 \end{bmatrix}$$
$$d_{\nu+1,\nu+1}(\tau) \triangleq K_{u,\nu+1}^T(\tau)R_u(\tau)K_{u,\nu+1}(\tau) + K_{d,\nu+1}^T(\tau)(\delta S_d)^{-1}(\tau)K_{d,\nu+1}(\tau)$$
$$+ K_{f,\nu+1}^T(\tau)(\delta S_f)^{-1}(\tau)K_{f,\nu+1}(\tau)$$

and

$$R_\nu(\tau) \triangleq \begin{bmatrix} R_{u,\nu}(\tau; K_{u,\nu}, K_{u,\nu+1}) + R_{f,\nu}(\tau; K_{f,\nu}, K_{f,\nu+1}) + R_{f,\nu}(\tau; K_{f,\nu}, K_{f,\nu+1}) & 0 \\ 0 & 0 \end{bmatrix}$$

and, finally

$$O(\tau) \triangleq \begin{bmatrix} Q_z(\tau) & Q_z(\tau) \\ Q_z(\tau) & Q_z(\tau) \end{bmatrix}.$$

In view of such expressions (4.65)–(4.68), the iterate differential equation (4.78) can be further rearranged as follows

$$\frac{d}{d\tau}\Lambda_{\nu+1}(\tau) = - F_{,m\nu+1}^T(\tau)\Lambda_{\nu+1}(\tau) - \Lambda_{\nu+1}(\tau)F_{m,\nu+1}(\tau) - N_{\nu+1}(\tau)$$
$$- R_\nu(\tau) - M_{\nu+1}(\tau). \quad (4.79)$$

Moreover, the iterate differential equation (4.71) can be written in terms of the $(\nu+2)$-th iteration

$$\frac{d}{d\tau}\Lambda_{\nu+2}(\tau) = -F^T_{m,\nu+1}(\tau)\Lambda_{\nu+2}(\tau) - \Lambda_{\nu+2}(\tau)F_{m,\nu+1}(\tau) - N_{\nu+1}(\tau) - M_{\nu+2}(\tau). \quad (4.80)$$

Subtracting Eq. (4.80) from that of (4.79) and letting $\delta\Lambda_{\nu+1}(\tau) \triangleq \Lambda_{\nu+1}(\tau) - \Lambda_{\nu+2}(\tau)$, it follows that

$$\frac{d}{d\tau}\delta\Lambda_{\nu+1}(\tau) = -F^T_{m,\nu+1}(\tau)\delta\Lambda_{\nu+1}(\tau) - \delta\Lambda_{\nu+1}(\tau)F_{m,\nu+1}(\tau) - R_\nu(\tau) - [M_{\nu+1}(\tau) - M_{\nu+2}(\tau)], \quad (4.81)$$

with the terminal-value condition $\delta\Lambda_{\nu+1}(t_f) = \Lambda_{\nu+1}(t_f) - \Lambda_{\nu+2}(t_f) = N_f - N_f = 0$.

It can be verified that the solution to Eq. (4.81) takes the form

$$\delta\Lambda_{\nu+1}(\tau) = \int_\tau^{t_f} \Phi^T_{\nu+1}(\sigma,\tau)[R_\nu(\tau) + M_{\nu+1}(\tau) - M_{\nu+2}(\tau)]\Phi_{\nu+1}(\sigma,\tau)d\sigma \quad (4.82)$$

in which $\Phi_{\nu+1}(\sigma,\tau)$ is the state transition matrix associated with $F_{m,\nu+1}(\tau)$.

Because $R_\nu(\tau)$ and $M_{\nu+1}(\tau) - M_{\nu+2}(\tau)$ are positive semidefinite on the interval $[t_0, t_f]$, the expression shown in (4.82) yields that $\delta\Lambda_{\nu+1}(\tau)$ is positive semidefinite. In other words, it is equivalently rewritten as

$$\Lambda_{\nu+1}(\tau) \geq \Lambda_{\nu+2}(\tau), \qquad \forall\, \tau \in [t_0, t_f],$$

which completes the proof for Part (a).

To prove Part (b), for any given nontrivial vector $z \in \mathbb{R}^n$ and $\tau \in [t_0, t_f]$, it is then easy to see that

$$z^T\Lambda_\nu(\tau)z \geq 0.$$

By the result of Part (a), it follows that

$$\{z^T\Lambda_\nu(\tau)z\}_{\nu=0}^{\infty}$$

is monotone decreasing sequence and bounded below by zero for all z. Therefore,

$$\lim_{\nu\to\infty} z^T\Lambda_\nu(\tau)\,z$$

exists. Equivalently, it implies that the limit

$$\lim_{\nu\to\infty}\Lambda_\nu(\tau) = \underline{\Lambda}(\tau)$$

also exists.

Finally, taking limit both sides of the iterate equation (4.71) as $\nu \to \infty$ yields

$$\lim_{\nu\to\infty}\left\{\frac{d}{d\alpha}\Lambda_{\nu+1}(\tau)\right\} = \lim_{\nu\to\infty}\left\{-F^T_{m,\nu}(\tau)\Lambda_{\nu+1}(\tau) - \Lambda_{\nu+1}(\tau)F_{m,\nu}(\tau) - N_\nu(\tau) - M_{\nu+1}(\tau)\right\}. \tag{4.83}$$

Interchanging limiting and differentiation operators in Eq. (4.83) results in

$$\frac{d}{d\tau}\underline{\Lambda}(\tau) = -\underline{F}^T_m(\tau)\underline{\Lambda}(\tau) - \underline{\Lambda}(\tau)\underline{F}_m(\tau) - \underline{N}(\tau) - \underline{M}(\tau).$$

Now rewriting $\underline{\Lambda}(\tau)$ in terms of

$$\underline{\Lambda}(\tau) = \hat{\mu}_1\,\underline{\mathscr{H}}_1(\tau) + \hat{\mu}_2\,\underline{\mathscr{H}}_2(\tau) + \cdots + \hat{\mu}_k\,\underline{\mathscr{H}}^*_k(\tau),$$

yields the limiting differential equations by (4.73)–(4.74) as in the Part (b).

4.5 Chapter Summary

A class of disturbance attenuation problems with unknown input time-delays in which the risk-averse controller together with persistent disturbances and faults operate is investigated. The dynamical estimates of the current states and feedback measurement time delays are generated by the use of a Kalman-like filter. The stochastic and adversarial elements of the overall system are analyzed through the zero-sum game-theoretic framework whereby the risk-averse control gains enabling a saddle-point equilibrium for the controller, persistent disturbances and faults are obtained by solving the set of backward-in-time matrix differential equations that are used to compute the mathematical statistics associated with performance uncertainty; e.g., mean, variance, skewness, etc. These higher-order statistics succinctly convey the goals of this chapter. Namely, performance variations of the underlying chi-squared random costs are sought to quantify for feedback information, while being primarily interested in influence mechanisms for risk-averse controls.

References

1. Stahlhut, J.W., Browne, T.J., Heydt, G.T., Vittal, V.: Latency viewed as a stochastic process and its impact on wide area power system control signals. IEEE Trans. Power Syst. **23**(1), 84–91 (2008)
2. Rhee, I., Speyer, J.L.: A game theoretic approach to a finite-time disturbance attenuation problem. IEEE Trans. Autom. Control **36**(9), 1021–1032 (1991)

3. Pham, K.D., Sain, M.K., Liberty, S.R.: Cost cumulant control: state-feedback, finite-horizon paradigm with application to seismic protection. In: Miele, A. (ed.) Spec. Issue J. Optim. Theory Appl. **115**(3), 685–710. Kluwer Academic/Plenum, New York (2002)
4. Pham, K.D.: New risk-averse control paradigm for stochastic two-time-scale systems and performance robustness. In: Miele, A. (ed.) J. Optim. Theory Appl. **146**(2), 511–537 (2010)
5. Dotta, D., Silva, A.S., Decker, I.C.: Wide-area measurement-based two-level control design considering signal transmission delay. IEEE Trans. Power Syst. **24**(1), 208–216 (2009)
6. Wu, H.X., Ni, H., Heydt, G.T.: The impact of time delay on robust control design in power systems. IEEE Power Eng. **2**(2), 1511–1516 (2002)
7. Brockett, R.W.: Finite Dimensional Linear Systems. Wiley, New York (1970)
8. Basar, T., Olsder, G.J.: Dynamic Non-cooperative Game Theory. Society for Industrial and Applied Mathematics, Philadelphia, PA (1998)
9. Fleming, W.H., Rishel, R.W.: Deterministic and Stochastic Optimal Control. Springer, New York (1975)

Chapter 5
Performance Risk Management in Weakly Coupled Bilinear Stochastic Systems

5.1 Introduction

The study of bilinear systems is both unique and necessary. It provides basic understanding and critical analysis of historical background to the development of nonlinear systems and its more recent approximation by means of bilinear systems and it is unique by virtue of objective expositions of key problems currently impacting on the installation of bilinear systems into a position where they can eventually approximate any nonlinear input-output maps. It is a necessary report in the sense that bilinear systems are embarking on present and future applications as can be seen from a variety of physical, chemical, biological and nuclear systems [1]. Additionally, it is especially rewarding to review some detailed findings and technical issues investigated by several authors, see for examples [2,3] and the likes.

By contrast, remarkably little has been published about post-design performance analysis going beyond the traditional expected-value approach wherein the conventional measure of closed-loop performance is by and large centered on expected values. Aspects of asymmetry or skewness in the probability distribution functions of the non-Gaussian random utilities are matters of critical importance, as succinctly pointed out by [4]. Less citation of research evidence and comparable literature on resilient controls of bilinear stochastic systems with risk consequences occurs than is expected. Therefore, it is against such a background that this chapter presents the first conspectus on ordering uncertain prospects in weakly coupled bilinear stochastic systems which have perfect state-feedback measurements.

The rest of the chapter is organized as follows. Section 5.2 concerns about particular aspects of the class of weakly coupled bilinear stochastic systems. A major step in the difficult progression of characterization for all the mathematical statistics pertaining to the restrictive family of chi-squared performance measures is reported in Sect. 5.3. Moreover, Sect. 5.4 touches on the initiation of resilient controls with performance reliability which is manifested in the problem statements

K.D. Pham, *Resilient Controls for Ordering Uncertain Prospects*, Springer Optimization and Its Applications 98, DOI 10.1007/978-3-319-08705-4_5

and solution method. Yet, active responsiveness to change led by the procedural mechanism for risk-averse control designs is subject of discussion in Sect. 5.5. Finally, some conclusions are further included in Sect. 5.6.

5.2 Meeting the Problem

Throughout the chapter assumes a fixed probability space $(\Omega, \mathbb{F}, \{\mathbb{F}_{t_0,t} : t \in [t_0, t_f]\}, \mathbb{P})$ with filtration satisfying the usual conditions. All the filtrations are right continuous and complete and $\mathbb{F}_{t_f} \triangleq \{\mathbb{F}_{t_0,t} : t \in [t_0, t_f]\}$. In addition, let $\mathcal{L}^2_{\mathbb{F}_{t_f}}([t_0, t_f]; \mathbb{R}^n)$ denote the space of $\mathbb{F}_{t_f}$-adapted random processes $\{\hbar(t) : t \in [t_0, t_f]$ such that $E\{\int_{t_0}^{t_f} ||\hbar(t)||^2 \, dt\} < \infty\}$.

Small wonder, then, that parallels with the reports on complex structures of nonlinear weakly coupled stochastic control systems are hardly subtle. Agreement was reached that bilinearization is one of enabling tools for analysis. Figure 5.1 illustrates a particular emphasis on the following class of time-varying weakly coupled bilinear stochastic systems on $[t_0, t_f]$ governed by

$$\begin{bmatrix} \dot{x}_1(t) \\ \dot{x}_2(t) \end{bmatrix} = \begin{bmatrix} A_1(t) & \epsilon A_2(t) \\ \epsilon A_3(t) & A_4(t) \end{bmatrix} \begin{bmatrix} x_1(t) \\ x_2(t) \end{bmatrix} + \begin{bmatrix} B_1(t) & \epsilon B_2(t) \\ \epsilon B_3(t) & B_4(t) \end{bmatrix} \begin{bmatrix} u_1(t) \\ u_2(t) \end{bmatrix}$$
$$+ \begin{bmatrix} x_1(t) \\ x_2(t) \end{bmatrix} \begin{bmatrix} M_a(t) & \epsilon M_b(t) \\ \epsilon M_c(t) & M_d(t) \end{bmatrix} \begin{bmatrix} u_1(t) \\ u_2(t) \end{bmatrix} + \begin{bmatrix} G_1(t) & \epsilon G_2(t) \\ \epsilon G_3(t) & G_4(t) \end{bmatrix} \begin{bmatrix} w_1(t) \\ w_2(t) \end{bmatrix} \tag{5.1}$$

with the initial known conditions

$$\begin{bmatrix} x_1(t_0) \\ x_2(t_0) \end{bmatrix} = \begin{bmatrix} x_1^0 \\ x_2^0 \end{bmatrix}$$

whereupon for $i = 1, 2$, $x_i(t)$ is the controlled state process valued in $\mathbb{R}^{n_i}$ of the reduced-order subsystem i; $u_i(t)$ is the control process valued in a closed convex subset of $\mathbb{R}^{m_i}$; and ϵ is a small weak coupling parameter. Exogenous inputs modeled

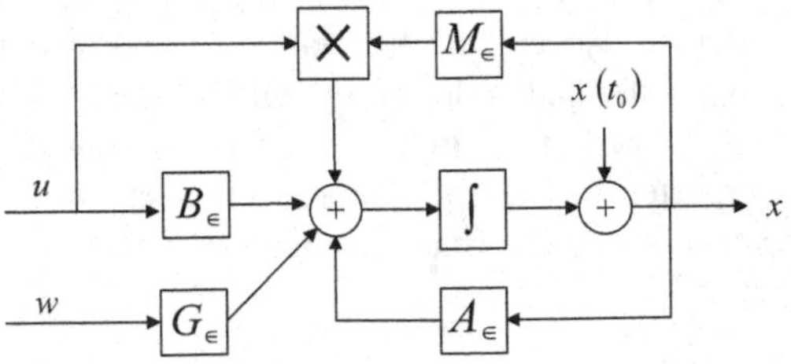

Fig. 5.1 A description of stochastic bilinear controlled systems

by white noises $w_i(t) \in \mathbb{R}^{p_i}$ assume the form of mutually uncorrelated stationary $\mathbb{F}_{t_f}$-adapted Gaussian processes with means $E\{w_i(t)\} = m_{w_i}(t)$ and covariances $cov\{w_i(t_1), w_i(t_2)\} = W_i\delta(t_1 - t_2)$ for all $t_1, t_2 \in [t_0, t_f]$ and $W_i > 0$. As is expected, the system coefficients A_j, B_j, G_j for $j = 1, 2, 3, 4$ as well as M_a, M_b, M_c, and M_d are of appropriate dimensions, weakly coupled, and continuous-time matrix functions.

Not surprisingly, the restricted class of performance measures whose ranges of variability are bounded below from zero, is of primarily interest. Henceforth, it is assumed that

$$
\begin{aligned}
J = &\begin{bmatrix} x_1(t_f) \\ x_2(t_f) \end{bmatrix}^T \begin{bmatrix} Q_1^f & \epsilon Q_2^f \\ \epsilon (Q_2^f)^T & Q_3^f \end{bmatrix} \begin{bmatrix} x_1(t_f) \\ x_2(t_f) \end{bmatrix} \\
&+ \int_{t_0}^{t_f} \begin{bmatrix} x_1(\tau) \\ x_2(\tau) \end{bmatrix}^T \begin{bmatrix} Q_1(\tau) & \epsilon Q_2(\tau) \\ \epsilon Q_2^T(\tau) & Q_3(\tau) \end{bmatrix} \begin{bmatrix} x_1(\tau) \\ x_2(\tau) \end{bmatrix} d\tau \\
&+ \int_{t_0}^{t_f} \begin{bmatrix} u_1(\tau) \\ u_2(\tau) \end{bmatrix}^T \begin{bmatrix} R_1(\tau) & 0 \\ 0 & R_2(\tau) \end{bmatrix} \begin{bmatrix} u_1(\tau) \\ u_2(\tau) \end{bmatrix} d\tau. \qquad (5.2)
\end{aligned}
$$

As seen from the notational simplicity, the change of variables that pertains to the system description governed by (5.1) and (5.2) is inescapably necessary; e.g.,

$$
x(t) \triangleq \begin{bmatrix} x_1(t) \\ x_2(t) \end{bmatrix}; \quad u(t) \triangleq \begin{bmatrix} u_1(t) \\ u_2(t) \end{bmatrix}; \quad w(t) \triangleq \begin{bmatrix} w_1(t) \\ w_2(t) \end{bmatrix}; \quad x(t_0) \triangleq x_0 = \begin{bmatrix} x_1^0 \\ x_2^0 \end{bmatrix}
$$

$$
A_\epsilon(t) \triangleq \begin{bmatrix} A_1(t) & \epsilon A_2(t) \\ \epsilon A_3(t) & A_4(t) \end{bmatrix}; \; B_\epsilon(t) \triangleq \begin{bmatrix} B_1(t) & \epsilon B_2(t) \\ \epsilon B_3(t) & B_4(t) \end{bmatrix}; \; Q_\epsilon(t) \triangleq \begin{bmatrix} Q_1(\tau) & \epsilon Q_2(\tau) \\ \epsilon Q_2^T(\tau) & Q_3(\tau) \end{bmatrix}
$$

$$
M_\epsilon(t) \triangleq \begin{bmatrix} M_a(t) & \epsilon M_b(t) \\ \epsilon M_c(t) & M_d(t) \end{bmatrix}; \quad G_\epsilon(t) \triangleq \begin{bmatrix} G_1(t) & \epsilon G_2(t) \\ \epsilon G_3(t) & G_4(t) \end{bmatrix}; \quad Q_\epsilon^f \triangleq \begin{bmatrix} Q_1^f & \epsilon Q_2^f \\ \epsilon (Q_2^f)^T & Q_3^f \end{bmatrix}
$$

and

$$
cov\{w(t_1), w(t_2)\} = W\delta(t_1 - t_2); \quad W \triangleq \begin{bmatrix} W_1 & 0 \\ 0 & W_2 \end{bmatrix}; \quad R(\tau) \triangleq \begin{bmatrix} R_1(\tau) & 0 \\ 0 & R_2(\tau) \end{bmatrix}.
$$

Beyond this, the weakly coupled bilinear stochastic control system under consideration is further rewritten as

$$
\dot{x}(t) = A_\epsilon(t)x(t) + (B_\epsilon(t) + x(t)M_\epsilon(t))u(t) + G_\epsilon(t)w(t), \quad x(t_0). \qquad (5.3)
$$

Understandably, there are always elements of uncertainty in asserting what would have happened if this or that had been done in the past. Such an assessment finds ample corroboration in the σ-algebras

$$\mathbb{F}_{t_0,t} \triangleq \sigma\{w(\tau) : t_0 \le \tau \le t\}$$
$$\mathscr{G}^y_{t_0,t} \triangleq \sigma\{x(\tau) : t_0 \le \tau \le t\}, \quad t \in [t_0, t_f].$$

That said, there is reason to believe that the admissible set of feedback control laws should make efficient use of the information available $\mathscr{G}^y_{t_f} \triangleq \{\mathscr{G}^y_{t_0,t} : t \in [t_0, t_f]\} \subset \{\mathbb{F}_{t_0,t} : t \in [t_0, t_f]\}$; e.g.,

$$\mathbb{U}^y[t_0, t_f] \triangleq \{u \in \mathscr{L}^2_{\mathscr{G}^y_{t_f}}([t_0, t_f], \mathbb{R}^{m_1+m_2}), \mathbb{P} - a.s.\}$$

where $\mathbb{U}^y[t_0, t_f]$ is a closed convex subset of $\mathscr{L}^2_{\mathbb{F}_{t_f}}([t_0, t_f], \mathbb{R}^{m_1+m_2})$.

Associated with an admissible initial condition $(x(\cdot), u(\cdot))$ is an integral-quadratic form cost $J : \mathbb{R}^{n_1+n_2} \times \mathbb{U}^y[t_0, t_f] \mapsto \mathbb{R}^+$ on $[t_0, t_f]$

$$J(x_0, u(\cdot)) = x^T(t_f) Q^f_\epsilon x(t_f) + \int_0^{t_f} [x^T(\tau) Q_\epsilon(\tau) x(\tau) + u^T(\tau) R(\tau) u(\tau)] d\tau \tag{5.4}$$

In view of the linear structure of the system (5.3)–(5.4) and the information structure $\mathscr{G}^y_{t_f}$, non-randomized control strategies are therefore, restricted to the linear mapping $\gamma : [t_0, t_f] \times \mathbb{R}^{n_1+n_2} \mapsto \mathbb{U}^y[t_0, t_f]$ with the rule of actions

$$u(t) = \gamma(t, x(t)) \triangleq K(t)x(t) + l(t) \tag{5.5}$$

where the deterministic matrix-valued functions $K \in \mathscr{C}([t_0, t_f]; \mathbb{R}^{(m_1+m_2)\times(n_1+n_2)})$ and vector-valued functions $l \in \mathscr{C}([t_0, t_f]; \mathbb{R}^{(m_1+m_2)})$ are the admissible feedback gains and feedforward inputs are yet to be defined.

Consolidation of the system dynamics (5.3) is now equivalently treated with Brownian motions $dw(t)$ instead of their nonexistent time derivatives $w(t)$; e.g.,

$$\begin{aligned} dx(t) &= (A_\epsilon(t)x(t) + (B_\epsilon(t) + x(t)M_\epsilon(t))u(t) + G_\epsilon(t)m_w(t))dt \\ &\quad + G_\epsilon(t)U_W \Lambda_W^{1/2} dw(t), \quad x(t_0) \end{aligned} \tag{5.6}$$

where $dw(t) \triangleq \tilde{w}(t)dt = (w(t) - m_w(t))dt$ and $m_w(t) \triangleq E\{w(t)\}$ together with the corresponding singular-value decomposition of the intensity $W \triangleq U_W \Lambda_W^{1/2} U_W^T$.

In the sequel, the assimilation of the control policy (5.5) into the controlled system (5.6) is governed by the Ito stochastic differential equation

$$dx(t) = ((A_\epsilon(t) + B_\epsilon(t, x)K(t))x(t) + B_\epsilon(t, x)l(t) + G_\epsilon(t)m_w(t))dt \tag{5.7}$$
$$+ G_\epsilon(t)U_W \Lambda_W^{1/2} dw(t), \quad x(t_0) \tag{5.8}$$

in which $B_\epsilon(t, x) \triangleq B_\epsilon(t) + x(t)M_\epsilon(t)$.

As it turns out, the finite-horizon integral-quadratic-form cost further becomes

$$J(K,l) = x^T(t_f)Q_f x(t_f) + \int_{t_0}^{t_f} [x^T(\tau)(Q_\epsilon(\tau) + K^T(\tau)R(\tau)K(\tau))x(\tau) + 2x^T(\tau)K^T(\tau)R(\tau)l(\tau) + l^T(\tau)R(\tau)l(\tau)]d\tau. \tag{5.9}$$

5.3 Responses to Performance Uncertainties

A rather innovative approach of coping with the uncertain cost (5.9) of the chi-squared type is to use its mathematical statistics in a responsive mean-risk aware performance index for ordering uncertain prospects as appropriately required in statistical optimal control hereafter. The focus aims to provide a procedural mechanism involved modeling and characterization of the mathematical statistics associated with (5.9). In particular, the initial condition (t_0, x_0) is replaced by any arbitrary pair (τ, x_τ). Then, for the given affine input l and admissible feedback gain K, the cost functional (5.9) is seen as the "cost-to-go", $J(\tau, x_\tau)$. The correspondent moment-generating function is given by the definition

$$\varphi(\tau, x_\tau, \theta) \triangleq E\{\exp(\theta J(\tau, x_\tau))\} \tag{5.10}$$

where the scalar $\theta \in \mathbb{R}^+$ is a small parameter. Thus, the cumulant-generating function immediately follows

$$\psi(\tau, x_\tau, \theta) \triangleq \ln\{\varphi(\tau, x_\tau, \theta)\} \tag{5.11}$$

in which $\ln\{\cdot\}$ denotes the natural logarithmic transformation of an enclosed entity.

Theorem 5.3.1 (Cumulant-Generating Function). *For all $\tau \in [t_0, t_f]$ and $\theta \in \mathbb{R}^+$, let $\varphi(\tau, x_\tau, \theta) \triangleq \varrho(\tau, \theta)\exp\{x_\tau^T \Upsilon(\tau, \theta)x_\tau + 2x_\tau^T \eta(\tau, \theta)\}$ and further let $\upsilon(\tau, \theta) \triangleq \ln\{\varrho(\tau, \theta)\}$. Then, the cumulant-generating function is given by*

$$\psi(\tau, x_\tau, \theta) = x_\tau^T \Upsilon(\tau, \theta)x_\tau + 2x_\tau^T \eta(\tau, \theta) + \upsilon(\tau, \theta) \tag{5.12}$$

where the cumulant building variables $\Upsilon(\tau, \theta)$, $\eta(\tau, \theta)$ and $\upsilon(\tau, \theta)$ solve the backward-in-time differential equations

$$\frac{d}{d\tau}\Upsilon(\tau, \theta) = -(A_\epsilon(\tau) + B_\epsilon(\tau, x)K(\tau))^T \Upsilon(\tau, \theta) - \Upsilon(\tau, \theta)(A_\epsilon(\tau) + B_\epsilon(\tau, x)K(\tau)) - 2\Upsilon(\tau, \theta)G_\epsilon(\tau)WG_\epsilon^T(\tau)\Upsilon(\tau, \theta) - \theta(Q_\epsilon(\tau) + K^T(\tau)R(\tau)K(\tau)) \tag{5.13}$$

$$\frac{d}{d\tau}\eta(\tau,\theta) = -(A_\epsilon(\tau) + B_\epsilon(\tau,x)K(\tau))^T\eta(\tau,\theta) - \Upsilon(\tau,\theta)(B_\epsilon(\tau,x)l(\tau) + G_\epsilon(\tau)m_w(\tau)) - \theta K^T(\tau)R(\tau)l(\tau) \tag{5.14}$$

$$\frac{d}{d\tau}\upsilon(\tau,\theta) = -\operatorname{Tr}\{\Upsilon(\tau,\theta)G_\epsilon(\tau)WG_\epsilon^T(\tau)\} - 2\eta^T(\tau,\theta)(B_\epsilon(\tau,x)l(\tau) + G_\epsilon(\tau)l(\tau)) - \theta l^T(\tau)R(\tau)l(\tau) \tag{5.15}$$

together with the terminal-value conditions $\Upsilon(t_f,\theta) = \theta Q_\epsilon^f$, $\eta(t_f,\theta) = 0$ *and* $\upsilon(t_f,\theta) = 0$.

Proof. For any $\theta \in \mathbb{R}^+$ given and $\varpi(\tau,x_\tau,\theta) \triangleq \exp\{\theta J(\tau,x_\tau)\}$, the moment-generating function therefore becomes

$$\varphi(\tau,x_\tau,\theta) = E\{\varpi(\tau,x_\tau,\theta)\}$$

with the time derivative of

$$\frac{d}{d\tau}\varphi(\tau,x_\tau,\theta) = -\theta\Big\{x_\tau^T[Q_\epsilon(\tau) + K^T(\tau)R(\tau)K(\tau)]x_\tau + 2x_\tau^T K^T(\tau)R(\tau)l(\tau) + l^T(\tau)R(\tau)l(\tau)\Big\}\varphi(\tau,x_\tau,\theta). \tag{5.16}$$

Using the standard Ito's stochastic differential formula, it yields

$$\begin{aligned}
d\varphi(\tau,x_\tau,\theta) &= E\{d\varpi(\tau,x_\tau,\theta)\} \\
&= E\Big\{\varpi_\tau(\tau,x_\tau,\theta)\,d\tau + \varpi_{x_\tau}(\tau,x_\tau,\theta)\,dx_\tau \\
&\quad + \frac{1}{2}\operatorname{Tr}\{\varpi_{x_\tau x_\tau}(\tau,x_\tau,\theta)G_\epsilon(\tau)WG_\epsilon^T(\tau)\}\,d\tau\Big\} \\
&= \varphi_\tau(\tau,x_\tau,\theta)\,d\tau + \varphi_{x_\tau}(\tau,x_\tau,\theta)\,[(A_\epsilon(\tau) + B_\epsilon(\tau,x)K(\tau))x_\tau \\
&\quad + B_\epsilon(\tau,x)l(\tau) + G_\epsilon(\tau)m_w(\tau)]d\tau \\
&\quad + \frac{1}{2}\operatorname{Tr}\{\varphi_{x_\tau x_\tau}(\tau,x_\tau,\theta)\,G_\epsilon(\tau)\,WG_\epsilon^T(\tau)\}\,d\tau
\end{aligned}$$

from which, by the definition

$$\varphi(\tau,x_\tau,\theta) \triangleq \varrho(\tau,\theta)\exp\{x_\tau^T\Upsilon(\tau,\theta)x_\tau + 2x_\tau^T\eta(\tau,\theta)\}$$

and its partial derivatives

$$\varphi_\tau(\tau, x_\tau, \theta) = \Big[\frac{\frac{d}{d\tau}\varrho(\tau,\theta)}{\varrho(\tau,\theta)} + x_\tau^T \frac{d}{d\tau}\Upsilon(\tau,\theta)x_\tau + 2x_\tau^T \frac{d}{d\tau}\eta(\tau,\theta)\Big]\varphi(\tau, x_\tau, \theta)$$

$$\varphi_{x_\tau}(\tau, x_\tau, \theta) = \{x_\tau^T\left[\Upsilon(\tau,\theta) + \Upsilon^T(\tau,\theta)\right] + 2\eta^T(\tau,\theta)\}\,\varphi(\tau, x_\tau, \theta)$$

$$\varphi_{x_\tau x_\tau}(\tau, x_\tau, \theta) = \left[\Upsilon(\tau,\theta) + \Upsilon^T(\tau,\theta)\right]\varphi(\tau, x_\tau, \theta)$$
$$+ \left[\Upsilon(\tau,\theta) + \Upsilon^T(\tau,\theta)\right]x_\tau x_\tau^T\left[\Upsilon(\tau,\theta) + \Upsilon^T(\tau,\theta)\right]\varphi(\tau, x_\tau, \theta)$$

leads to

$$\frac{d}{d\tau}\varphi(\tau, x_\tau, \theta) = \Bigg\{\frac{\frac{d}{d\tau}\varrho(\tau,\theta)}{\varrho(\tau,\theta)} + 2x_\tau^T\Upsilon(\tau,\theta)G_\epsilon(\tau)WG_\epsilon^T(\tau)\Upsilon(\tau,\theta)x_\tau$$
$$+ x_\tau^T\Bigg[\frac{d}{d\tau}\Upsilon(\tau,\theta) + (A_\epsilon(\tau)$$
$$+ B_\epsilon(\tau,x)K(\tau))^T\Upsilon(\tau,\theta) + \Upsilon(\tau,\theta)(A_\epsilon(\tau)$$
$$+ B_\epsilon(\tau,x)K(\tau))\Bigg]x_\tau + 2x_\tau^T\Bigg[\frac{d}{d\tau}\eta(\tau,\theta) + (A_\epsilon(\tau)$$
$$+ B_\epsilon(\tau,x)K(\tau))^T\eta(\tau,\theta)$$
$$+ \Upsilon(\tau,\theta)(B_\epsilon(\tau,x)l(\tau) + G_\epsilon(\tau)m_w(\tau))\Bigg]$$
$$+ \mathrm{Tr}\left\{\Upsilon(\tau,\theta)G_\epsilon(\tau)WG_\epsilon^T(\tau)\right\}$$
$$+ 2\eta^T(\tau,\theta)(B_\epsilon(\tau,x)l(\tau) + G_\epsilon(\tau)m_w(\tau))\Bigg\}\varphi(\tau, x_\tau; \theta). \tag{5.17}$$

By applying the result (5.16) to this finding (5.17) and having both linear and quadratic terms independent of the arbitrary x_τ, it requires that the results (5.13)–(5.15) hold true. At last, the terminal-value conditions: $\Upsilon(t_f,\theta) = \theta Q_\epsilon^f$, $\eta(t_f,\theta) = 0$, $\varrho(t_f,\theta) = 1$ and $\upsilon(t_f,\theta) = 0$.

By definition, the mathematical statistics associated with the chi-squared random performance measure (5.9) are now generated by the use of a Maclaurin series expansion for the cumulant-generating function

$$\psi(\tau, x_\tau, \theta) \triangleq \sum_{r=1}^{\infty}\kappa_r\frac{\theta^r}{r!} = \sum_{r=1}^{\infty}\frac{\partial^r}{\partial\theta^r}\psi(\tau, x_\tau, \theta)\Big|_{\theta=0}\frac{\theta^r}{r!} \tag{5.18}$$

in which κ_r are called the rth performance-measure statistics.

In addition, the coefficients of the Maclaurin series expansion are computed by using the result (5.12)

$$\left.\frac{\partial^r}{\partial\theta^r}\psi(\tau,x_\tau,\theta)\right|_{\theta=0} = x_\tau^T \left.\frac{\partial^r}{\partial\theta^r}\Upsilon(\tau,\theta)\right|_{\theta=0} x_\tau + 2x_\tau^T \left.\frac{\partial^r}{\partial\theta^r}\eta(\tau,\theta)\right|_{\theta=0} + \left.\frac{\partial^r}{\partial\theta^r}\upsilon(\tau,\theta)\right|_{\theta=0}. \tag{5.19}$$

In view of the results (5.18) and (5.19), the performance-measure statistics associated with the weakly coupled bilinear stochastic systems are then described as follows

$$\kappa_r = x_\tau^T \left.\frac{\partial^r}{\partial\theta^r}\Upsilon(\tau,\theta)\right|_{\theta=0} x_\tau + 2x_\tau^T \left.\frac{\partial^r}{\partial\theta^r}\eta(\tau,\theta)\right|_{\theta=0} + \left.\frac{\partial^r}{\partial\theta^r}\upsilon(\tau,\theta)\right|_{\theta=0} \tag{5.20}$$

for any finite $1 \le r < \infty$.

For notational convenience, the definition that follows is essential, e.g.,

$$H(\tau,r) \triangleq \left.\frac{\partial^r}{\partial\theta^r}\Upsilon(\tau,\theta)\right|_{\theta=0},$$

$$\breve{D}(\tau,i) \triangleq \left.\frac{\partial^r}{\partial\theta^r}\eta(\tau,\theta)\right|_{\theta=0},$$

$$D(\tau,r) \triangleq \left.\frac{\partial^r}{\partial\theta^r}\upsilon(\tau,\theta)\right|_{\theta=0}.$$

Then, the performance-measure statistics obtained in the sequel exhibit a familiar pattern of quadratic affine in arbitrary initial system state x_τ as being illustrated here.

Theorem 5.3.2 (Performance-Measure Statistics). *With the particular interest in the bilinear stochastic system governed by (5.8)–(5.9), the kth performance-measure statistic considered for performance risk aversion is given by*

$$\kappa_k = x_0^T H(t_0,k)x_0 + 2x_0^T \breve{D}(t_0,k) + D(t_0,k), \quad k \in \mathbb{Z}^+ \tag{5.21}$$

in which the building variables $\{H(\tau,r)\}_{r=1}^k$, $\{\breve{D}(\tau,r)\}_{r=1}^k$ and $\{D(\tau,r)\}_{r=1}^k$ evaluated at $\tau = t_0$ satisfy the time-backward differential equations (with the dependence of $H(\tau,r)$, $\breve{D}(\tau,r)$ and $D(\tau,r)$ upon the admissible l and K suppressed)

$$\frac{d}{d\tau}H(\tau,1) = -(A_\epsilon(\tau)+B_\epsilon(\tau,x)K(\tau))^T H(\tau,1) - H(\tau,1)(A_\epsilon(\tau)+B_\epsilon(\tau,x)K(\tau)) - Q_\epsilon(\tau) - K^T(\tau)R(\tau)K(\tau) \tag{5.22}$$

$$\frac{d}{d\tau}H(\tau,r) = -(A_\epsilon(\tau)+B_\epsilon(\tau,x)K(\tau))^T H(\tau,r) - H(\tau,r)(A_\epsilon(\tau)+B_\epsilon(\tau,x)K(\tau))$$
$$-\sum_{s=1}^{r-1}\frac{2r!}{s!(r-s)!}H(\tau,s)G_\epsilon(\tau)WG_\epsilon^T(\tau)H(\tau,r-s), \quad 2 \le r \le k \tag{5.23}$$

$$\frac{d}{d\tau}\breve{D}(\tau,1) = -(A_\epsilon(\tau)+B_\epsilon(\tau,x)K(\tau))^T \breve{D}(\tau,1)$$
$$- H(\tau,1)(B(\tau,x)l(\tau) + G_\epsilon(\tau)m_w(\tau)) - K^T(\tau)R(\tau)l(\tau) \tag{5.24}$$

$$\frac{d}{d\tau}\breve{D}(\tau,r) = -(A_\epsilon(\tau)+B_\epsilon(\tau,x)K(\tau))^T \breve{D}(\tau,r)$$
$$- H(\tau,r)(B_\epsilon(\tau,x)l(\tau) + G_\epsilon(\tau)m_w(\tau)), \quad 2 \le r \le k \tag{5.25}$$

and

$$\frac{d}{d\tau}D(\tau,1) = -\operatorname{Tr}\left\{H(\tau,1)G_\epsilon(\tau)WG_\epsilon^T(\tau)\right\}$$
$$- 2\breve{D}^T(\tau,1)(B_\epsilon(\tau,x)l(\tau) + G_\epsilon(\tau)m_w(\tau)) - l^T(\tau)R(\tau)l(\tau) \tag{5.26}$$

$$\frac{d}{d\tau}D(\tau,r) = -\operatorname{Tr}\left\{H(\tau,r)G_\epsilon(\tau)WG_\epsilon^T(\tau)\right\}$$
$$- 2\breve{D}^T(\tau,r)(B_\epsilon(\tau,x)l(\tau) + G_\epsilon(\tau)m_w(\tau)), \quad 2 \le r \le k \tag{5.27}$$

where the terminal-value conditions $H(t_f,1) = Q_\epsilon^f$, $H(t_f,r) = 0$ *for* $2 \le r \le k$, $\breve{D}(t_f,r) = 0$ *for* $1 \le r \le k$ *and* $D(t_f,r) = 0$ *for* $1 \le r \le k$.

Proof. The proof is a straight forward repetition of the work [5], hence it is omitted.

5.4 Statements of Mayer Problem with Performance Risk

All of the associations which have been discussed in the previous section are contending for adherence among uncertain prospects of the chi-squared performance measure. A serious attempt to ascertain which of these associations will be most influential over the resilient control problem in determining what performance appraisal provides what information for possible direct intervention, when risk-averse attitudes get screened and how revision of control as corrective actions gets done for performance risk aversion – requires a more deeply cutting analysis of the underlying sources of (5.22)–(5.27). Such basic analysis is the sequel development.

It is hardly surprising, therefore, that the dynamics (5.22)–(5.27) of performance-measure statistics (5.21) are the focus of attention. In preparing for the statements of resilient controls for bilinear stochastic systems, the k-tuple state variables $\mathscr{H}$, $\breve{\mathscr{D}}$ and $\mathscr{D}$ are defined as follows

$$\mathscr{H}(\cdot) \triangleq (\mathscr{H}_1(\cdot), \ldots, \mathscr{H}_k(\cdot)),$$

$$\breve{\mathscr{D}}(\cdot) \triangleq \left(\breve{\mathscr{D}}_1(\cdot), \ldots, \breve{\mathscr{D}}_k(\cdot)\right),$$

$$\mathscr{D}(\cdot) \triangleq (\mathscr{D}_1(\cdot), \ldots, \mathscr{D}_k(\cdot)),$$

where each elements $\mathscr{H}_r \in \mathscr{C}^1([t_0, t_f]; \mathbb{R}^{(n_1+n_2)\times(n_1+n_2)})$ of $\mathscr{H}$, $\breve{\mathscr{D}}_r \in \mathscr{C}^1([t_0, t_f]; \mathbb{R}^{n_1+n_2})$ of $\breve{\mathscr{D}}$ and finally, $\mathscr{D}_r \in \mathscr{C}^1([t_0, t_f]; \mathbb{R})$ of $\mathscr{D}$ have the representations

$$\mathscr{H}_r(\cdot) = H(\cdot, r), \qquad \breve{\mathscr{D}}_r(\cdot) = \breve{D}(\cdot, r), \qquad \mathscr{D}_r(\cdot) = D(\cdot, r)$$

with the right members satisfying the dynamic equations (5.22)–(5.27) on the finite horizon $[t_0, t_f]$. The problem formulation can be considerably simplified if the bounded and Lipschitz continuous mappings are introduced

$$\mathscr{F}_r : [t_0, t_f] \times (\mathbb{R}^{(n_1+n_2)\times(n_1+n_2)})^k \times \mathbb{R}^{(m_1+m_2)\times(n_1+n_2)} \mapsto \mathbb{R}^{(n_1+n_2)\times(n_1+n_2)}$$

$$\breve{\mathscr{G}}_r : [t_0, t_f] \times (\mathbb{R}^{(n_1+n_2)\times(n_1+n_2)})^k \times (\mathbb{R}^{n_1+n_2})^k \times \mathbb{R}^{(m_1+m_2)\times(n_1+n_2)} \times \mathbb{R}^{m_1+m_2} \mapsto \mathbb{R}^{n_1+n_2}$$

$$\mathscr{G}_r : [t_0, t_f] \times (\mathbb{R}^{(n_1+n_2)\times(n_1+n_2)})^k \times (\mathbb{R}^{n_1+n_2})^k \times \mathbb{R}^{m_1+m_2} \mapsto \mathbb{R}$$

where the actions are given by

$$\mathscr{F}_1(\tau, \mathscr{H}, K) = -(A_\epsilon(\tau) + B_\epsilon(\tau, x)K(\tau))^T \mathscr{H}_1(\tau) - \mathscr{H}_1(\tau)(A_\epsilon(\tau) + B_\epsilon(\tau, x)K(\tau)) - Q_\epsilon(\tau) - K^T(\tau)R(\tau)K(\tau)$$

$$\mathscr{F}_r(\tau, \mathscr{H}, K) = -(A_\epsilon(\tau) + B_\epsilon(\tau, x)K(\tau))^T \mathscr{H}_r(\tau) - \mathscr{H}_r(\tau)(A_\epsilon(\tau) + B_\epsilon(\tau, x)K(\tau)) - \sum_{s=1}^{r-1} \frac{2r!}{s!(r-s)!} \mathscr{H}_s(\tau)G_\epsilon(\tau)WG_\epsilon^T(\tau)\mathscr{H}_{r-s}(\tau), \quad 2 \le r \le k$$

$$\breve{\mathscr{G}}_1\left(\tau, \mathscr{H}, \breve{\mathscr{D}}, K, l\right) = -(A_\epsilon(\tau) + B_\epsilon(\tau, x)K(\tau))^T \breve{\mathscr{D}}_1(\tau) - \mathscr{H}_1(\tau)(B_\epsilon(\tau, x)l(\tau) + G_\epsilon(\tau)m_w(\tau)) - K^T(\tau)R(\tau)l(\tau)$$

$$\breve{\mathscr{G}}_r\left(\tau, \mathscr{H}, \breve{\mathscr{D}}, K, l\right) = -(A_\epsilon(\tau) + B_\epsilon(\tau, x)K(\tau))^T \breve{\mathscr{D}}_r(\tau) - \mathscr{H}_r(\tau)(B_\epsilon(\tau, x)l(\tau) + G_\epsilon(\tau)m_w(\tau)), \quad 2 \le r \le k$$

$$\mathscr{G}_1\left(\tau, \mathscr{H}, \breve{\mathscr{D}}, l\right) = -\operatorname{Tr}\left\{\mathscr{H}_1(\tau)G_\epsilon(\tau)WG_\epsilon^T(\tau)\right\} - 2\breve{\mathscr{D}}_1^T(\tau)(B_\epsilon(\tau, x)l(\tau) + G_\epsilon(\tau)m_w(\tau)) - l^T(\tau)R(\tau)l(\tau)$$

$$\mathscr{G}_r\left(\tau, \mathscr{H}, \breve{\mathscr{D}}, l\right) = -\operatorname{Tr}\{\mathscr{H}_r(\tau) G_\epsilon(\tau) W G_\epsilon^T(\tau)\}$$
$$\qquad - 2\breve{\mathscr{D}}_r^T(\tau)(B_\epsilon(\tau,x) l(\tau) + G_\epsilon(\tau) m_w(\tau)), \quad 2 \le r \le k.$$

Next the synergistic development in the product mappings is also possible due to

$$\mathscr{F} \triangleq \mathscr{F}_1 \times \cdots \times \mathscr{F}_k,$$
$$\breve{\mathscr{G}} \triangleq \breve{\mathscr{G}}_1 \times \cdots \times \breve{\mathscr{G}}_k,$$
$$\mathscr{G} \triangleq \mathscr{G}_1 \times \cdots \times \mathscr{G}_k.$$

Thus, the dynamic equations of motion (5.22)–(5.27) can be rewritten as

$$\frac{d}{d\tau}\mathscr{H}(\tau) = \mathscr{F}(\tau, \mathscr{H}(\tau), K(\tau)), \quad \mathscr{H}(t_f) = \mathscr{H}_f \tag{5.28}$$
$$\frac{d}{d\tau}\breve{\mathscr{D}}(\tau) = \breve{\mathscr{G}}\left(\tau, \mathscr{H}(\tau), \breve{\mathscr{D}}(\tau), K(\tau), l(\tau)\right), \quad \breve{\mathscr{D}}(t_f) = \breve{\mathscr{D}}_f \tag{5.29}$$
$$\frac{d}{d\tau}\mathscr{D}(\tau) = \mathscr{G}\left(\tau, \mathscr{H}(\tau), \breve{\mathscr{D}}(\tau), l(\tau)\right), \quad \mathscr{D}(t_f) = \mathscr{D}_f \tag{5.30}$$

in which the terminal-value conditions $\mathscr{H}_f = \left(Q_\epsilon^f, 0, \ldots, 0\right)$, $\breve{\mathscr{D}}_f = (0, 0, \ldots, 0)$ and $\mathscr{D}_f = (0, 0, \ldots, 0)$.

Understandably, the product system (5.28)–(5.30) uniquely determines $\mathscr{H}$, $\breve{\mathscr{D}}$ and $\mathscr{D}$ once the admissible affine input l and feedback gain K are specified. Henceforth, they are considered as $\mathscr{H} = \mathscr{H}(\cdot, K)$, $\breve{\mathscr{D}} = \breve{\mathscr{D}}(\cdot, K, l)$ and $\mathscr{D} = \mathscr{D}(\cdot, K, l)$.

Stimulated by various degrees of performance uncertainty, performance-measure statistics are now utilized in cognizant selection rules to order uncertain prospects, such as mean-value aware performance indexes, to improve their reporting and information dissemination on performance values and risks for performance appraisal.

Definition 5.4.1 (Mean-Risk Aware Performance Index). Let $k \in \mathbb{Z}^+$ and the sequence $\mu = \{\mu_r \ge 0\}_{r=1}^k$ with $\mu_1 > 0$. Then, the performance appraisal with mean and risk awareness

$$\phi_0 : [t_0, t_f] \times (\mathbb{R}^{(n_1+n_2)\times(n_1+n_2)})^k \times (\mathbb{R}^{n_1+n_2})^k \times \mathbb{R}^k \mapsto \mathbb{R}^+$$

in statistical optimal control on $[t_0, t_f]$ is of Mayer type and defined by

$$\phi_0\left(t_0, \mathscr{H}(t_0), \breve{\mathscr{D}}(t_0), \mathscr{D}(t_0)\right) \triangleq \underbrace{\mu_1 \kappa_1}_{\text{Mean Measure}} + \underbrace{\mu_2 \kappa_2 + \cdots + \mu_k \kappa_k}_{\text{Risk Measures}}$$
$$= \sum_{r=1}^k \mu_r \left[x_0^T \mathscr{H}_r(t_0) x_0 + 2 x_0^T \breve{\mathscr{D}}_r(t_0) + \mathscr{D}_r(t_0)\right] \tag{5.31}$$

where the scalar, real constants μ_r represent parametric design of freedom and the unique solutions $\{\mathscr{H}_r(t_0)\}_{r=1}^k$, $\left\{\breve{\mathscr{D}}_r(t_0)\right\}_{r=1}^k$ and $\{\mathscr{D}_r(t_0)\}_{r=1}^k$ evaluated at $\tau = t_0$ satisfy the dynamical equations (5.28)–(5.30).

For the given terminal data $(t_f, \mathscr{H}_f, \breve{\mathscr{D}}_f, \mathscr{D}_f)$, the classes of admissible feedforward inputs and feedback gains that can be adjusted optimally to cope with the chi-squared performance measure are defined as follows.

Definition 5.4.2 (Admissible Affine Inputs and Feedback Gains). Let compact subsets $\overline{L} \subset \mathbb{R}^m$ and $\overline{K} \subset \mathbb{R}^{(m_1+m_2)\times(n_1+n_2)}$ be the sets of allowable vector and matrix values. For the given $k \in \mathbb{Z}^+$ and the sequence $\mu = \{\mu_r \geq 0\}_{r=1}^k$ with $\mu_1 > 0$, the admissible feedforward inputs $\mathscr{L}_{t_f,\mathscr{H}_f,\breve{\mathscr{D}}_f,\mathscr{D}_f;\mu}$ and feedback gains $\mathscr{K}_{t_f,\mathscr{H}_f,\breve{\mathscr{D}}_f,\mathscr{D}_f;\mu}$ respectively resume the classes of $\mathscr{C}([t_0, t_f]; \mathbb{R}^{m_1+m_2})$ and $\mathscr{C}([t_0, t_f]; \mathbb{R}^{(m_1+m_2)\times(n_1+n_2)})$ with values $l(\cdot) \in \overline{L}$ and $K(\cdot) \in \overline{K}$ for which the solutions to the dynamic equations (5.28)–(5.30) exist on $[t_0, t_f]$.

Likewise, the optimization statements for the working of weakly coupled bilinear stochastic systems of interest, especially one conducive to appropriate risk-averse behavior over a finite horizon are stated in the sequel.

Definition 5.4.3 (Optimization Problem of Mayer Type). Fix $k \in \mathbb{Z}^+$ and the sequence $\mu = \{\mu_r \geq 0\}_{r=1}^k$ with $\mu_1 > 0$. Then, the Mayer optimization problem over $[t_0, t_f]$ is given by the minimization of the risk-value aware performance index (5.31) over $l(\cdot) \in \mathscr{L}_{t_f,\mathscr{H}_f,\breve{\mathscr{D}}_f,\mathscr{D}_f;\mu}$, $K(\cdot) \in \mathscr{K}_{t_f,\mathscr{H}_f,\breve{\mathscr{D}}_f,\mathscr{D}_f;\mu}$ and subject to the dynamic equations of motion (5.28)–(5.30) for $\tau \in [t_0, t_f]$.

Construction of scalar-valued functions which are the candidates for the value function plays a key role in the dynamic programming approach and leads directly to the concept of a reachable set in the latter half of this chapter.

Definition 5.4.4 (Reachable Set). Let $\mathscr{Q}$ be defined $\mathscr{Q} \triangleq \left\{\left(\varepsilon, \mathscr{Y}, \breve{\mathscr{Z}}, \mathscr{Z}\right) \in [t_0, t_f] \times (\mathbb{R}^{n\times n})^k \times (\mathbb{R}^n)^k \times \mathbb{R}^k\right\}$ such that $\mathscr{L}_{\varepsilon,\mathscr{Y},\breve{\mathscr{Z}},\mathscr{Z};\mu} \times \mathscr{K}_{\varepsilon,\mathscr{Y},\breve{\mathscr{Z}},\mathscr{Z};\mu} \neq \emptyset$.

By adapting to the initial cost problem and the terminologies present in the statistical optimal control here, the Hamilton-Jacobi-Bellman (HJB) equation satisfied by the value function $\mathscr{V}\left(\varepsilon, \mathscr{Y}, \breve{\mathscr{Z}}, \mathscr{Z}\right)$ is then given as follows.

Theorem 5.4.1 (HJB Equation for Mayer Problem). *Let $\left(\varepsilon, \mathscr{Y}, \breve{\mathscr{Z}}, \mathscr{Z}\right)$ be any interior point of the reachable set $\mathscr{Q}$ at which the value function $\mathscr{V}\left(\varepsilon, \mathscr{Y}, \breve{\mathscr{Z}}, \mathscr{Z}\right)$ is differentiable. If there exist optimal feedforward $l^* \in \mathscr{L}_{\varepsilon,\mathscr{Y},\breve{\mathscr{Z}},\mathscr{Z};\mu}$ and feedback $K^* \in \mathscr{K}_{\varepsilon,\mathscr{Y},\breve{\mathscr{Z}},\mathscr{Z};\mu}$, then the partial differential equation of dynamic programming*

$$
\begin{aligned}
0 = \min_{l \in \overline{L},\, K \in \overline{K}} \Big\{ &\frac{\partial}{\partial \varepsilon} \mathscr{V}\left(\varepsilon, \mathscr{Y}, \breve{\mathscr{Z}}, \mathscr{Z}\right) + \frac{\partial}{\partial \operatorname{vec}(\mathscr{Y})} \mathscr{V}\left(\varepsilon, \mathscr{Y}, \breve{\mathscr{Z}}, \mathscr{Z}\right) \operatorname{vec}\left(\mathscr{F}\left(\varepsilon, \mathscr{Y}, K\right)\right) \\
&+ \frac{\partial}{\partial \operatorname{vec}\left(\breve{\mathscr{Z}}\right)} \mathscr{V}\left(\varepsilon, \mathscr{Y}, \breve{\mathscr{Z}}, \mathscr{Z}\right) \operatorname{vec}\left(\breve{\mathscr{G}}\left(\varepsilon, \mathscr{Y}, \breve{\mathscr{Z}}, K, l\right)\right) \\
&+ \frac{\partial}{\partial \operatorname{vec}(\mathscr{Z})} \mathscr{V}\left(\varepsilon, \mathscr{Y}, \breve{\mathscr{Z}}, \mathscr{Z}\right) \operatorname{vec}\left(\mathscr{G}\left(\varepsilon, \mathscr{Y}, \breve{\mathscr{Z}}, l\right)\right) \Big\}
\end{aligned} \tag{5.32}
$$

is satisfied when the boundary condition is given by

$$
\mathscr{V}\left(t_0, \mathscr{H}(t_0), \breve{\mathscr{D}}(t_0), \mathscr{D}(t_0)\right) = \phi_0\left(t_0, \mathscr{H}(t_0), \breve{\mathscr{D}}(t_0), \mathscr{D}(t_0)\right).
$$

Proof. The proof can be obtained by adapting the results from [6] with the aid of the isomorphic mapping $\operatorname{vec}(\cdot)$ from the whole vector space of $\mathbb{R}^{s_1 \times s_2}$ to the entire vector space of $\mathbb{R}^{s_1 s_2}$ as shown in [7].

Such a dynamic programming framework and the results aforementioned support builds on the sufficient condition for the optimality as follows.

Theorem 5.4.2 (Verification Theorem). *Fix $k \in \mathbb{Z}^+$ and let $\mathscr{W}\left(\varepsilon, \mathscr{Y}, \breve{\mathscr{Z}}, \mathscr{Z}\right)$ be a continuously differentiable solution of the HJB equation (5.32) which satisfies the boundary condition*

$$
\mathscr{W}\left(t_0, \mathscr{H}(t_0), \breve{\mathscr{D}}(t_0), \mathscr{D}(t_0)\right) = \phi_0\left(t_0, \mathscr{H}(t_0), \breve{\mathscr{D}}(t_0), \mathscr{D}(t_0)\right). \tag{5.33}
$$

Let $(t_f, \mathscr{H}_f, \breve{\mathscr{D}}_f, \mathscr{D}_f)$ be in $\mathscr{Q}$; (l, K) in $\mathscr{L}_{t_f, \mathscr{H}_f, \breve{\mathscr{D}}_f, \mathscr{D}_f; \mu} \times \mathscr{K}_{t_f, \mathscr{H}_f, \breve{\mathscr{D}}_f, \mathscr{D}_f; \mu}$; $\mathscr{H}$, $\breve{\mathscr{D}}$ and $\mathscr{D}$ the corresponding solutions of (5.28)–(5.30). Then $\mathscr{W}(\tau, \mathscr{H}(\tau), \breve{\mathscr{D}}(\tau), \mathscr{D}(\tau))$ is a non-increasing function of τ. If the 2-tuple (l^*, K^*) is in $\mathscr{L}_{t_f, \mathscr{H}_f, \breve{\mathscr{D}}_f, \mathscr{D}_f; \mu} \times \mathscr{K}_{t_f, \mathscr{H}_f, \breve{\mathscr{D}}_f, \mathscr{D}_f; \mu}$ defined on $[t_0, t_f]$ with corresponding solutions, $\mathscr{H}^*$, $\breve{\mathscr{D}}^*$ and $\mathscr{D}^*$ of the dynamical equations (5.28)–(5.30) such that for $\tau \in [t_0, t_f]$

$$
\begin{aligned}
0 = &\frac{\partial}{\partial \varepsilon} \mathscr{W}\left(\tau, \mathscr{H}^*(\tau), \breve{\mathscr{D}}^*(\tau), \mathscr{D}^*(\tau)\right) \\
&+ \frac{\partial}{\partial \operatorname{vec}(\mathscr{Y})} \mathscr{W}\left(\tau, \mathscr{H}^*(\tau), \breve{\mathscr{D}}^*(\tau), \mathscr{D}^*(\tau)\right) \operatorname{vec}\left(\mathscr{F}\left(\tau, \mathscr{H}^*(\tau), K^*(\tau)\right)\right) \\
&+ \frac{\partial}{\partial \operatorname{vec}(\breve{\mathscr{Z}})} \mathscr{W}\left(\tau, \mathscr{H}^*(\tau), \breve{\mathscr{D}}^*(\tau), \mathscr{D}^*(\tau)\right) \operatorname{vec}\left(\breve{\mathscr{G}}\left(\tau, \mathscr{H}^*(\tau), \breve{\mathscr{D}}^*(\tau), K^*(\tau), l^*(\tau)\right)\right) \\
&+ \frac{\partial}{\partial \operatorname{vec}(\mathscr{Z})} \mathscr{W}\left(\tau, \mathscr{H}^*(\tau), \breve{\mathscr{D}}^*(\tau), \mathscr{D}^*(\tau)\right) \operatorname{vec}\left(\mathscr{G}\left(\tau, \mathscr{H}^*(\tau), \breve{\mathscr{D}}^*(\tau), l^*(\tau)\right)\right)
\end{aligned} \tag{5.34}
$$

then l^* and K^* are optimal. Moreover,

$$\mathscr{W}\left(\varepsilon, \mathscr{Y}, \breve{\mathscr{Z}}, \mathscr{Z}\right) = \mathscr{V}\left(\varepsilon, \mathscr{Y}, \breve{\mathscr{Z}}, \mathscr{Z}\right) \tag{5.35}$$

where $\mathscr{V}\left(\varepsilon, \mathscr{Y}, \breve{\mathscr{Z}}, \mathscr{Z}\right)$ is the value function.

Proof. The proof is relegated to the adaptation of the Mayer-form verification theorem in [6]. And the technical details can be found in [7].

5.5 Risk-Averse Control as Adaptive Behavior

While recognizing that the dynamic programming approach is applicable to the control optimization of Mayer type, the states of the dynamical system (5.28)–(5.30) defined on the interval $[t_0, \varepsilon]$ have the terminal values denoted by $\mathscr{H}(\varepsilon) = \mathscr{Y}$, $\breve{\mathscr{D}}(\varepsilon) = \breve{\mathscr{Z}}$, and $\mathscr{D}(\varepsilon) = \mathscr{Z}$, the quadratic-affine property of the mean-risk aware performance index (5.31) is equally insistent that a candidate solution to the HJB equation (5.32) takes the form of

$$\mathscr{W}\left(\varepsilon, \mathscr{Y}, \breve{\mathscr{Z}}, \mathscr{Z}\right) = x_0^T \sum_{r=1}^{k} \mu_r \left(\mathscr{Y}_r + \mathscr{E}_r(\varepsilon)\right) x_0 + 2x_0^T \sum_{r=1}^{k} \mu_r \left(\breve{\mathscr{Z}}_r + \breve{\mathscr{T}}_r(\varepsilon)\right) + \sum_{r=1}^{k} \mu_r \left(\mathscr{Z}_r + \mathscr{T}_r(\varepsilon)\right) \tag{5.36}$$

whereupon the time-parametric functions $\mathscr{E}_r \in \mathscr{C}^1([t_0, t_f]; \mathbb{R}^{(n_1+n_2)\times(n_1+n_2)})$, $\breve{\mathscr{T}}_r \in \mathscr{C}^1([t_0, t_f]; \mathbb{R}^{n_1+n_2})$ and $\mathscr{T}_r \in \mathscr{C}^1([t_0, t_f]; \mathbb{R})$ are to be determined.

Using the isomorphic vec mapping, there is no difficulty to verify the following result: Fix $k \in \mathbb{Z}^+$ and let $\left(\varepsilon, \mathscr{Y}, \breve{\mathscr{Z}}, \mathscr{Z}\right)$ be any interior point of the reachable set $\mathscr{Q}$ at which the real-valued function $\mathscr{W}\left(\varepsilon, \mathscr{Y}, \breve{\mathscr{Z}}, \mathscr{Z}\right)$ of the form (5.36) is differentiable. The derivative of $\mathscr{W}\left(\varepsilon, \mathscr{Y}, \breve{\mathscr{Z}}, \mathscr{Z}\right)$ with respect to ε is given

$$\frac{d}{d\varepsilon}\mathscr{W}\left(\varepsilon, \mathscr{Y}, \breve{\mathscr{Z}}, \mathscr{Z}\right) = x_0^T \sum_{r=1}^{k} \mu_r (\mathscr{F}_r(\varepsilon, \mathscr{Y}, K) + \frac{d}{d\varepsilon}\mathscr{E}_r(\varepsilon))x_0 + 2x_0^T \sum_{r=1}^{k} \mu_r (\breve{\mathscr{G}}_r\left(\varepsilon, \mathscr{Y}, \breve{\mathscr{Z}}, K, l\right) + \frac{d}{d\varepsilon}\breve{\mathscr{T}}_r(\varepsilon)) + \sum_{r=1}^{k} \mu_r (\mathscr{G}_r\left(\varepsilon, \mathscr{Y}, \breve{\mathscr{Z}}, l\right) + \frac{d}{d\varepsilon}\mathscr{T}_r(\varepsilon)) \tag{5.37}$$

provided $l \in \overline{L}$ and $K \in \overline{K}$.

Responding to the guess solution (5.36) and the result (5.37), the HJB equation for the Mayer problem here can therefore be rewritten as follows

$$0 = \min_{l \in \overline{L}, K \in \overline{K}} \Big\{ x_0^T \sum_{r=1}^{k} \mu_r (\mathscr{F}_r(\varepsilon, \mathscr{Y}, K) + \frac{d}{d\varepsilon} \mathscr{E}_r(\varepsilon)) x_0$$

$$+ 2x_0^T \sum_{r=1}^{k} \mu_r (\breve{\mathscr{G}}_r \left(\varepsilon, \mathscr{Y}, \breve{\mathscr{Z}}, K, l\right) + \frac{d}{d\varepsilon} \breve{\mathscr{T}}_r(\varepsilon))$$

$$+ \sum_{r=1}^{k} \mu_r (\mathscr{G}_r \left(\varepsilon, \mathscr{Y}, \breve{\mathscr{Z}}, l\right) + \frac{d}{d\varepsilon} \mathscr{T}_r(\varepsilon)) \Big\} . \tag{5.38}$$

Due to the fact the initial condition x_0 is an arbitrary vector, the necessary condition for an extremum of (5.38) on $[t_0, \varepsilon]$ is obtained by differentiating the expression within the bracket of (5.38) with respect to l and K as

$$l(\varepsilon, \breve{\mathscr{Z}}) = -R^{-1}(\varepsilon) B_\epsilon^T(\varepsilon, x) \sum_{r=1}^{k} \hat{\mu}_r \breve{\mathscr{Z}}_r \tag{5.39}$$

$$K(\varepsilon, \mathscr{Y}) = -R^{-1}(\varepsilon) B_\epsilon^T(\varepsilon, x) \sum_{r=1}^{k} \hat{\mu}_r \mathscr{Y}_r \tag{5.40}$$

where the adjusted degrees of freedom $\hat{\mu}_r \triangleq \mu_r / \mu_1$ and $\mu_1 > 0$.

Recently the relevance of the time-dependent functions $\{\mathscr{E}_r(\cdot)\}_{r=1}^k$, $\left\{\breve{\mathscr{T}}_r(\cdot)\right\}_{r=1}^k$ and $\{\mathscr{T}_r(\cdot)\}_{r=1}^k$ to the sufficient condition of (5.38) is further emphasized by the need for the left-hand side of (5.38) being zero for any $\varepsilon \in [t_0, t_f]$. In this regard, the selection of the time-dependent functions $\{\mathscr{E}_r(\cdot)\}_{r=1}^k$, $\left\{\breve{\mathscr{T}}_r(\cdot)\right\}_{r=1}^k$ and $\{\mathscr{T}_r(\cdot)\}_{r=1}^k$ whose derivation is similar to those from [5] is therefore briefly presented as follows

$$\frac{d}{d\varepsilon} \mathscr{E}_1(\varepsilon) = (A_\epsilon(\varepsilon) + B_\epsilon(\varepsilon, x) K(\varepsilon))^T \mathscr{H}_1(\varepsilon) + \mathscr{H}_1(\varepsilon)(A_\epsilon(\varepsilon) + B_\epsilon(\varepsilon, x) K(\varepsilon))$$
$$+ Q_\epsilon(\varepsilon) + K^T(\varepsilon) R(\varepsilon) K(\varepsilon) \tag{5.41}$$

$$\frac{d}{d\varepsilon} \mathscr{E}_r(\varepsilon) = (A_\epsilon(\varepsilon) + B_\epsilon(\varepsilon, x) K(\varepsilon))^T \mathscr{H}_r(\varepsilon) + \mathscr{H}_r(\varepsilon)(A_\epsilon(\varepsilon) + B_\epsilon(\varepsilon, x) K(\varepsilon))$$
$$+ \sum_{s=1}^{r-1} \frac{2r!}{s!(r-s)!} \mathscr{H}_s(\varepsilon) G_\epsilon(\varepsilon) W G_\epsilon^T(\varepsilon) \mathscr{H}_{r-s}(\varepsilon), \quad 2 \le r \le k \tag{5.42}$$

$$\frac{d}{d\varepsilon}\breve{\mathscr{T}}_1(\varepsilon) = (A_\epsilon(\varepsilon) + B_\epsilon(\varepsilon, x)K(\varepsilon))^T \breve{\mathscr{D}}_1(\varepsilon) + \mathscr{H}_1(\varepsilon)(B_\epsilon(\varepsilon, x)l(\varepsilon) + G_\epsilon(\varepsilon)m_w(\varepsilon)) + K^T(\varepsilon)R(\varepsilon)l(\varepsilon) \quad (5.43)$$

$$\frac{d}{d\varepsilon}\breve{\mathscr{T}}_r(\varepsilon) = (A_\epsilon(\varepsilon) + B_\epsilon(\varepsilon, x)K(\varepsilon))^T \breve{\mathscr{D}}_r(\varepsilon) + \mathscr{H}_r(\varepsilon)(B_\epsilon(\varepsilon, x)l(\varepsilon) + G_\epsilon(\varepsilon)m_w(\varepsilon)), \quad 2 \le r \le k \quad (5.44)$$

$$\frac{d}{d\varepsilon}\mathscr{T}_1(\varepsilon) = \mathrm{Tr}\left\{\mathscr{H}_1(\varepsilon)G_\epsilon(\varepsilon)WG_\epsilon^T(\varepsilon)\right\} + 2\breve{\mathscr{D}}_1^T(\varepsilon)(B_\epsilon(\varepsilon, x)l(\varepsilon) + G_\epsilon(\varepsilon)m_w(\varepsilon)) + l^T(\varepsilon)R(\varepsilon)l(\varepsilon) \quad (5.45)$$

$$\frac{d}{d\varepsilon}\mathscr{T}_r = \mathrm{Tr}\left\{\mathscr{H}_r(\varepsilon)G_\epsilon(\varepsilon)WG_\epsilon^T(\varepsilon)\right\} + 2\breve{\mathscr{D}}_r^T(\varepsilon)(B_\epsilon(\varepsilon, x)l(\varepsilon) + G_\epsilon(\tau)m_w(\varepsilon)), \quad 2 \le r \le k. \quad (5.46)$$

In effect, the boundary condition of $\mathscr{W}(\varepsilon, \mathscr{Y}, \breve{\mathscr{Z}}, \mathscr{Z})$ implies that

$$x_0^T \sum_{r=1}^{k} \mu_r \left(\mathscr{H}_r(t_0) + \mathscr{E}_r(t_0)\right) x_0 + 2x_0^T \sum_{r=1}^{k} \mu_r \left(\breve{\mathscr{D}}_r(t_0) + \breve{\mathscr{T}}_r(t_0)\right)$$

$$+ \sum_{r=1}^{k} \mu_r \left(\mathscr{D}_r(t_0) + \mathscr{T}_r(t_0)\right) = x_0^T \sum_{r=1}^{k} \mu_r \mathscr{H}_r(t_0) x_0 + 2x_0^T \sum_{r=1}^{k} \mu_r \breve{\mathscr{D}}_r(t_0) + \sum_{r=1}^{k} \mu_r \mathscr{D}_r(t_0).$$

In the light of the result aforementioned, one can easily obtain $\mathscr{E}_r(t_0) = 0$, $\breve{\mathscr{T}}_r(t_0) = 0$ and $\mathscr{T}_r(t_0) = 0$. Finally, the optimal feedforward input (5.39) and feedback gain (5.40) minimizing the mean-risk aware performance index (5.31) become

$$l^*(\varepsilon) = -\, R^{-1}(\varepsilon)B_\epsilon^T(\varepsilon, x)\sum_{r=1}^{k} \hat{\mu}_r \breve{\mathscr{D}}_r^*(\varepsilon)$$

$$K^*(\varepsilon) = -\, R^{-1}(\varepsilon)B_\epsilon^T(\varepsilon, x)\sum_{r=1}^{k} \hat{\mu}_r \mathscr{H}_r^*(\varepsilon).$$

A brief description on control analysis and the algorithmic development on risk-averse control as adaptive behavior for the class of weakly coupled bilinear stochastic systems is now captured in the following presentation, which is further depicted in Fig. 5.2.

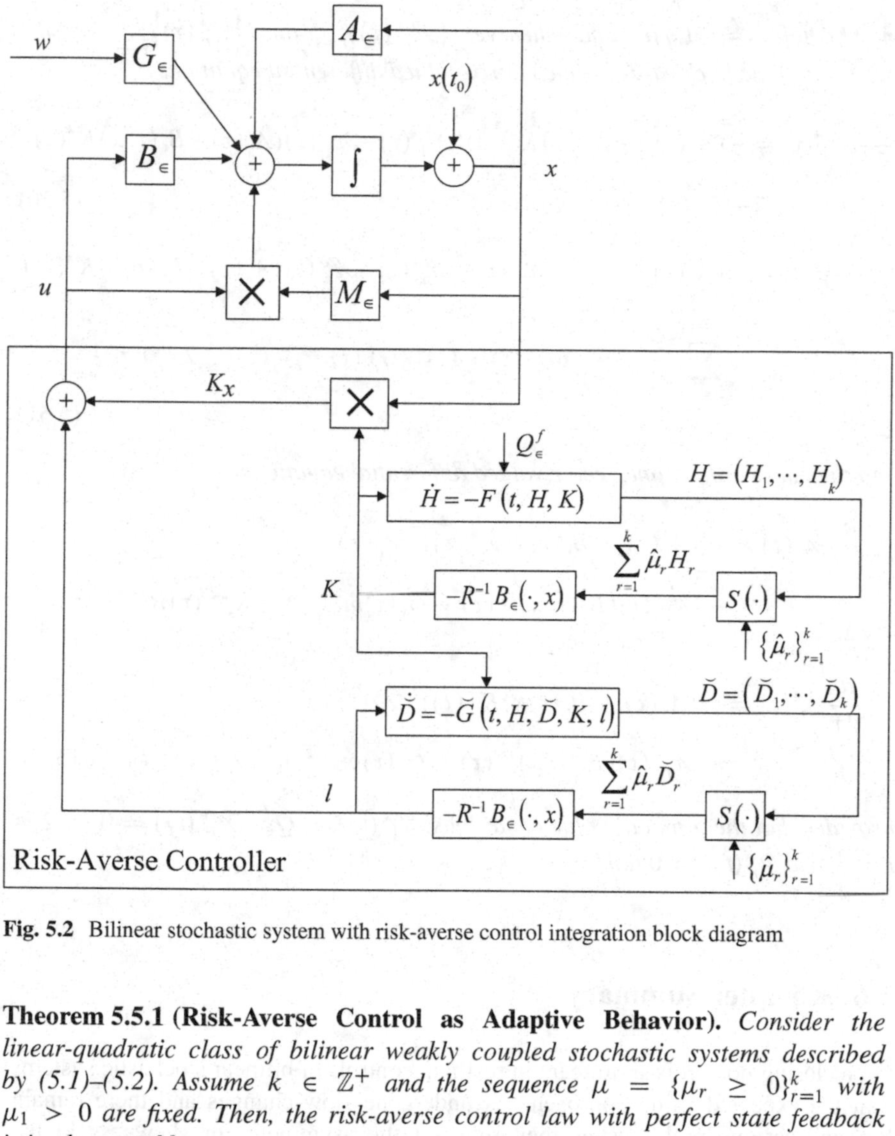

Fig. 5.2 Bilinear stochastic system with risk-averse control integration block diagram

Theorem 5.5.1 (Risk-Averse Control as Adaptive Behavior). *Consider the linear-quadratic class of bilinear weakly coupled stochastic systems described by (5.1)–(5.2). Assume* $k \in \mathbb{Z}^+$ *and the sequence* $\mu = \{\mu_r \geq 0\}_{r=1}^{k}$ *with* $\mu_1 > 0$ *are fixed. Then, the risk-averse control law with perfect state feedback is implemented by*

$$u^*(t) = K^*(t)x^*(t) + l^*(t), \quad t = t_0 + t_f - \tau, \forall \tau \in [t_0, t_f] \tag{5.47}$$

$$K^*(\tau) = -R^{-1}(\tau)B_\epsilon^T(\tau, x)\sum_{r=1}^{k}\hat{\mu}_r \mathscr{H}_r^*(\tau) \tag{5.48}$$

$$l^*(\tau) = -R^{-1}(\tau)B_\epsilon^T(\tau, x)\sum_{r=1}^{k}\hat{\mu}_r \breve{\mathscr{D}}_r^*(\tau) \tag{5.49}$$

in which $\hat{\mu}_r \triangleq \mu_r/\mu_1$ *and whenever* $\{\mathcal{H}_r^*(\tau)\}_{r=1}^k$ *and* $\left\{\breve{\mathcal{D}}_r^*(\tau)\right\}_{r=1}^k$ *are the solutions of the backward-in-time matrix-valued differential equations*

$$\frac{d}{d\tau}\mathcal{H}_1^*(\tau) = -(A_\epsilon(\tau)+B_\epsilon(\tau,x)K^*(\tau))^T\mathcal{H}_1^*(\tau) - \mathcal{H}_1^*(\tau)(A_\epsilon(\tau)+B_\epsilon(\tau,x)K^*(\tau)) - Q_\epsilon(\tau) - K^{*T}(\tau)R(\tau)K^*(\tau) \quad (5.50)$$

$$\frac{d}{d\tau}\mathcal{H}_r^*(\tau) = -(A_\epsilon(\tau)+B_\epsilon(\tau,x)K^*(\tau))^T\mathcal{H}_r^*(\tau) - \mathcal{H}_r^*(\tau)(A_\epsilon(\tau)+B_\epsilon(\tau,x)K^*(\tau)) - \sum_{s=1}^{r-1}\frac{2r!}{s!(r-s)!}\mathcal{H}_s^*(\tau)G_\epsilon(\tau)WG_\epsilon^T(\tau)\mathcal{H}_{r-s}^*(\tau), \quad 2 \le r \le k \quad (5.51)$$

and the backward-in-time vector-valued differential equations

$$\frac{d}{d\tau}\breve{\mathcal{D}}_1^*(\tau) = -\,(A_\epsilon(\tau) + B_\epsilon(\tau,x)K^*(\tau))^T\breve{\mathcal{D}}_1^*(\tau) - \mathcal{H}_1^*(\tau)(B_\epsilon(\tau,x)l^*(\tau) + G_\epsilon(\tau)m_w(\tau)) - K^{*T}(\tau)R(\tau)l^*(\tau) \quad (5.52)$$

$$\frac{d}{d\tau}\breve{\mathcal{D}}_r^*(\tau) = -\,(A_\epsilon(\tau) + B_\epsilon(\tau,x)K^*(\tau))^T\breve{\mathcal{D}}_r^*(\tau) - \mathcal{H}_r^*(\tau)(B_\epsilon(\tau,x)l^*(\tau) + G_\epsilon(\tau)m_w(\tau)), \quad 2 \le r \le k \quad (5.53)$$

provided that the terminal-value conditions $\mathcal{H}_1^*(t_f) = Q_\epsilon^f$, $\mathcal{H}_r^*(t_f) = 0$ *for* $2 \le r \le k$ *and* $\breve{\mathcal{D}}_r^*(t_f) = 0$ *for* $1 \le r \le k$.

5.6 Chapter Summary

On a closing note, it is significant in resilient controls of bilinear stochastic systems that the expected value approach has undergone slow changes and there remain big and serious problems as they rule out the asymmetry or skewness in the probabilistic performance distributions of the chi-squared random costs. In order to do away with the fact that no distinct progress has been made in determining optimal selection rules for uncertain prospects beyond the expected values, the research findings as proposed here lay stress on the key aspects of the quadratic class of weakly coupled bilinear stochastic systems together with its characterization of performance-measure statistics so as to set up the corresponding statistical optimal control problem in which, from recently opened doors and also from windows, new perspectives on the procedural mechanism for resilient controller designs endowed with perfect state-feedback measurements and subject to performance risk aversion are revealed.

References

1. Mohler, R.R.: Nonlinear Systems: Cybernetics Applications to Bilinear Control. Prentice-Hall, Englewood Cliffs (1991)
2. Desai, U.: Realization of bilinear stochastic systems. IEEE Trans. Autom. Control **AC-31**, 189–192 (1986)
3. Yaz, E.: Full and reduced-ordered observer design for discrete stochastic bilinear systems. IEEE Trans. Autom. Control **AC-37**, 503–505 (1992)
4. Pollatsek, A., Tversky, A.: Theory of risk. J. Math. Psychol. **7**, 540–553 (1970)
5. Pham, K.D.: Linear-Quadratic Controls in Risk-Averse Decision Making: Performance-Measure Statistics and Control Decision Optimization. Springer Briefs in Optimization. Springer, New York (2012). ISBN:978-1-4614-5078-8
6. Fleming W.H., Rishel, R.W.: Deterministic and Stochastic Optimal Control. Springer, New York (1975)
7. Pham, K.D.: Performance-reliability-aided decision-making in multiperson quadratic decision games against jamming and estimation confrontations. In: Giannessi, F. (ed.) J. Optim. Theory Appl. **149**(1), 599–629 (2011)

Chapter 6
Risk-Averse Control of Weakly Coupled Bilinear Stochastic Systems

6.1 Introduction

Bilinear systems are key systems to comprehensive understanding of linear and nonlinear systems with far-reaching impact on a variety of physical, chemical, biological and nuclear systems [1]. In essence, the product of state and control variables which arise from bilinear systems can provide means for approximating any nonlinear input-output maps as closely as desired provided that the nonlinear maps satisfy certain continuity and causality conditions. Thus, bilinear systems are contended as the cross-fertilization of ideas; e.g., Are all nonlinear systems bilinear?

There is an abundance of scholarly literature about stochastic bilinear control systems including such [2,3] and others. Additionally, concepts relating to trajectory regulatory and sensitivity reductions seem to be the winner. About these matters much has been written [4] and [5]. One common result is that the performance appraisal of such research is largely seen as rather simple and incomplete, where the conventional measure of performance variations is largely centered on expected values and thereby, may not be accurate enough to describe complex behavior of closed-loop performance distributions of today's real world problems.

The changes with which this chapter is grappling and which are propelling performance uncertainty analysis and quantification of the chi-squared random costs for the restrictive class of weakly coupled bilinear stochastic systems into previously unexperienced stochastic control and development process certainly involve many levels of statistical measures for performance risk. Understandably, changes in optimal selection rules for ordering uncertain prospects due to the inherent asymmetry or skewness in the probability distribution functions are visibly influencing the innovative technical approach as being proposed and thereby, overcoming the limitations of traditional evaluation schemes which are mainly based on just the expected outcome and outcome variance and thus would necessarily imply indifference between some courses of action as pointed out in [6].

K.D. Pham, *Resilient Controls for Ordering Uncertain Prospects*, Springer Optimization and Its Applications 98, DOI 10.1007/978-3-319-08705-4_6

The structure of this chapter is as follows. Section 6.2 contains the development of all the mathematical statistics for performance robustness for the underlying family of chi-squared random costs. Positive attitudes towards the concept of resilient controls for low sensitivity and performance risk aversion is quite tangible in the problem statements and solution method of Mayer type as shown in Sect. 6.3. In addition, Sect. 6.4 gives an account of the research attention focused on the control algorithm. All in all, Sect. 6.5 offers some final remarks.

6.2 Assessing Performance Statistics for Robustness

Throughout the chapter, a fixed probability space $(\Omega, \mathbb{F}, \{\mathbb{F}_{t_0,t} : t \in [t_0, t_f]\}, \mathbb{P})$ with filtration is satisfying the usual conditions. All the filtrations are right continuous and complete and $\mathbb{F}_{t_f} \triangleq \{\mathbb{F}_{t_0,t} : t \in [t_0, t_f]\}$. In addition, let $\mathcal{L}^2_{\mathbb{F}_{t_f}}([t_0, t_f]; \mathbb{R}^n)$ denote the space of $\mathbb{F}_{t_f}$-adapted random processes $\{\hbar(t) : t \in [t_0, t_f]$ such that $E\{\int_{t_0}^{t_f} ||\hbar(t)||^2 \, dt\} < \infty\}$.

Due to very complex structures of nonlinear weakly coupled stochastic control systems, the presence of small weak coupling and constant parameters as well as the bilinearization procedure will be exploited to result in better understanding and approximation. Very often in the literature on weakly coupled bilinear stochastic control systems, the mathematical model for time-varying weakly coupled bilinear stochastic systems on a finite horizon $[t_0, t_f]$ is recorded as

$$\begin{bmatrix} \dot{x}_1(t) \\ \dot{x}_2(t) \end{bmatrix} = \begin{bmatrix} A_1(t,\zeta) & \epsilon A_2(t,\zeta) \\ \epsilon A_3(t,\zeta) & A_4(t,\zeta) \end{bmatrix} \begin{bmatrix} x_1(t) \\ x_2(t) \end{bmatrix} + \begin{bmatrix} B_1(t,\zeta) & \epsilon B_2(t,\zeta) \\ \epsilon B_3(t,\zeta) & B_4(t,\zeta) \end{bmatrix} \begin{bmatrix} u_1(t) \\ u_2(t) \end{bmatrix}$$
$$+ \begin{bmatrix} x_1(t) \\ x_2(t) \end{bmatrix} \begin{bmatrix} M_a(t,\zeta) & \epsilon M_b(t,\zeta) \\ \epsilon M_c(t,\zeta) & M_d(t,\zeta) \end{bmatrix} \begin{bmatrix} u_1(t) \\ u_2(t) \end{bmatrix} + \begin{bmatrix} G_1(t) & \epsilon G_2(t) \\ \epsilon G_3(t) & G_4(t) \end{bmatrix} \begin{bmatrix} w_1(t) \\ w_2(t) \end{bmatrix} \tag{6.1}$$

with the initial known conditions

$$\begin{bmatrix} x_1(t_0) \\ x_2(t_0) \end{bmatrix} = \begin{bmatrix} x_1^0 \\ x_2^0 \end{bmatrix}$$

where for $i = 1, 2$, $x_i(t)$ is the controlled state process valued in $\mathbb{R}^{n_i}$ of the reduced-order subsystem i; $u_i(t)$ is the control process valued in a closed convex subset of $\mathbb{R}^{m_i}$; ϵ is a small weak coupling parameter; and ζ is a constant parameter. Exogenous process noises $w_i(t) \in \mathbb{R}^{p_i}$ take the form of mutually uncorrelated stationary $\mathbb{F}_{t_f}$-adapted Gaussian processes with means $E\{w_i(t)\} = m_{w_i}(t)$ and covariances $cov\{w_i(t_1), w_i(t_2)\} = W_i \delta(t_1 - t_2)$ for all $t_1, t_2 \in [t_0, t_f]$ and $W_i > 0$.

In the problem description (6.1), the system coefficients A_j, B_j, G_j for $j = 1, 2, 3, 4$ as well as M_a, M_b, M_c, and M_d are of appropriate dimensions, weakly coupled, and continuous-time matrix functions of the constant parameter ζ while the disturbance distribution coefficients G_j are independent of ζ.

The following notation is used in order to relate (6.1) and (6.4)

$$x(t) \triangleq \begin{bmatrix} x_1(t) \\ x_2(t) \end{bmatrix}; \quad u(t) \triangleq \begin{bmatrix} u_1(t) \\ u_2(t) \end{bmatrix}; \quad w(t) \triangleq \begin{bmatrix} w_1(t) \\ w_2(t) \end{bmatrix}; \quad x(t_0) \triangleq x_0 = \begin{bmatrix} x_1^0 \\ x_2^0 \end{bmatrix}$$

$$A_\epsilon(t,\zeta) \triangleq \begin{bmatrix} A_1(t,\zeta) & \epsilon A_2(t,\zeta) \\ \epsilon A_3(t,\zeta) & A_4(t,\zeta) \end{bmatrix}; \quad B_\epsilon(t,\zeta) \triangleq \begin{bmatrix} B_1(t,\zeta) & \epsilon B_2(t,\zeta) \\ \epsilon B_3(t,\zeta) & B_4(t,\zeta) \end{bmatrix}$$

$$M_\epsilon(t,\zeta) \triangleq \begin{bmatrix} M_a(t,\zeta) & \epsilon M_b(t,\zeta) \\ \epsilon M_c(t,\zeta) & M_d(t,\zeta) \end{bmatrix}; \quad G_\epsilon(t) \triangleq \begin{bmatrix} G_1(t) & \epsilon G_2(t) \\ \epsilon G_3(t) & G_4(t) \end{bmatrix}.$$

The weakly coupled bilinear stochastic control system under consideration is represented compactly by

$$\dot{x}(t) = A_\epsilon(t,\zeta)x(t) + (B_\epsilon(t,\zeta) + x(t)M_\epsilon(t,\zeta))u(t) + G_\epsilon(t)w(t), \quad x(t_0) \tag{6.2}$$

where the aggregate process noise $w(t) \in \mathbb{R}^{p_1+p_2}$ is modeled by a stationary $\mathbb{F}_{t_f}$-adapted Gaussian process with its mean $E\{w(t)\} = m_w(t)$ and intensity matrix $cov\{w(t_1), w(t_2)\} = W\delta(t_1 - t_2)$ for all t_1, $t_2 \in [t_0, t_f]$ and $W \triangleq \begin{bmatrix} W_1 & 0 \\ 0 & W_2 \end{bmatrix} > 0$.

Note here that the state $x(t)$ is considered as the function of the constant parameter variable ζ, i.e., $x(t) \equiv \pi(t, x_0, \zeta)$. The traditional approach to sensitivity analysis for the system trajectory has been to define a sensitivity variable by

$$\sigma(t) \triangleq \frac{\partial}{\partial\zeta}\pi(t, x_0, \zeta)\bigg|_{\zeta=\zeta_0}. \tag{6.3}$$

As usual, the restricted class of performance costs having its ranges of variability bounded below from zero, is associated with the stochastic controlled system (6.2)

$$J(u) = x^T(t_f)Q_\epsilon^f x(t_f) + \int_{t_0}^{t_f} \left[x^T(\tau)Q_\epsilon(\tau)x(\tau) + u^T(\tau)R(\tau)u(\tau) + \sigma^T(\tau)S(\tau)\sigma(\tau)\right] d\tau \tag{6.4}$$

in which the terminal penalty weighting Q_ϵ^f, the state weighting $Q_\epsilon(\tau)$, the trajectory sensitivity weighting $S(\tau)$ and control weighting $R(\tau)$ with invertibility are continuous-time matrix functions with symmetry and positive semi-definiteness; whereas the other change of variables is given by

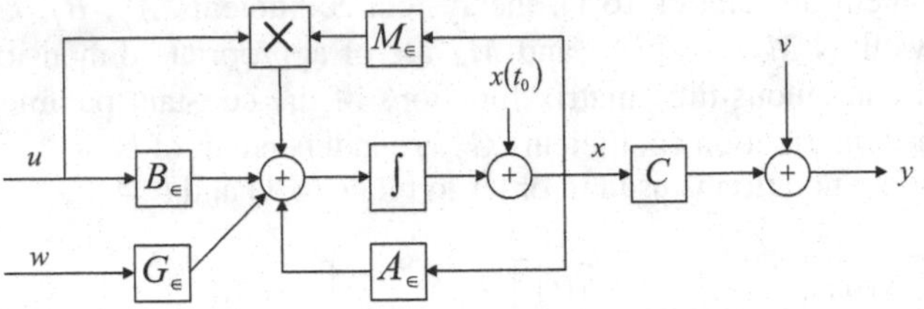

Fig. 6.1 Structure of bilinear stochastic system with some noisy observations

$$Q_\epsilon^f \triangleq \begin{bmatrix} Q_1^f & \epsilon Q_2^f \\ \epsilon (Q_2^f)^T & Q_3^f \end{bmatrix}; \; Q_\epsilon(\tau) \triangleq \begin{bmatrix} Q_1(\tau) & \epsilon Q_2(\tau) \\ \epsilon Q_2^T(\tau) & Q_3(\tau) \end{bmatrix}; \; R(\tau) \triangleq \begin{bmatrix} R_1(\tau) & 0 \\ 0 & R_2(\tau) \end{bmatrix}.$$

Towards output feedback control, $y(t)$ is the measured output of the manipulated process $x(t)$ to which, the control design has available

$$y(t) = C(t)x(t) + v(t) \tag{6.5}$$

wherein $v(t) \in \mathbb{R}^q$ is an exogenous measurement noise modeled by a mutually uncorrelated stationary $\mathbb{F}_{t_f}$-adapted Gaussian process with its mean $E\{v(t)\} = m_v(t)$ and covariance $cov\{v(t_1), v(t_2)\} = N\delta(t_1 - t_2)$ for all t_1, $t_2 \in [t_0, t_f]$. At this point, Fig. 6.1 gives a possible arrangement of the basic diagram for the system description by (6.2) and (6.5).

When interpreting the mathematical model (6.2) and (6.5) in Ito stochastic differentials, one has a way to treat white noises on a mathematically rigorous basis

$$dx(t) = (A_\epsilon(t,\zeta)x(t) + (B_\epsilon(t,\zeta) + x(t)M_\epsilon(t,\zeta))u(t) + G_\epsilon(t)m_w(t))dt$$
$$+ G_\epsilon(t)U_W \Lambda_W^{1/2} dw(t), \quad x(t_0) = x_0 \tag{6.6}$$
$$dy(t) = (C(t)x(t) + m_v(t))dt + dv(t) \tag{6.7}$$

where for each $t \in [t_0, t_f]$, U_W and Λ_W correspond to the eigen-decomposition of the intensity W of the white noise $w(t)$ such that $W = U_W \Lambda_W U_W^T$, from which the incremental stationary Wiener processes $dw(t)$ and $dv(t)$ are defined as $dw(t) \triangleq (w(t) - m_w(t))dt$ and $dv(t) \triangleq (v(t) - m_v(t))dt$, respectively.

Furthermore, the development hereafter continually probes beneath the surface of the following σ-algebras

$$\mathbb{F}_{t_0,t} \triangleq \sigma\{(w(\tau), v(\tau)) : t_0 \le \tau \le t\}$$
$$\mathcal{G}_{t_0,t}^y \triangleq \sigma\{y(\tau) : t_0 \le \tau \le t\}, \quad t \in [t_0, t_f]$$

to determine their underlying causes; e.g., the information structure available defined by $\mathscr{G}^y_{t_f} \triangleq \{\mathscr{G}^y_{t_0,t} : t \in [t_0, t_f]\} \subset \{\mathbb{F}_{t_0,t} : t \in [t_0, t_f]\}$.

To cope with the information structure $\mathscr{G}^y_{t_f}$ which is specifically supported by the stochastic differential equation (6.5), the admissible set of feedback control laws therefore reduces to

$$\mathbb{U}^y[t_0, t_f] \triangleq \{u \in \mathscr{L}^2_{\mathscr{G}^y_{t_f}}([t_0, t_f], \mathbb{R}^{m_1+m_2}), \quad t \in [t_0, t_f], \ \mathbb{P} - a.s.\}$$

whereupon $\mathbb{U}^y[t_0, t_f]$ is a closed convex subset of $\mathscr{L}^2_{\mathscr{G}^y_{t_f}}([t_0, t_f], \mathbb{R}^{m_1+m_2})$.

Concurrent with the incomplete information pattern by $\mathscr{G}^y_{t_0,t}$ has been the set of admissible control policies to take the form, for some $\mathfrak{d}(\cdot, \cdot)$

$$u(t) = \mathfrak{d}(t, y(\tau)), \quad \tau \in [t_0, t]. \tag{6.8}$$

In furtherance of the development of conditional probability density $p(x(t)|\mathscr{G}^y_{t_0,t})$, the probability density of $x(t)$ conditioned on $\mathscr{G}^y_{t_0,t}$ represents the sufficient statistics to describe the conditional stochastic effects of future control actions. Under the Gaussian assumption, the conditional probability density $p(x(t)|\mathscr{G}^y_{t_0,t})$ is parameterized by its conditional mean $\hat{x}(t) \triangleq E\{x(t)|\mathscr{G}^y_{t_0,t}\}$ and state-estimate error covariance $\Sigma(t) \triangleq E\{[x(t) - \hat{x}(t)][x(t) - \hat{x}(t)]^T|\mathscr{G}^y_{t_0,t}\}$. In the context of the linear-Gaussian structure, it is further reported that $\Sigma(t)$ is independent of controls and feedback observations. Therefore, to look for optimal control laws of the aforementioned form, it is only required that $u(t) = \gamma(t, \hat{x}(t))$.

Of particular interest, the search for optimal control policies is now consistently and productively restricted to linear time-varying feedback laws generated from the accessible states $\hat{x}(t)$ by

$$u(t) = K(t)\hat{x}(t) + l(t), \quad t \in [t_0, t_f] \tag{6.9}$$

where both $K(\cdot)$ and $l(\cdot)$ which are of appropriate dimensions and continuous-time matrix and vector functions, will be formally defined shortly.

Then, for the admissible $K(\cdot)$, $l(\cdot)$, and the pair (t_0, x_0), it gives a sufficient condition for the existence of $x(t)$ in (6.6). In view of (6.9), the controlled system (6.6) is rewritten accordingly

$$\begin{aligned} dx(t) &= (A_\epsilon(t,\zeta)x(t) + \Delta_\epsilon(t,x,\zeta)K(t)\hat{x}(t) + \Delta_\epsilon(t,x,\zeta)l(t) \\ &\quad + G_\epsilon(t)m_w(t))dt + G_\epsilon(t)U_W \Lambda_W^{1/2} dw(t), \quad x(t_0) = x_0 \end{aligned} \tag{6.10}$$

where for convenience, it is necessary to have $\Delta_\epsilon(t,x,\zeta) \triangleq B_\epsilon(t,\zeta) + x(t)M_\epsilon(t,\zeta)$.

In addition, the variation of the parameter ζ from the nominal value ζ_0 was assumed to be small, such that the Taylor's expansion of the state and control functions can be approximated by retaining only the first two terms.

Differentiating (6.10) with respect to ζ and evaluating at $\zeta = \zeta_0$ yields the stochastic differential equation

$$d\sigma(t) = (A_\epsilon^\zeta(t,\zeta_0)x(t) + \Delta_\epsilon^\zeta(t,x,\zeta_0)K(t)\hat{x}(t)A_\epsilon(t,\zeta_0)\sigma(t) \\ +\Delta_\epsilon(t,x,\zeta_0)K(t)\hat{\sigma}(t) + \Delta_\epsilon^\zeta(t,x,\zeta_0)l(t))dt, \quad \sigma(t_0) = 0 \quad (6.11)$$

where $A_\epsilon^\zeta(\cdot,\zeta_0) \triangleq \frac{\partial A_\epsilon(\cdot,\zeta)}{\partial\zeta}\Big|_{\zeta=\zeta_0}$, $\Delta_\epsilon^\zeta(\cdot,\cdot,\zeta_0) \triangleq \frac{\partial \Delta_\epsilon(\cdot,\cdot,\zeta)}{\partial\zeta}\Big|_{\zeta=\zeta_0}$, and $\hat{\sigma}(\cdot) \triangleq \frac{\partial \hat{x}(\cdot)}{\partial\zeta}\Big|_{\zeta=\zeta_0}$.

As noted earlier, the linear-Gaussian assumptions imply that the conditional mean $\hat{x}(t)$ and the state-estimate error covariance $\Sigma(t)$ can be generated by the Kalman-like filter

$$d\hat{x}(t) = ((A_\epsilon(t,\zeta) + \Delta_\epsilon(t,x,\zeta)K(t))\hat{x}(t) + \Delta_\epsilon(t,x,\zeta)l(t) + G_\epsilon(t)m_w(t))dt \\ + L(t)(dy(t) - (C(t)\hat{x}(t) + m_v(t))dt), \quad \hat{x}(t_0) = x_0 \quad (6.12)$$

and its state-estimate error covariance equation

$$L(t) = \Sigma(t)C^T(t)N^{-1} \quad (6.13)$$

$$\frac{d}{dt}\Sigma(t) = A_\epsilon(t,\zeta)\Sigma(t) + \Sigma(t)A_\epsilon^T(t,\zeta) + G_\epsilon(t)WG_\epsilon^T(t) \\ - \Sigma(t)C^T(t)N^{-1}C(t)\Sigma(t), \quad \Sigma(t_0) = 0. \quad (6.14)$$

Figure 6.2 shows the observer as described in Eq. (6.12) for finding the conditional mean estimates of the current states.

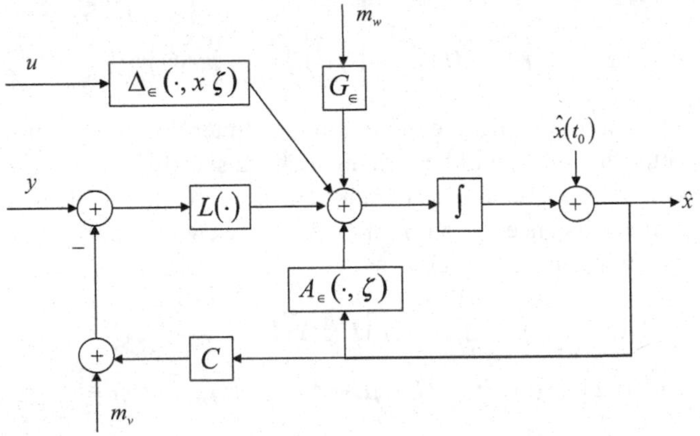

Fig. 6.2 Block diagram of Kalman-like observer for the current system states

It can be argued that such a partial of Eq. (6.12) with respect to ζ is

$$\begin{aligned} d\hat{\sigma}(t) &= (A_\epsilon^\zeta(t,\zeta_0) + \Delta_\epsilon^\zeta(t,x,\zeta_0)K(t))\hat{x}(t)dt + (A_\epsilon(t,\zeta_0) \\ &\quad + \Delta_\epsilon(t,x,\zeta_0)K(t))\hat{\sigma}(t)dt + L_\zeta(t)C(t)\tilde{x}(t)dt + \Delta_\epsilon^\zeta(t,x,\zeta_0)l(t)dt \\ &\quad + L(t)C(t)\tilde{\sigma}(t)dt + L_\zeta(t)dv(t), \quad \hat{\sigma}(t_0) = 0 \end{aligned} \tag{6.15}$$

where the state-estimate errors $\tilde{x}(t) \triangleq x(t) - \hat{x}(t)$, $\tilde{\sigma}(t) \triangleq \frac{\partial \tilde{x}}{\partial \zeta}\Big|_{\zeta=\zeta_0}$ and $L_\zeta \triangleq \frac{\partial L}{\partial \zeta}\Big|_{\zeta=\zeta_0}$. The detailed structure of the corresponding observer is described in Fig. 6.3

Using the definitions for $\tilde{x}(t)$ and $\tilde{\sigma}(t)$ together with the initial-value condition $\tilde{x}(t_0) = 0$, Figs. 6.4 and 6.5 respectively illustrate the block diagram interpretations of the errors described in Eqs. (6.16) and (6.17) of the estimates (6.12) and (6.15)

$$d\tilde{x}(t) = (A_\epsilon(t,\zeta) - L(t)C(t))\tilde{x}(t)dt - L(t)dv(t) + G_\epsilon(t)U_W \Lambda_W^{1/2} dw(t) \tag{6.16}$$

and

$$\begin{aligned} d\tilde{\sigma}(t) &= ((A_\epsilon^\zeta(t,\zeta_0) - L_\zeta(t)C(t))\tilde{x}(t) + (A_\epsilon(t,\zeta_0) - L(t)C(t))\tilde{\sigma}(t))dt \\ &\quad - L_\zeta(t)dv(t), \quad \tilde{\sigma}(t_0) = 0 \end{aligned} \tag{6.17}$$

whereupon the gain $L_\zeta(t) = \Sigma_\zeta(t)C^T(t)N^{-1}$ and the state-estimate error covariances $\Sigma_\zeta \triangleq \frac{\partial \Sigma}{\partial \zeta}\Big|_{\zeta=\zeta_0}$ associated with the state estimates are satisfying

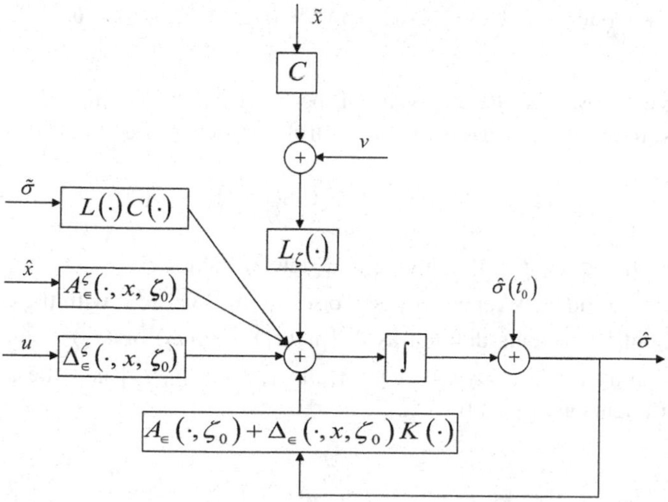

Fig. 6.3 Observer diagram for trajectory sensitivity variables

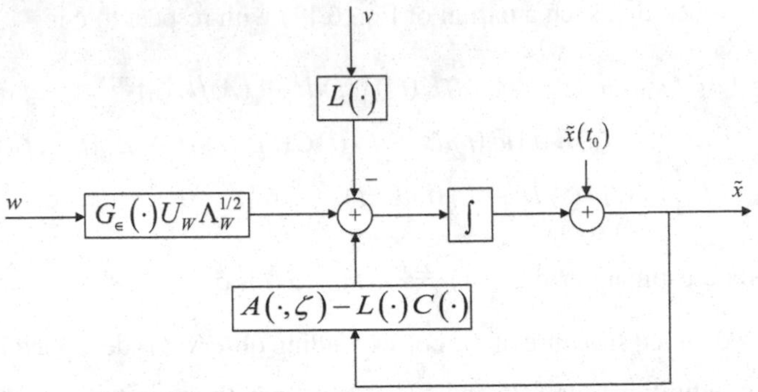

Fig. 6.4 Block diagram interpretation of estimate errors of the trajectory sensitivity variables

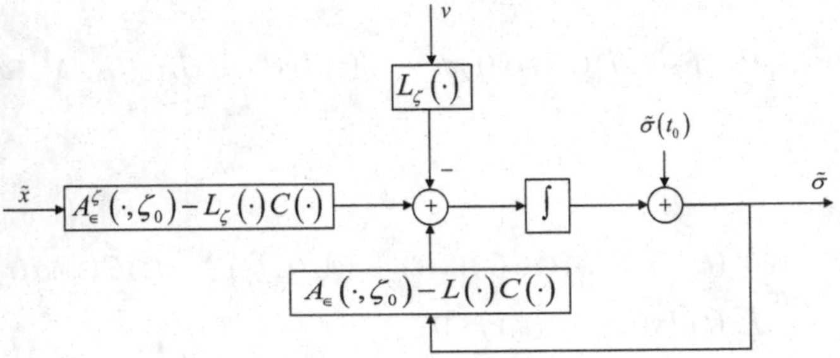

Fig. 6.5 Block diagram of estimate errors of the system states

$$
\begin{aligned}
\frac{d}{dt}\Sigma_\zeta(t) &= A_\epsilon^\zeta(t,\zeta_0)\Sigma(t) + A_\epsilon(t,\zeta_0)\Sigma_\zeta(t) + \Sigma_\zeta(t)A_\epsilon^T(t,\zeta_0) - \Sigma(t)C(t)N^{-1}C(t)\Sigma_\zeta(t) \\
&\quad + \Sigma(t)(A_\epsilon^\zeta)^T(t,\zeta_0) - \Sigma_\zeta(t)C^T(t)N^{-1}C(t)\Sigma(t), \quad \Sigma_\zeta(t_0) = 0. \qquad (6.18)
\end{aligned}
$$

To progress towards the analysis of performance reliability, the aggregate dynamics is, therefore, governed by the controlled stochastic differential equation

$$
dz(t) = (A_z(t)z(t) + b_z(t))dt + G_z(t)dw_z(t), \quad z(t_0) = z_0 \qquad (6.19)
$$

where for each $t \in [t_0, t_f]$, the augmented state variables $z^T = \left[\hat{x}^T \; \tilde{x}^T \; \hat{\sigma}^T \; \tilde{\sigma}^T\right]$, the aggregate stationary Wiener process noises $w_z^T = \left[w^T \; v^T\right]$ with the correlation of independent increments defined as $E\left\{[w_z(\tau_1) - w_z(\tau_2)][w_z(\tau_1) - w_z(\tau_2)]^T\right\} = W_z|\tau_1 - \tau_2|$ and $W_z \triangleq diag(W, N)$ for all $\tau_1, \tau_2 \in [t_0, t_f]$ and the augmented system coefficients are given by

$$A_z(t) = \begin{bmatrix} A_z^{11} & A_z^{12} & A_z^{13} & A_z^{14} \\ A_z^{21} & A_z^{22} & A_z^{23} & A_z^{24} \\ A_z^{31} & A_z^{32} & A_z^{33} & A_z^{34} \\ A_z^{41} & A_z^{42} & A_z^{43} & A_z^{44} \end{bmatrix}(t); \quad z_0 = \begin{bmatrix} x_0 \\ 0 \\ 0 \\ 0 \end{bmatrix}$$

$$b_z(t) = \begin{bmatrix} \Delta_\epsilon(t,x,\zeta)l(t) + G_\epsilon(t)m_W(t) \\ 0 \\ \Delta_\epsilon^\zeta(t,x,\zeta_0)l(t) \\ 0 \end{bmatrix}; \quad G_z(t) = \begin{bmatrix} 0 & L(t) \\ G_\epsilon(t)U_W\Lambda_W^{1/2} & -L(t) \\ 0 & L_\zeta(t) \\ 0 & -L_\zeta(t) \end{bmatrix}$$

provided $A_z^{11}(t) = A_\epsilon(t,\zeta) + \Delta_\epsilon(t,x,\zeta)K(t)$, $A_z^{12}(t) = L(t)C(t)$, $A_z^{13}(t) = A_z^{14}(t) = 0$; $A_z^{21}(t) = A_z^{23}(t) = A_z^{24}(t) = 0$, $A_z^{22}(t) = A_\epsilon(t,\zeta) - L(t)C(t)$; $A_z^{31}(t) = A_\epsilon^\zeta(t,\zeta_0) + \Delta_\epsilon^\zeta(t,x,\zeta_0)K(t)$, $A_z^{32}(t) = L_\zeta(t)C(t)$, $A_z^{33}(t) = A_\epsilon(t,\zeta_0) + \Delta_\epsilon(t,x,\zeta_0)K(t)$, $A_z^{34}(t) = L(t)C(t)$; and $A_z^{41}(t) = A_z^{43}(t) = 0$, $A_z^{42}(t) = A_\epsilon^\zeta(t,\zeta_0) - L_\zeta(t)C(t)$, and $A_z^{44}(t) = A_\epsilon(t,\zeta_0) - L(t)C(t)$ for all $t \in [t_0,t_f]$ and some ζ and ζ_0.

Additionally, the performance measure (6.4) of finite-horizon integral-quadratic-form cost is now rewritten as follows

$$J(K,l) = z^T(t_f)N_f z(t_f) + \int_{t_0}^{t_f} [z^T(\tau)N_z(\tau)z(\tau) + 2z^T(\tau)M_z(\tau) + l^T(\tau)R(\tau)l(\tau)]d\tau \tag{6.20}$$

where the continuous-time weightings are given by

$$N_f = \begin{bmatrix} Q_f & Q_f & 0 & 0 \\ Q_f & Q_f & 0 & 0 \\ 0 & 0 & 0 & 0 \\ 0 & 0 & 0 & 0 \end{bmatrix}; \quad M_z(\cdot) \triangleq \begin{bmatrix} K^T R l \\ 0 \\ 0 \\ 0 \end{bmatrix}(\cdot); \quad N_z(\cdot) = \begin{bmatrix} Q + K^T R K & Q & 0 & 0 \\ Q & Q & 0 & 0 \\ 0 & 0 & S & S \\ 0 & 0 & S & S \end{bmatrix}(\cdot).$$

One of the best ways to understand a concept of performance risks is to study the history of its development. Understanding the degrees of performance uncertainty facing control designers during the development of performance reliability helps one better understand the concept. The purpose here is to provide an understanding of performance risk measures. Henceforth, the performance risk measures are supposedly a major improvement over traditional performance averages.

After proposing the performance risk measures herein, the cumulates or semi-invariances of (6.20) are therefore of interest for two reasons: firstly, they are relevant to performance riskiness and secondly, performance measure (6.20) is not a normal random variable. Therefore, the cumulates will help control engineers make proper feedback control decisions when faced with non-normal performance measure distributions. As expected, these semi-invariances of the chi-squared random variable (6.20) are generated by the Maclaurin series expansion of the cumulant-generating function of (6.20).

Theorem 6.2.1 (Cumulant-Generating Function). *Let $z(\cdot)$ be a state variable of the stochastic dynamics concerning sensitivity (6.19) with initial values $z(\tau) \equiv z_\tau$ and $\tau \in [t_0, t_f]$. Further let the moment-generating function be defined by*

$$\varphi(\tau, z_\tau, \theta) \triangleq \varrho(\tau, \theta) \exp\left\{z_\tau^T \Upsilon(\tau, \theta) z_\tau + 2 z_\tau^T \eta(\tau, \theta)\right\} \tag{6.21}$$

$$\upsilon(\tau, \theta) \triangleq \ln\{\varrho(\tau, \theta)\}, \qquad \theta \in \mathbb{R}^+. \tag{6.22}$$

Then, the cumulant-generating function has the form of quadratic affine

$$\psi(\tau, z_\tau, \theta) = z_\tau^T \Upsilon(\tau, \theta) z_\tau + 2 z_\tau^T \eta(\tau, \theta) + \upsilon(\tau, \theta) \tag{6.23}$$

where the scalar solution $\upsilon(\tau, \theta)$ solves the backward-in-time differential equation

$$\begin{aligned} \frac{d}{d\tau}\upsilon(\tau, \theta) = &- \operatorname{Tr}\left\{\Upsilon(\tau, \theta) G_z(\tau) W_z G_z^T(\tau)\right\} \\ &- 2\eta^T(\tau, \theta) b_z(\tau) - \theta l^T(\tau) R(\tau) l(\tau), \quad \upsilon\left(t_f, \theta\right) = 0 \end{aligned} \tag{6.24}$$

the matrix solution $\Upsilon(\tau, \theta)$ satisfies the backward-in-time differential equation

$$\begin{aligned} \frac{d}{d\tau}\Upsilon(\tau, \theta) = &- A_z^T(\tau) \Upsilon(\tau, \theta) - \Upsilon(\tau, \theta) A_z(\tau) \\ &- 2\Upsilon(\tau, \theta) G_z(\tau) W_z G_z^T(\tau) \Upsilon(\tau, \theta) - \theta N_z(\tau), \quad \Upsilon(t_f, \theta) = \theta N_f \end{aligned} \tag{6.25}$$

and the vector solution $\eta(\tau, \theta)$ satisfies the backward-in-time differential equation

$$\frac{d}{d\tau}\eta(\tau, \theta) = -A_z^T(\tau) \eta(\tau, \theta) - \Upsilon(\tau, \theta) b_z(\tau) - \theta M_z(\tau), \quad \eta(t_f, \theta) = 0. \tag{6.26}$$

Also, the scalar solution $\varrho(\tau)$ satisfies the backward-in-time differential equation

$$\begin{aligned} \frac{d}{d\tau}\varrho(\tau, \theta) = &\ \varrho(\tau, \theta) \left[-\operatorname{Tr}\left\{\Upsilon(\tau, \theta) G_z(\tau) W_z G_z^T(\tau)\right\}\right. \\ &\left. - 2\eta^T(\tau, \theta) b_z(\tau) - \theta l^T(\tau) R(\tau) l(\tau)\right], \quad \varrho\left(t_f, \theta\right) = 1. \end{aligned} \tag{6.27}$$

Proof. For notional simplicity, it is convenient to have $\varpi(\tau, z_\tau, \theta) \triangleq \exp\{\theta J(\tau, z_\tau)\}$ in which the performance measure (6.20) is rewritten as the cost-to-go function from an arbitrary state z_τ at a running time $\tau \in [t_0, t_f]$, that is,

$$J(\tau,z_\tau)=z^T(t_f)N_f z(t_f)+\int_\tau^{t_f}[z^T(t)N_z(t)z(t)+2z^T(t)M_z(t)+l^T(t)R(t)l(t)]dt \tag{6.28}$$

subject to the stochastic differential equation

$$dz(t)=(A_z(t)z(t)+b_z(t))dt+G_z(t)dw_z(t),\quad z(\tau)=z_\tau. \tag{6.29}$$

By definition, the moment-generating function is

$$\varphi(\tau,z_\tau,\theta)\triangleq E\{\varpi(\tau,z_\tau,\theta)\}.$$

Therefore, the total time derivative of $\varphi(\tau,z_\tau,\theta)$ is obtained as

$$\frac{d}{d\tau}\varphi(\tau,z_\tau,\theta)=-\theta[z_\tau^T N_a(\tau)z_\tau+2z_\tau^T M_z(\tau)+l^T(\tau)R(\tau)l(\tau)]\varphi(\tau,z_\tau,\theta).$$

Using the standard Ito's stochastic differential formula, it follows

$$\begin{aligned}d\varphi(\tau,z_\tau,\theta)&=E\{d\varpi(\tau,z_\tau,\theta)\}\\&=E\left\{\varpi_\tau(\tau,z_\tau,\theta)\,d\tau+\varpi_{z_\tau}(\tau,z_\tau,\theta)\,dz_\tau+\frac{1}{2}\mathrm{Tr}\left\{\varpi_{z_\tau z_\tau}(\tau,z_\tau,\theta)G_z(\tau)W_zG_z^T(\tau)\right\}d\tau\right\},\\&=\varphi_\tau(\tau,z_\tau,\theta)d\tau+\varphi_{z_\tau}(\tau,z_\tau,\theta)(A_z(\tau)z_\tau+b_z(\tau))d\tau\\&+\frac{1}{2}\mathrm{Tr}\left\{\varphi_{z_\tau z_\tau}(\tau,z_\tau,\theta)G_z(\tau)W_zG_z^T(\tau)\right\}d\tau\end{aligned}$$

which under the hypothesis of $\varphi(\tau,z_\tau,\theta)=\varrho(\tau,\theta)\exp\left\{z_\tau^T\Upsilon(\tau,\theta)z_\tau+2z_\tau^T\eta(\tau,\theta)\right\}$, its partial derivatives and some algebraic manipulations yield

$$\begin{aligned}&-\theta[z_\tau^T N(\tau)z_\tau+2z_\tau^T M_z(\tau)+l^T(\tau)R(\tau)l(\tau)]\varphi(\tau,z_\tau,\theta)\\&=\left\{\frac{\frac{d}{d\tau}\varrho(\tau,\theta)}{\varrho(\tau,\theta)}+z_\tau^T\frac{d}{d\tau}\Upsilon(\tau,\theta)z_\tau+2z_\tau^T\frac{d}{d\tau}\eta(\tau,\theta)+z_\tau^T[A_z^T(\tau)\Upsilon(\tau,\theta)+\Upsilon(\tau,\theta)A_z(\tau)]z_\tau\right.\\&\qquad+2z_\tau^T\Upsilon(\tau,\theta)b_z(\tau)+2z_\tau^T A_z(\tau)\eta(\tau,\theta)+2\eta^T(\tau,\theta)b_z(\tau)\\&\left.+2z_\tau^T\Upsilon(\tau,\theta)G_z(\tau)W_zG_z^T(\tau)\Upsilon(\tau,\theta)z_\tau+\mathrm{Tr}\left\{\Upsilon(\tau,\theta)G_z(\tau)W_zG_z^T(\tau)\right\}\right\}\varphi(\tau,z_\tau,\theta).\end{aligned}$$

To have constant, linear and quadratic terms independent of arbitrary z_τ, it requires that the following expressions hold

$$\frac{d}{d\tau}\Upsilon(\tau,\theta)=-A_z^T(\tau)\Upsilon(\tau,\theta)-\Upsilon(\tau,\theta)A_z(\tau)-2\Upsilon(\tau,\theta)G_z(\tau)W_zG_z^T(\tau)\Upsilon(\tau,\theta)-\theta N_z(\tau)$$

$$\frac{d}{d\tau}\eta(\tau,\theta)=-A_z^T(\tau)\eta(\tau,\theta)-\Upsilon(\tau,\theta)b_z(\tau)-\theta M_z(\tau)$$

with the terminal-value conditions $\Upsilon(t_f,\theta) = \theta N_f$ and $\eta(t_f,\theta) = 0$. Finally, the backward-in-time differential equation satisfied by $\upsilon(\tau,\theta)$ is obtained as follows

$$\frac{d}{d\tau}\upsilon(\tau,\theta) = -\operatorname{Tr}\{\Upsilon(\tau,\theta)G_z(\tau)\,W_z G_z^T(\tau)\}$$
$$-2\eta^T(\tau,\theta)b_z(\tau) - \theta l^T(\tau)R(\tau)l(\tau), \quad \upsilon(t_f,\theta) = 0$$

which completes the proof.

Believing the higher-order statistics associated with (6.20) to be the appropriate measures of risks, the determination of these mathematical statistics underpinning complex behavior of (6.20) is technically feasible because of a Maclaurin series expansion of the cumulant-generating function (6.23), e.g.,

$$\psi(\tau,z_\tau,\theta) = \sum_{r=1}^{\infty} \frac{\partial^{(r)}}{\partial\theta^{(r)}}\psi(\tau,z_\tau,\theta)\bigg|_{\theta=0}\frac{\theta^r}{r!} \tag{6.30}$$

in which all $\kappa_r \triangleq \frac{\partial^{(r)}}{\partial\theta^{(r)}}\psi(\tau,z_\tau,\theta)\big|_{\theta=0}$ are called performance-measure statistics. Moreover, the series expansion coefficients are computed by using the cumulant-generating function (6.23)

$$\frac{\partial^{(r)}}{\partial\theta^{(r)}}\psi(\tau,z_\tau,\theta)\bigg|_{\theta=0} = z_\tau^T\frac{\partial^{(r)}}{\partial\theta^{(r)}}\Upsilon(\tau,\theta)\bigg|_{\theta=0} z_\tau$$
$$+ 2z_\tau^T\frac{\partial^{(r)}}{\partial\theta^{(r)}}\eta(\tau,\theta)\bigg|_{\theta=0} + \frac{\partial^{(r)}}{\partial\theta^{(r)}}\upsilon(\tau,\theta)\bigg|_{\theta=0}. \tag{6.31}$$

In view of the definition (6.30), the rth-performance-measure statistic follows

$$\kappa_r = z_\tau^T\frac{\partial^{(r)}}{\partial\theta^{(r)}}\Upsilon(\tau,\theta)\bigg|_{\theta=0} z_\tau + 2z_\tau^T\frac{\partial^{(r)}}{\partial\theta^{(r)}}\eta(\tau,\theta)\bigg|_{\theta=0} + \frac{\partial^{(r)}}{\partial\theta^{(r)}}\upsilon(\tau,\theta)\bigg|_{\theta=0} \tag{6.32}$$

for any finite $1 \le r < \infty$. For notational convenience, the change of notations

$$H_r(\tau) \triangleq \frac{\partial^{(r)}\Upsilon(\tau,\theta)}{\partial\theta^{(r)}}\bigg|_{\theta=0}; \breve{D}_r(\tau) \triangleq \frac{\partial^{(r)}\eta(\tau,\theta)}{\partial\theta^{(r)}}\bigg|_{\theta=0}; D_r(\tau) \triangleq \frac{\partial^{(r)}\upsilon(\tau,\theta)}{\partial\theta^{(r)}}\bigg|_{\theta=0}$$

is therefore introduced to summarize the underlying performance variations. The emerging problem of risk-averse optimal control design being formulated then simply requires a new model concept for performance benefits and risks, which essentially consists of tradeoffs between performance values and risks for reliable control decisions. It therefore underlies the concept of risk aversion in closed-loop performance assessment and feedback design under uncertainty.

Theorem 6.2.2 (Performance-Measure Statistics). *Let the bilinear-quadratic stochastic system be described by (6.19) and (6.20). For $k \in \mathbb{N}$ fixed, the kth-cumulant of performance measure (6.20) is*

$$\kappa_k = z_0^T H_k(t_0) z_0 + 2 z_0^T \breve{D}_k(t_0) + D_k(t_0) \tag{6.33}$$

where the supporting variables $\{H_r(\tau)\}_{r=1}^k$, $\{\breve{D}_r(\tau)\}_{r=1}^k$, and $\{D_r(\tau)\}_{r=1}^k$ evaluated at $\tau = t_0$ satisfy the differential equations (with the dependence of $H_r(\tau)$, $\breve{D}_r(\tau)$, and $D_r(\tau)$ upon the admissible $K(\tau)$ and $l(\tau)$ suppressed)

$$\frac{d}{d\tau} H_1(\tau) = -A_z^T(\tau) H_1(\tau) - H_1(\tau) A_z(\tau) - N_z(\tau) \tag{6.34}$$

$$\frac{d}{d\tau} H_r(\tau) = -A_z^T(\tau) H_r(\tau) - H_r(\tau) A_z(\tau) - \sum_{s=1}^{r-1} \frac{2r!}{s!(r-s)!} H_s(\tau) G_z(\tau) W_z G_z^T(\tau) H_{r-s}(\tau), \quad 2 \le r \le k \tag{6.35}$$

and

$$\frac{d}{d\tau} \breve{D}_1(\tau) = -A_z^T(\tau) \breve{D}_1(\tau) - H_1(\tau) b_z(\tau) - M_z(\tau) \tag{6.36}$$

$$\frac{d}{d\tau} \breve{D}_r(\tau) = -A_z^T(\tau) \breve{D}_r(\tau) - H_r(\tau) b_z(\tau), \quad 2 \le r \le k \tag{6.37}$$

and, finally

$$\frac{d}{d\tau} D_1(\tau) = -\operatorname{Tr}\{H_1(\tau) G_z(\tau) W_z G_z^T(\tau)\} - 2\breve{D}_1^T(\tau) b_z(\tau) - l^T(\tau) R(\tau) l(\tau) \tag{6.38}$$

$$\frac{d}{d\tau} D_r(\tau) = -\operatorname{Tr}\{H_r(\tau) G_z(\tau) W_z G_z^T(\tau)\} - 2\breve{D}_r^T(\tau) b_z(\tau), \quad 2 \le r \le k \tag{6.39}$$

where the terminal-value conditions $H_1(t_f) = N_f$, $H_r(t_f) = 0$ for $2 \le r \le k$, $\breve{D}_r(t_f) = 0$ for $1 \le r \le k$, and $D_r(t_f) = 0$ for $1 \le r \le k$.

Proof. The expression of statistical measures of risks (6.33) is readily justified by using the result (6.32). What remains is to show that the solutions $H_r(\tau)$, $\breve{D}_r(\tau)$, and $D_r(\tau)$ for $1 \le r \le k$ indeed satisfy the backward-in-time dynamical equations (6.34)–(6.39). Notice that the dynamical equations (6.34)–(6.39) satisfied by the solutions $H_r(\tau)$, $\breve{D}_r(\tau)$, and $D_r(\tau)$ are therefore obtained by successively taking derivatives with respect to θ of the supporting equations (6.24)–(6.25) under the assumptions of $(A_\epsilon, \Delta_\epsilon)$ and (A_ϵ, C) being uniformly stabilizable and detectable on $[t_0, t_f]$.

6.3 Depicting the Problem Statements

Next the recent successes and key findings in Sect. 6.2 have demonstrated the significant roles played by statistical measures of risks as described in (6.20). To formulate in precise terms the feedback control optimization problem of interest, it is important to note that all the performance-measure statistics are functions of time-backward evolutions and do not depend on intermediate interaction values $z(t)$. Henceforth, the time-backward trajectories (6.34)–(6.39) are therefore considered as the new dynamical equations with the associated state variables $H_r(\tau)$, $\breve{D}_r(\tau)$, and $D_r(\tau)$, not the traditional system states $z(t)$.

For such risk-averse optimal control problems, the importance of a compact statement is advocated to aid mathematical manipulations. To make this more precise, one may think of the k-tuple state variables

$$\mathscr{H}(\cdot) \triangleq (\mathscr{H}_1(\cdot), \ldots, \mathscr{H}_k(\cdot));\ \breve{\mathscr{D}}(\cdot) \triangleq (\breve{\mathscr{D}}_1(\cdot), \ldots, \breve{\mathscr{D}}_k(\cdot));\ \mathscr{D}(\cdot) \triangleq (\mathscr{D}_1(\cdot), \ldots, \mathscr{D}_k(\cdot))$$

whose the state variables $\mathscr{H}_r \in \mathscr{C}^1([t_0, t_f]; \mathbb{R}^{4n \times 4n})$, $\breve{\mathscr{D}}_r \in \mathscr{C}^1([t_0, t_f]; \mathbb{R}^{4n})$, and $\mathscr{D}_r \in \mathscr{C}^1([t_0, t_f]; \mathbb{R})$ for $n = n_1 + n_2$ having the representations $\mathscr{H}_r(\cdot) \triangleq H_r(\cdot)$, $\breve{\mathscr{D}}_r(\cdot) \triangleq \breve{D}_r(\cdot)$ and $\mathscr{D}_r(\cdot) \triangleq D_r(\cdot)$ with the right members satisfying the dynamics (6.34)–(6.39) are defined on $[t_0, t_f]$.

In the remainder of the development, the convenient mappings with the rules of action are given by

$$\mathscr{F}_1(\tau, \mathscr{H}) \triangleq -A_z^T(\tau)\mathscr{H}_1(\tau) - \mathscr{H}_1(\tau)A_z(\tau) - N_z(\tau)$$

$$\mathscr{F}_r(\tau, \mathscr{H}) \triangleq -A_z^T(\tau)\mathscr{H}_r(\tau) - \mathscr{H}_r(\tau)A_z(\tau) - \sum_{s=1}^{r-1} \frac{2r!}{s!(r-s)!} \mathscr{H}_s(\tau)G_z(\tau)W_z G_z^T(\tau)\mathscr{H}_{r-s}(\tau)$$

and

$$\breve{\mathscr{G}}_1(\tau, \mathscr{H}, \breve{\mathscr{D}}) \triangleq -A_z^T(\tau)\breve{\mathscr{D}}_1(\tau) - \mathscr{H}_1(\tau)b_z(\tau) - M_z(\tau)$$

$$\breve{\mathscr{G}}_r(\tau, \mathscr{H}, \breve{\mathscr{D}}) \triangleq -A_z^T(\tau)\breve{\mathscr{D}}_r(\tau) - \mathscr{H}_r(\tau)b_z(\tau)$$

and, finally

$$\mathscr{G}_1(\tau, \mathscr{H}, \breve{\mathscr{D}}) \triangleq -\mathrm{Tr}\left\{\mathscr{H}_1(\tau)G_z(\tau)W_z G_z^T(\tau)\right\} - 2\breve{\mathscr{D}}_1^T(\tau)b_z(\tau) - l^T(\tau)R(\tau)l(\tau)$$

$$\mathscr{G}_r(\tau, \mathscr{H}, \breve{\mathscr{D}}) \triangleq -\mathrm{Tr}\left\{\mathscr{H}_r(\tau)G_z(\tau)W_z G_z^T(\tau)\right\} - 2\breve{\mathscr{D}}_r^T(\tau)b_z(\tau).$$

For a compact formulation, the product mappings

$$\mathscr{F} \triangleq \mathscr{F}_1 \times \cdots \times \mathscr{F}_k; \quad \breve{\mathscr{G}} \triangleq \breve{\mathscr{G}}_1 \times \cdots \times \breve{\mathscr{G}}_k; \quad \mathscr{G} \triangleq \mathscr{G}_1 \times \cdots \times \mathscr{G}_k$$

are necessary. Thus, the dynamical equations (6.34)–(6.39) in the statistical optimal control become

$$\frac{d}{d\tau}\mathscr{H}(\tau) = \mathscr{F}(\tau, \mathscr{H}(\tau)), \quad \mathscr{H}(t_f) \equiv \mathscr{H}_f \tag{6.40}$$

$$\frac{d}{d\tau}\breve{\mathscr{D}}(\tau) = \breve{\mathscr{G}}(\tau, \mathscr{H}(\tau), \breve{\mathscr{D}}(\tau)), \quad \breve{\mathscr{D}}(t_f) \equiv \breve{\mathscr{D}}_f \tag{6.41}$$

$$\frac{d}{d\tau}\mathscr{D}(\tau) = \mathscr{G}(\tau, \mathscr{H}(\tau), \breve{\mathscr{D}}(\tau)), \quad \mathscr{D}(t_f) \equiv \mathscr{D}_f \tag{6.42}$$

where the terminal-value conditions $\mathscr{H}_f \triangleq (N_f, 0, \ldots, 0)$, $\breve{\mathscr{D}}_f = (0, \ldots, 0)$, and $\mathscr{D}_f = (0, \ldots, 0)$.

Of note, understanding of the connections between the state variables $\mathscr{H}$, $\breve{\mathscr{D}}$, and $\mathscr{D}$; and the admissible feedback control parameters K and l is framed by the dynamical system (6.40)–(6.42). In this regard, the interpretation $\mathscr{H}(\cdot) \equiv \mathscr{H}(\cdot, K)$, $\breve{\mathscr{D}}(\cdot) \equiv \breve{\mathscr{D}}(\cdot, K, l)$ and $\mathscr{D}(\cdot) \equiv \mathscr{D}(\cdot, l)$ is authoritatively understood. For the given terminal data $(t_f, \mathscr{H}_f, \breve{\mathscr{D}}_f, \mathscr{D}_f)$, the classes of admissible feedback parameters are next defined.

Definition 6.3.1 (Admissible Feedback Parameters). Let compact subset $\overline{K} \subset \mathbb{R}^{m \times n}$ and $\overline{L} \subset \mathbb{R}^m$ for $m = m_1 + m_2$ be the set of allowable feedback gain and affine values. For the given $k \in \mathbb{N}$ and the sequence $\mu = \{\mu_r \geq 0\}_{r=1}^k$ with $\mu_1 > 0$, the set of feedback gains $\mathscr{K}_{t_f, \mathscr{H}_f, \breve{\mathscr{D}}_f, \mathscr{D}_f; \mu}$ and $\mathscr{L}_{t_f, \mathscr{H}_f, \breve{\mathscr{D}}_f, \mathscr{D}_f; \mu}$ are the classes of $\mathscr{C}([t_0, t_f]; \mathbb{R}^{m \times n})$ and $\mathscr{C}([t_0, t_f]; \mathbb{R}^m)$ with values $K(\cdot) \in \overline{K}$ and $L(\cdot) \in \overline{L}$ for which solutions to the dynamical equations (6.40)–(6.42) with the terminal-value conditions $\mathscr{H}(t_f) = \mathscr{H}_f$, $\breve{\mathscr{D}}(t_f) = \breve{\mathscr{D}}_f$, and $\mathscr{D}(t_f) = \mathscr{D}_f$ exist on $[t_0, t_f]$.

What follows is a defining development in reliable feedback control research that clarifies all issues and gives the control researcher a comprehensive view. Such an event in the research on statistical measures of risks occurred with the development of the performance-measure statistics in Sect. 6.2. These measures liberate the controller from the constraint of only one utility function, which is fine if the controller utility is best represented by expected performance. Yet, it is now important to progress from the traditional approach of risk neutral to the birth of a whole rainbow of utility functions. In particular, a mean-risk aware performance index is considered as an example source of superiority and thus represents the whole gamut of control designer behavior from risk seeking to risk neutral to risk aversion.

On $\mathscr{K}_{t_f, \mathscr{H}_f, \breve{\mathscr{D}}_f, \mathscr{D}_f; \mu}$ the performance index with mean-risk considerations in the risk-averse optimal control follows.

Definition 6.3.2 (Mean-Risk Aware Performance Index). Fix $k \in \mathbb{N}$ and the sequence of scalar coefficients $\mu = \{\mu_r \geq 0\}_{r=1}^k$ with $\mu_1 > 0$. Then for the given z_0, the risk-value aware performance index pertaining to the statistical optimal control of the stochastic system with low sensitivity over $[t_0, t_f]$ is defined by

$$\phi_0(t_0, \mathscr{H}(t_0), \breve{\mathscr{D}}(t_0), \mathscr{D}(t_0)) \triangleq \underbrace{\mu_1\kappa_1}_{\text{Mean}} + \underbrace{\mu_2\kappa_2 + \cdots + \mu_k\kappa_k}_{\text{Risks}}$$
$$= \sum_{r=1}^{k} \mu_r [z_0^T \mathscr{H}_r(t_0) z_0 + 2 z_0^T \breve{\mathscr{D}}_r(t_0) + \mathscr{D}_r(t_0)] \tag{6.43}$$

where additional design freedom by means of μ_r's utilized by the control designer with risk-averse attitudes are sufficient to meet and exceed different levels of performance-based reliability requirements, for instance, mean (i.e., the average of performance measure), variance (i.e., the dispersion of values of performance measure around its mean), skewness (i.e., the anti-symmetry of the density of performance measure), kurtosis (i.e., the heaviness in the density tails of performance measure), etc., pertaining to closed-loop performance variations and uncertainties while the supporting solutions $\{\mathscr{H}_r(\tau)\}_{r=1}^k$, $\{\breve{\mathscr{D}}_r(\tau)\}_{r=1}^k$, and $\{\mathscr{D}_r(\tau)\}_{r=1}^k$ evaluated at $\tau = t_0$ satisfy the dynamical equations (6.40)–(6.42).

In effect, the optimization problem that follows is to minimize the mean-risk aware performance index (6.43) over all admissible feedback matrices $K = K(\cdot)$ in $\mathscr{K}_{t_f, \mathscr{H}_f, \breve{\mathscr{D}}_f, \mathscr{D}_f; \mu}$ and feedforward vectors $l = l(\cdot)$ in $\mathscr{L}_{t_f, \mathscr{H}_f, \breve{\mathscr{D}}_f, \mathscr{D}_f; \mu}$.

Definition 6.3.3 (Optimization Problem of Mayer Type). Fix $k \in \mathbb{N}$ and the sequence of scalar coefficients $\mu = \{\mu_r \geq 0\}_{r=1}^k$ with $\mu_1 > 0$. Then, the optimization problem of the statistical control over $[t_0, t_f]$ is given by

$$\min_{K(\cdot) \in \mathscr{K}_{t_f, \mathscr{H}_f, \breve{\mathscr{D}}_f, \mathscr{D}_f; \mu};\, l(\cdot) \in \mathscr{L}_{t_f, \mathscr{H}_f, \breve{\mathscr{D}}_f, \mathscr{D}_f; \mu}} \phi_0(t_0, \mathscr{H}(t_0), \breve{\mathscr{D}}(t_0), \mathscr{D}(t_0)) \tag{6.44}$$

subject to the dynamical equations (6.40)–(6.42) on $[t_0, t_f]$.

As already mentioned, the optimization considered here is in Mayer form and can be solved by applying an adaptation of the Mayer-form verification theorem of dynamic programming as given in [7]. To embed the aforementioned optimization into a larger optimal control problem, the terminal time and states $(t_f, \mathscr{H}_f, \breve{\mathscr{D}}, \mathscr{D}_f)$ are parameterized as $(\varepsilon, \mathscr{Y}, \breve{\mathscr{Z}}, \mathscr{Z})$. Thus, the value function now depends on the terminal-value condition parameterizations.

Definition 6.3.4 (Value Function). Suppose $(\varepsilon, \mathscr{Y}, \breve{\mathscr{Z}}, \mathscr{Z})$ is given and fixed. Then, the value function $\mathscr{V}(\varepsilon, \mathscr{Y}, \breve{\mathscr{Z}}, \mathscr{Z})$ is defined as the infimum of the performance index (6.43) over the admissible $K(\cdot) \in \mathscr{K}_{\varepsilon, \mathscr{Y}, \breve{\mathscr{Z}}, \mathscr{Z}; \mu}$ and $l(\cdot) \in \mathscr{L}_{\varepsilon, \mathscr{Y}, \breve{\mathscr{Z}}, \mathscr{Z}; \mu}$.

For convention, $\mathscr{V}(\varepsilon, \mathscr{Y}, \breve{\mathscr{Z}}, \mathscr{Z}) \triangleq \infty$ when $\mathscr{K}_{\varepsilon, \mathscr{Y}, \breve{\mathscr{Z}}, \mathscr{Z}; \mu} \times \mathscr{L}_{\varepsilon, \mathscr{Y}, \breve{\mathscr{Z}}, \mathscr{Z}; \mu}$ is empty. To avoid cumbersome notation, the dependence of trajectory solutions on $K(\cdot)$ and $l(\cdot)$ is suppressed. Next, some candidates for the value function are constructed with the help of the concept of a reachable set.

Definition 6.3.5 (Reachable Set). Let the reachable set $\hat{\mathscr{Q}}$ be

$$\hat{\mathscr{Q}} \triangleq \left\{(\varepsilon, \mathscr{Y}, \breve{\mathscr{Z}}, \mathscr{Z}) : \mathscr{K}_{\varepsilon,\mathscr{Y},\breve{\mathscr{Z}},\mathscr{Z};\mu} \times \mathscr{L}_{\varepsilon,\mathscr{Y},\breve{\mathscr{Z}},\mathscr{Z};\mu} \neq \emptyset\right\}.$$

Furthermore, the value function must satisfy both partial differential inequality and equation at each interior point of the reachable set, at which it is differentiable.

Theorem 6.3.1 (Hamilton-Jacobi-Bellman (HJB) Equation). *Let $(\varepsilon, \mathscr{Y}, \breve{\mathscr{Z}}, \mathscr{Z})$ be any interior point of $\hat{\mathscr{Q}}$, at which the scalar-valued function $\mathscr{V}(\varepsilon, \mathscr{Y}, \breve{\mathscr{Z}}, \mathscr{Z})$ is differentiable. Then $\mathscr{V}(\varepsilon, \mathscr{Y}, \breve{\mathscr{Z}}, \mathscr{Z})$ satisfies the partial differential inequality*

$$\begin{aligned}
0 \geq{} & \frac{\partial}{\partial \varepsilon}\mathscr{V}(\varepsilon, \mathscr{Y}, \breve{\mathscr{Z}}, \mathscr{Z}) + \frac{\partial}{\partial \operatorname{vec}(\mathscr{Y})}\mathscr{V}(\varepsilon, \mathscr{Y}, \breve{\mathscr{Z}}, \mathscr{Z})\operatorname{vec}(\mathscr{F}(\varepsilon, \mathscr{Y}, K)) \\
& + \frac{\partial}{\partial \operatorname{vec}(\breve{\mathscr{Z}})}\mathscr{V}(\varepsilon, \mathscr{Y}, \breve{\mathscr{Z}}, \mathscr{Z})\operatorname{vec}(\breve{\mathscr{G}}(\varepsilon, \mathscr{Y}, \breve{\mathscr{Z}}, K, l)) \\
& + \frac{\partial}{\partial \operatorname{vec}(\mathscr{Z})}\mathscr{V}(\varepsilon, \mathscr{Y}, \breve{\mathscr{Z}}, \mathscr{Z})\operatorname{vec}(\mathscr{G}(\varepsilon, \mathscr{Y}, \breve{\mathscr{Z}}, l))
\end{aligned} \tag{6.45}$$

for all $K \in \overline{\mathscr{K}}$, $l \in \overline{\mathscr{L}}$, and vec(·) *the vectorizing operator of enclosed entities.*

If there is an optimal feedback law (K^*, l^*) in $\mathscr{K}_{\varepsilon,\mathscr{Y},\breve{\mathscr{Z}},\mathscr{Z};\mu} \times \mathscr{L}_{\varepsilon,\mathscr{Y},\breve{\mathscr{Z}},\mathscr{Z};\mu}$, then the partial differential equation

$$\begin{aligned}
0 = \min_{K \in \overline{K}, l \in \overline{L}} & \left\{\frac{\partial}{\partial \varepsilon}\mathscr{V}(\varepsilon, \mathscr{Y}, \breve{\mathscr{Z}}, \mathscr{Z}) + \frac{\partial}{\partial \operatorname{vec}(\mathscr{Y})}\mathscr{V}(\varepsilon, \mathscr{Y}, \breve{\mathscr{Z}}, \mathscr{Z})\operatorname{vec}(\mathscr{F}(\varepsilon, \mathscr{Y}, K))\right. \\
& + \frac{\partial}{\partial \operatorname{vec}(\breve{\mathscr{Z}})}\mathscr{V}(\varepsilon, \mathscr{Y}, \breve{\mathscr{Z}}, \mathscr{Z})\operatorname{vec}(\breve{\mathscr{G}}(\varepsilon, \mathscr{Y}, \breve{\mathscr{Z}}, K, l)) \\
& \left. + \frac{\partial}{\partial \operatorname{vec}(\mathscr{Z})}\mathscr{V}(\varepsilon, \mathscr{Y}, \breve{\mathscr{Z}}, \mathscr{Z})\operatorname{vec}(\mathscr{G}(\varepsilon, \mathscr{Y}, \breve{\mathscr{Z}}, l))\right\}
\end{aligned} \tag{6.46}$$

is satisfied. The minimum in (6.46) is achieved by the optimal feedback policy composed by $K^*(\varepsilon)$ and $l^*(\varepsilon)$ at ε.

Proof. The readers are referred to the proof in [8].

Yet from a sufficient-condition analysis of the control optimization of Mayer type in the dynamic programming framework emerges the verification theorem.

Theorem 6.3.2 (Verification Theorem). *Fix $k \in \mathbb{N}$ and let $\mathscr{W}(\varepsilon, \mathscr{Y}, \breve{\mathscr{Z}}, \mathscr{Z})$ be a continuously differentiable solution of the HJB equation (6.46), which satisfies the boundary $\mathscr{W}(t_0, \mathscr{H}(t_0), \breve{\mathscr{Z}}(t_0), \mathscr{D}(t_0)) = \phi_0(t_0, \mathscr{H}(t_0), \breve{\mathscr{Z}}(t_0), \mathscr{D}(t_0))$. Let $(t_f, \mathscr{H}_f, \breve{\mathscr{D}}_f, \mathscr{D}_f)$ be a point of $\hat{\mathscr{Q}}$, let (K, l) be a pair of feedback parameters in $\mathscr{K}_{t_f,\mathscr{H}_f,\breve{\mathscr{D}}_f,\mathscr{D}_f;\mu} \times \mathscr{L}_{t_f,\mathscr{H}_f,\breve{\mathscr{D}}_f,\mathscr{D}_f;\mu}$ and let $\mathscr{H}(\cdot)$, $\breve{\mathscr{D}}(\cdot)$, and $\mathscr{D}(\cdot)$*

be the corresponding solutions of the dynamical equations (6.40)–(6.42). Then, $\mathscr{W}(\tau, \mathscr{H}(\tau), \breve{\mathscr{D}}(\tau), \mathscr{D}(\tau))$ *is a non-increasing function of* τ*. If* (K^*, l^*) *is the 2-tuple feedback parameters in* $\mathscr{K}_{t_f, \mathscr{H}_f, \breve{\mathscr{D}}_f, \mathscr{D}_f;\mu} \times \mathscr{L}_{t_f, \mathscr{H}_f, \breve{\mathscr{D}}_f, \mathscr{D}_f;\mu}$ *defined on* $[t_0, t_f]$ *with the corresponding solutions* $\mathscr{H}^*(\cdot)$, $\breve{\mathscr{D}}^*(\cdot)$, *and* $\mathscr{D}^*(\cdot)$ *of the dynamics (6.40)–(6.42) such that, for* $\tau \in [t_0, t_f]$,

$$
\begin{aligned}
0 &= \frac{\partial}{\partial \varepsilon}\mathscr{W}(\tau, \mathscr{H}^*(\tau), \breve{\mathscr{D}}^*(\tau), \mathscr{D}^*(\tau)) \\
&+ \frac{\partial}{\partial \operatorname{vec}(\mathscr{Y})}\mathscr{W}(\tau, \mathscr{H}^*(\tau), \breve{\mathscr{D}}^*(\tau), \mathscr{D}^*(\tau))\operatorname{vec}(\mathscr{F}(\tau, \mathscr{H}^*(\tau), K^*(\tau))) \\
&+ \frac{\partial}{\partial \operatorname{vec}(\breve{\mathscr{Z}})}\mathscr{W}(\tau, \mathscr{H}^*(\tau), \breve{\mathscr{D}}^*(\tau), \mathscr{D}^*(\tau))\operatorname{vec}(\breve{\mathscr{G}}(\tau, \mathscr{H}^*(\tau), \breve{\mathscr{D}}^*(\tau), K^*(\tau), l^*(\tau))) \\
&+ \frac{\partial}{\partial \operatorname{vec}(\mathscr{Z})}\mathscr{W}(\tau, \mathscr{H}^*(\tau), \breve{\mathscr{D}}^*(\tau), \mathscr{D}^*(\tau))\operatorname{vec}(\mathscr{G}(\tau, \mathscr{H}^*(\tau), \breve{\mathscr{D}}^*(\tau), l^*(\tau)))
\end{aligned} \tag{6.47}
$$

then (K^*, l^*) *is an optimal feedback pair and* $\mathscr{W}(\varepsilon, \mathscr{Y}, \breve{\mathscr{Z}}, \mathscr{Z}) = \mathscr{V}(\varepsilon, \mathscr{Y}, \breve{\mathscr{Z}}, \mathscr{Z})$, *where* $\mathscr{V}(\varepsilon, \mathscr{Y}, \breve{\mathscr{Z}}, \mathscr{Z})$ *is the value function.*

Proof. The detailed analysis is found in [8].

6.4 Low Sensitivity Control with Risk Aversion

In conformity with the framework of dynamic programming, the terminal time and states $(t_f, \mathscr{H}_f, \breve{\mathscr{D}}_f, \mathscr{D}_f)$ are now parameterized as $(\varepsilon, \mathscr{Y}, \breve{\mathscr{Z}}, \mathscr{Z})$ for a family of optimization problems. For instance, the states (6.40)–(6.42) defined on the interval $[t_0, \varepsilon]$ now have terminal values denoted by $\mathscr{H}(\varepsilon) \equiv \mathscr{Y}$, $\breve{\mathscr{D}}(\varepsilon) \equiv \breve{\mathscr{Z}}$, and $\mathscr{D}(\varepsilon) \equiv \mathscr{Z}$, where $\varepsilon \in [t_0, t_f]$. Furthermore, with $k \in \mathbb{N}$ and $(\varepsilon, \mathscr{Y}, \breve{\mathscr{Z}}, \mathscr{Z})$ in $\hat{\mathscr{Q}}$, the following real-value candidate:

$$
\begin{aligned}
\mathscr{W}(\varepsilon, \mathscr{Y}, \breve{\mathscr{Z}}, \mathscr{Z}) &= z_0^T \sum_{r=1}^{k} \mu_r(\mathscr{Y}_r + \mathscr{E}_r(\varepsilon))z_0 \\
&+ 2z_0^T \sum_{r=1}^{k} \mu_r(\breve{\mathscr{Z}}_r + \breve{\mathscr{T}}_r(\varepsilon)) + \sum_{r=1}^{k} \mu_r(\mathscr{Z}_r + \mathscr{T}_r(\varepsilon))
\end{aligned} \tag{6.48}
$$

for the value function is therefore differentiable. The time derivative of $\mathscr{W}(\varepsilon, \mathscr{Y}, \mathscr{Z})$ can also be shown of the form

$$\frac{d}{d\varepsilon}\mathscr{W}(\varepsilon, \mathscr{Y}, \breve{\mathscr{Z}}, \mathscr{Z}) = z_0^T \sum_{r=1}^{k} \mu_r (\mathscr{F}_r(\varepsilon, \mathscr{Y}, K) + \frac{d}{d\varepsilon}\mathscr{E}_r(\varepsilon)) z_0$$

$$+2z_0^T \sum_{r=1}^{k} \mu_r (\breve{\mathscr{G}}_r(\varepsilon, \mathscr{Y}, \breve{\mathscr{Z}}, K, l) + \frac{d}{d\varepsilon}\breve{\mathscr{T}}_r(\varepsilon)) + \sum_{r=1}^{k} \mu_r (\mathscr{G}_r(\varepsilon, \mathscr{Y}, \breve{\mathscr{Z}}, l) + \frac{d}{d\varepsilon}\mathscr{T}_r(\varepsilon))$$

where $\mathscr{E}_r$, $\breve{\mathscr{T}}_r$, and $\mathscr{T}_r$ are the time-parameter functions to be determined. At the boundary, it requires $\mathscr{W}(t_0, \mathscr{H}(t_0), \breve{\mathscr{D}}(t_0), \mathscr{D}(t_0)) = \phi_0(t_0, \mathscr{H}(t_0), \breve{\mathscr{D}}(t_0), \mathscr{D}(t_0))$

$$z_0^T \sum_{r=1}^{k} \mu_r (\mathscr{H}_r(t_0) + \mathscr{E}_r(t_0)) z_0 + 2z_0^T \sum_{r=1}^{k} \mu_r (\breve{\mathscr{D}}_r(t_0) + \breve{\mathscr{T}}_r(t_0)) + \sum_{r=1}^{k} \mu_r (\mathscr{D}_r(t_0) + \mathscr{T}_r(t_0))$$
$$= z_0^T \sum_{r=1}^{k} \mu_r \mathscr{H}_r(t_0) z_0 + 2z_0^T \sum_{r=1}^{k} \mu_r \breve{\mathscr{D}}_r(t_0) + \sum_{r=1}^{k} \mu_r \mathscr{D}_r(t_0). \tag{6.49}$$

By matching the boundary condition (6.49), it yields that $\mathscr{E}_r(t_0) = 0$, $\breve{\mathscr{T}}_r(t_0) = 0$, and $\mathscr{T}_r(t_0) = 0$ for $1 \le r \le k$. Next it is necessary to verify that this candidate value function satisfies (6.47) along the corresponding trajectories produced by the feedback pair (K, l) resulting from the minimization in (6.46). Subsequently, it follows

$$0 = \min_{K \in \overline{K}, l \in \overline{L}} \Big\{ z_0^T \sum_{r=1}^{k} \mu_r \mathscr{F}_r(\varepsilon, \mathscr{Y}, K) z_0 + 2z_0^T \sum_{r=1}^{k} \mu_r \breve{\mathscr{G}}_r(\varepsilon, \mathscr{Y}, \breve{\mathscr{Z}}, K, l) + \sum_{r=1}^{k} \mu_r \frac{d}{d\varepsilon}\mathscr{T}_r(\varepsilon)$$
$$+ \sum_{r=1}^{k} \mu_r \mathscr{G}_r(\varepsilon, \mathscr{Y}, \breve{\mathscr{Z}}, l) + z_0^T \sum_{r=1}^{k} \mu_r \frac{d}{d\varepsilon}\mathscr{E}_r(\varepsilon) z_0 + 2z_0^T \sum_{r=1}^{k} \mu_r \frac{d}{d\varepsilon}\breve{\mathscr{T}}_r(\varepsilon) \Big\}. \tag{6.50}$$

Now the aggregate matrix coefficients $A_z(t)$, $b_z(t)$, $M_z(t)$ and $N_z(t)$ for $t \in [t_0, t_f]$ of the composite dynamics (6.19) with trajectory sensitivity concern are approximated and partitioned to conform with the n-dimensional structure of (6.1); e.g.,

$$I_0^T \triangleq [I\,0\,0\,0]; \quad I_1^T \triangleq [0\,I\,0\,0]; \quad I_2^T \triangleq [0\,0\,I\,0]; \quad I_3^T \triangleq [0\,0\,0\,I]$$

where I is an $n \times n$ identity matrix and

$$A_z(t) \approx I_0(A_\epsilon(t,\zeta) + \Delta_\epsilon(t,x,\zeta)K(t))I_0^T + I_2(A_\epsilon^\zeta(t,\zeta_0) + \Delta_\epsilon^\zeta(t,x,\zeta_0)K(t))I_0^T$$
$$+ I_2(A_\epsilon(t,\zeta_0) + \Delta_\epsilon(t,x,\zeta_0)K(t))I_2^T$$
$$b_z(t) \approx I_0\Delta_\epsilon(t,x,\zeta)l + I_2\Delta_\epsilon^\zeta(t,x,\zeta_0)l\,; \quad M_z(t) \approx I_0K^T(t)R(t)l$$
$$N_z(t) \approx I_0(Q(t) + K^T(t)R(t)K(t))I_0^T.$$

Therefore, the derivative of the expression in (6.50) with respect to the admissible feedback K and feedforward l yields the necessary conditions for an extremum of (6.46) on $[t_0, t_f]$,

$$K = - R^{-1}(\varepsilon)[\Delta_\epsilon^T(\varepsilon, x, \zeta)I_0^T + (\Delta_\epsilon^\xi(\varepsilon, x, \zeta_0))^T I_2^T]\sum_{s=1}^{k} \hat{\mu}_s \mathscr{Y}_s(I_0^T)^\dagger \tag{6.51}$$

$$l = - R^{-1}(\varepsilon)[\Delta_\epsilon^T(\varepsilon, x, \zeta)I_0^T + (\Delta_\epsilon^\xi(\varepsilon, x, \zeta_0))^T I_2^T]\sum_{s=1}^{k} \hat{\mu}_s \mathscr{Z}_s \tag{6.52}$$

where $\hat{\mu}_s = \frac{\mu_s}{\mu_1}$ with $\mu_1 > 0$ and $(I_0^T)^\dagger$ is an pseudo-inverse of I_0^T.

With the admissible feedback law (6.51)–(6.52) replaced in the expression of the bracket (6.50) and having $\{\mathscr{Y}_s\}_{s=1}^k$ evaluated on the solution trajectories (6.40)–(6.42), the time dependent functions $\mathscr{E}_r(\varepsilon)$, $\breve{\mathscr{T}}_r(\varepsilon)$ and $\mathscr{T}_r(\varepsilon)$ are therefore chosen such that the sufficient condition (6.47) in the verification theorem is satisfied in the presence of the arbitrary value of z_0; for example

$$\frac{d}{d\varepsilon}\mathscr{E}_1(\varepsilon) = A_z^T(\varepsilon)\mathscr{H}_1(\varepsilon) + \mathscr{H}_1(\varepsilon)A_z(\varepsilon) + N_a(\varepsilon)$$

$$\frac{d}{d\varepsilon}\mathscr{E}_r(\varepsilon) = A_z^T(\varepsilon)\mathscr{H}_r(\varepsilon) + \mathscr{H}_r(\varepsilon)A_z(\varepsilon)$$

$$+ \sum_{s=1}^{r-1} \frac{2r!}{s!(r-s)!}\mathscr{H}_s(\varepsilon)G_z(\varepsilon)W_z G_z^T(\varepsilon)\mathscr{H}_{r-s}(\varepsilon), \quad 2 \le r \le k$$

and

$$\frac{d}{d\varepsilon}\breve{\mathscr{T}}_1(\varepsilon) = A_z^T(\varepsilon)\breve{\mathscr{D}}_1(\varepsilon) + \mathscr{H}_1(\varepsilon)b_z(\varepsilon) + M_z(\varepsilon)$$

$$\frac{d}{d\varepsilon}\breve{\mathscr{T}}_r(\varepsilon) = A_z^T(\varepsilon)\breve{\mathscr{D}}_r(\varepsilon) + \mathscr{H}_r(\varepsilon)b_z(\varepsilon), \quad 2 \le r \le k$$

and, finally

$$\frac{d}{d\varepsilon}\mathscr{T}_r(\varepsilon) = \mathrm{Tr}\left\{\mathscr{H}_1(\varepsilon)G_z(\varepsilon)W_z G_z^T(\varepsilon)\right\} + 2\breve{\mathscr{D}}_1^T(\varepsilon)b_z(\varepsilon) + l^T(\varepsilon)R(\varepsilon)l(\varepsilon)$$

$$\frac{d}{d\varepsilon}\mathscr{T}_r(\varepsilon) = \mathrm{Tr}\left\{\mathscr{H}_r(\varepsilon)G_z(\varepsilon)W_z G_z^T(\varepsilon)\right\} + 2\breve{\mathscr{D}}_r^T(\varepsilon)b_z(\varepsilon), \quad 2 \le r \le k.$$

where the initial-value conditions $\mathscr{E}_r(t_0) = 0$, $\breve{\mathscr{T}}_r(t_0) = 0$ and $\mathscr{T}_r(t_0) = 0$ for $1 \le r \le k$. Therefore, the sufficient condition (6.47) of the verification theorem is satisfied so that the extremizing feedback law defined by (6.51)–(6.52) becomes optimal.

As the development has underscored, when the resilient control considered is concerned with performance riskiness, the design principles for weakly coupled bilinear stochastic systems with risk-averse output feedback and low sensitivity is the order of the day.

Theorem 6.4.1 (Statistical Control with Low Sensitivity). *Let $(A_\varepsilon, \Delta_\varepsilon)$ and (A_ε, C) be uniformly stabilizable and detectable. Consider $u(t) = K(t)\hat{x}(t) + l(t)$ for all $t \in [t_0, t_f]$, where the state estimates $\hat{x}(t)$ is governed by the dynamical observer (6.12). Then, the low sensitivity control strategy with risk aversion is supported by the output-feedback parameters $K^*(\cdot)$ and $l^*(\cdot)$*

$$K^*(t) = -R^{-1}(t)[\Delta_\epsilon^T(t,x,\zeta)I_0^T + (\Delta_\epsilon^\xi(t,x,\zeta_0))^T I_2^T]\sum_{s=1}^{k}\hat{\mu}_s\mathcal{H}_s^*(t)(I_0^T)^\dagger \tag{6.53}$$

$$l^*(t) = -R^{-1}(t)[\Delta_\epsilon^T(t,x,\zeta)I_0^T + (\Delta_\epsilon^\xi(t,x,\zeta_0))^T I_2^T]\sum_{s=1}^{k}\hat{\mu}_s\breve{\mathcal{D}}_s^*(t) \tag{6.54}$$

where the forward time $t \triangleq t_0 + t_f - \tau$ and the normalized parametric design of freedom $\hat{\mu}_s \triangleq \frac{\mu_r}{\mu_1}$. The optimal state solutions $\{\mathcal{H}_r^*(\cdot)\}_{r=1}^k$ are supporting the mathematical statistics for performance robustness and governed by the backward-in-time matrix valued differential equations

$$\frac{d}{d\tau}\mathcal{H}_1^*(\tau) = -(A_z^*)^T(\tau)\mathcal{H}_1^*(\tau) - \mathcal{H}_1^*(\tau)A_z^*(\tau) - N_z^*(\tau), \quad \mathcal{H}_1^*(t_f) = N_f \tag{6.55}$$

$$\frac{d}{d\tau}\mathcal{H}_r^*(\tau) = -(A_z^*)^T(\tau)\mathcal{H}_r^*(\tau) - \mathcal{H}_r^*(\tau)A_z^*(\tau)$$
$$-\sum_{s=1}^{r-1}\frac{2r!}{s!(r-s)!}\mathcal{H}_s^*(\tau)G_z(\tau)W_zG_z^T(\tau)\mathcal{H}_{r-s}^*(\tau), \quad \mathcal{H}_r^*(t_f) = 0, \quad 2 \le r \le k \tag{6.56}$$

and

$$\frac{d}{d\tau}\breve{\mathcal{D}}_1^*(\tau) = -(A_z^*)^T(\tau)\breve{\mathcal{D}}_1^*(\tau) - \mathcal{H}_1^*(\tau)b_z^*(\tau) - M_z^*(\tau), \quad \breve{\mathcal{D}}_1^*(t_f) = 0 \tag{6.57}$$

$$\frac{d}{d\tau}\breve{\mathcal{D}}_r^*(\tau) = -(A_z^*)^T(\tau)\breve{\mathcal{D}}_r^*(\tau) - \mathcal{H}_r^*(\tau)b_z^*(\tau), \quad \breve{\mathcal{D}}_r^*(t_f) = 0, \quad 2 \le r \le k \tag{6.58}$$

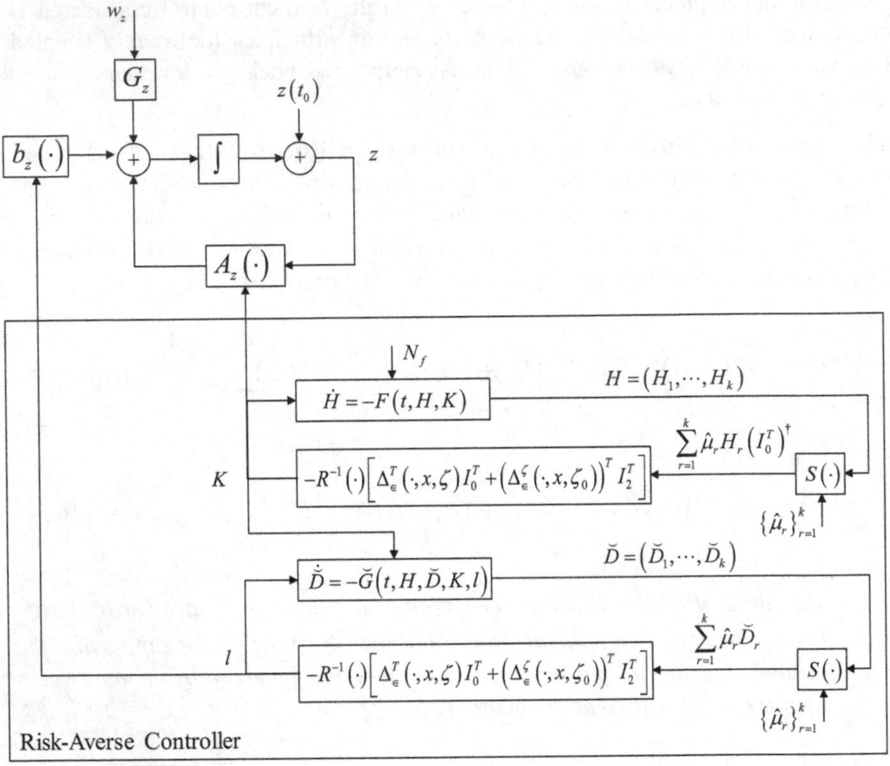

Fig. 6.6 Structure of risk-averse controlled bilinear system with noisy output observations

while the Kalman filter gain $L(t)$ and its partial $L_{\zeta}(t)$ with respect to the constant parameter ζ are governed by the estimate error-covariance equations (6.14) and (6.18) on $[t_0, t_f]$.

In summary, the feedback control system of Fig. 6.6 represents the effects of risk-averse optimization in performance robustness, sensitivity variations in response trajectory, and random errors in observing the output feedback measurements.

6.5 Chapter Summary

A summary of the conclusions is in order. In view of risk-averse control with low sensitivity and noisy feedback measurement, it is proper that the system state estimates are now taking a prominent place in participating cooperatively in the algorithmic development. Understandably, the particular mean-risk performance index provided by all appropriate statistical measures of performance expectation and risks is perhaps the most tangible reflection of the risk-averse attitudes that are now prevailing within those control designers. As for a future outlook, the technical scope here should be more fully extended and its results need thoroughly evaluated.

References

1. Mohler, R.R.: Nonlinear Systems: Cybernetics Applications to Bilinear Control. Prentice-Hall, Englewood Cliffs (1991)
2. Desai, U.: Realization of bilinear stochastic systems. IEEE Trans. Autom. Control **AC-31**, 189–192 (1986)
3. Yaz, E.: Full and reduced-ordered observer design for discrete stochastic bilinear systems. IEEE Trans. Autom. Control **AC-37**, 503–505 (1992)
4. D'Angelo, H., Moe, M.L., Hendricks, T.C.: Trajectory sensitivity of an optimal control systems. In: Proceedings of the 4th Allerton Conference on Circuit and Systems Theory, Monticello, pp. 489–498 (1966)
5. Kahne, S.J.: Low-sensitivity design of optimal linear control systems. IEEE Trans. Aerospace Electron. Syst. **4**(3), 374–379 (1968)
6. Pollatsek, A., Tversky, A.: Theory of risk. J. Math. Psychol. **7**, 540–553 (1970)
7. Fleming W.H., Rishel, R.W.: Deterministic and Stochastic Optimal Control. Springer, New York (1975)
8. Pham, K.D.: Performance-reliability-aided decision-making in multiperson quadratic decision games against jamming and estimation confrontations. In: Giannessi, F. (ed.) J. Optim. Theory Appl. **149**(1), 599–629 (2011)

Chapter 7
Resilient Control of A Class of Uncertain Time-Delay Systems

7.1 Introduction

Modern research on auxiliary signal design for fault detection and robust control of dynamic systems subject to certain ranges of perturbations or disturbances are well under way [1] and [2]. The principles of many reconfigurable control systems through various fault detection and diagnosis are eloquently described in [3] and [4]. In addition, a resilient control and filtering framework for a class of uncertain dynamical systems has been proposed in [5] which encompasses inherent time-delay model, parametric uncertainties and external disturbances.

Until now, most of the research in this field was satisfied with deterministic or expected value approaches to the problem of defining control objectives. From a different perspective, it is crucial to design highly reliable and robust controlled systems that are fault tolerant and non-fragile to both structured uncertainties and unexpected extreme and rare events. Building such a non-fragile and robust control orientation requires a new science which needs to bridge the gap between statistical measures of performance uncertainties and control decision optimization for control feedback synthesis. There resulted from the work of [6] a set of mathematical statistics of probabilistic performance distributions is adept at predicting performance behavior in the simple but important class of linear-quadratic stochastic control problems. The robust and resilient control design pivoted on the inherent system tradeoff between robustness for actuator failure accommodation and resilience for performance risk considerations is well summarized in the recent report of [7].

Chapter 7 represents an integrated, systematic attempt at applying decision theory to the wide class of resilient control situations in which: continuous-time stochastic dynamical systems are linear and subject to both norm-bounded and convex-bounded parameter uncertainties; and state and control time delays in executing the intended control inputs. Such physical processes typically induce dynamical behaviors in the closed-loop performance in response to various time

© Springer International Publishing Switzerland 2014   125
K.D. Pham, *Resilient Controls for Ordering Uncertain Prospects*, Springer
Optimization and Its Applications 98, DOI 10.1007/978-3-319-08705-4_7

delay effects and stochastic stimuli. Given all these technical challenges, the resilient control design can be solved in an elegant manner if the decision theory and control feedback optimization are resorted. The envisaged and proposed solutions and methodologies are to overcome the norm nowadays through developing and utilizing: on one hand, performance gathering analysis that proves expected-value performance specifications not fitting the reality well, is needed in order to properly characterize the generalized chi-squared type of performance behavior as will be established in Sect. 7.3; on the other hand, Sect. 7.4 will demonstrate the feasibility of mean-risk control decision laws that brings about robustness against structural and stochastic uncertainties as well as resiliency towards state and input time delay elements. Finally, Sect. 7.5 provides some concluding remarks for the interested readers to update their knowledge in this growing research topic.

7.2 Toward Performance Resiliency

The formulation presupposes a fixed probability space with filtration, $(\Omega, \mathbb{F}, \{\mathbb{F}_{0,t} : t \in [0, t_f]\}, \mathbb{P})$ satisfying the usual conditions. Throughout the chapter, all filtrations are right continuous and complete and $\mathbb{F}_{t_f} \triangleq \{\mathbb{F}_{0,t} : t \in [0, t_f]\}$. In addition, let $\mathscr{L}^2_{\mathbb{F}_{t_f}}([0, t_f]; \mathbb{R}^n)$ denote the space of $\mathbb{F}_{t_f}$-adapted random processes $\{\hbar(t) : t \in [0, t_f]$ such that $E\{\int_0^{t_f} ||\hbar(t)||^2\, dt\} < \infty\}$.

On the fixed probability space $(\Omega, \mathbb{F}, \{\mathbb{F}_{0,t} : t \in [0, t_f]\}, \mathbb{P})$, it is now of interest to investigate resilient controls of the problem class of stochastic linear systems, uncertainty, state and input time delays, risk aversion and finite time horizon; e.g.,

$$\begin{aligned}\dot{x}(t) = ((A + \Delta A(t))x(t) + (B + \Delta B(t))u(t) + (A_d + \Delta A_d(t))x(t - \tau) \\ + (B_h + \Delta B_h(t))u(t - \eta) + D(t)w(t), \quad x(0) = x_0 \qquad (7.1)\end{aligned}$$

where $x(t)$ is the controlled state process valued in $\mathbb{R}^n$ and $u(t)$ is the control process valued in a closed convex subset of an appropriate metric space; and $w(t)$ is an exogenous state noise taking the form of a mutually uncorrelated stationary $\mathbb{F}_{t_f}$-adapted Gaussian process with its mean $E\{w(t)\} = m_w(t)$ and covariance $cov\{w(t_1), w(t_2)\} = W\delta(t_1 - t_2)$ for all $t_1, t_2 \in [0, t_f]$ and $W > 0$. Of note, $x(t) = g(t)$ for all $t \in [-\max(\tau, \eta), 0]$ while $u(t) = h(t)$ for all $t \in [-\eta, 0]$. And τ and η stand for the amount of delay in the state and input of the uncertain time-delay system (7.1). This operation is illustrated in Fig. 7.1

The acquisition of information and its processing for which different control actions are based upon are governed by

$$y(t) = (C + \Delta C(t))x(t) + (F + \Delta F(t))u(t) + R(t)w(t) + T(t)v(t) \qquad (7.2)$$

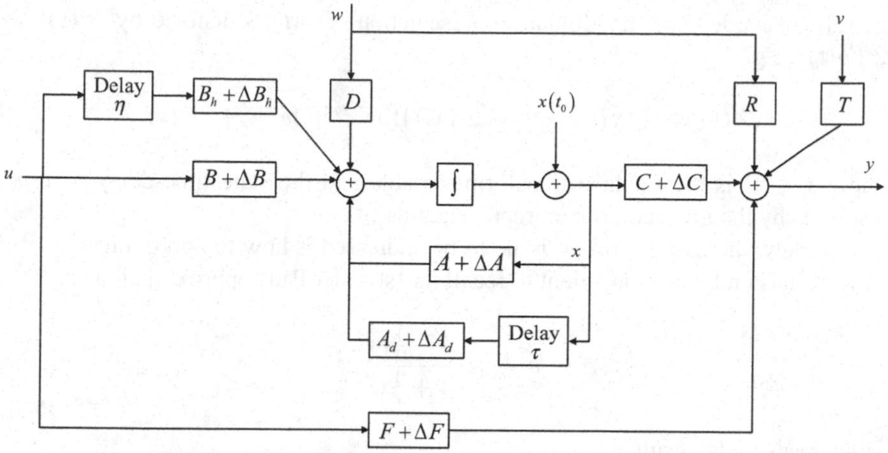

Fig. 7.1 Stochastic linear systems with uncertainty, state and input time delays, and risk aversion

where the exogenous measurement noise $v(t)$ is another mutually uncorrelated stationary $\mathbb{F}_{t_f}$-adapted Gaussian process with its mean $E\{v(t)\} = m_v(t)$ and covariance $cov\{v(t_1), v(t_2)\} = V\delta(t_1 - t_2)$ for all t_1, $t_2 \in [0, t_f]$ and $V > 0$.

The nominal system coefficients $A \in \mathbb{R}^{n \times n}$, $B \in \mathbb{R}^{n \times m}$, $C \in \mathbb{R}^{q \times n}$ and $F \in \mathbb{R}^{q \times m}$ are the constant matrices. In this part, the parameter perturbations $\Delta A(t)$, $\Delta B(t)$, $\Delta C(t)$ and $\Delta F(t)$ are real, time-varying matrix functions representing the norm-bounded parameter uncertainties and governed by

$$\begin{bmatrix} \Delta A(t) & \Delta B(t) \\ \Delta C(t) & \Delta F(t) \end{bmatrix} = \begin{bmatrix} H \\ H_c \end{bmatrix} \Delta(t) \begin{bmatrix} E \\ E_b \end{bmatrix} = \begin{bmatrix} H\Delta(t)E & H\Delta(t)E_b \\ H_c\Delta(t)E & H_c\Delta(t)E_b \end{bmatrix}$$

and

$$\Delta A_d(t) = H_d \Delta_1(t) E_d; \quad \Delta B_h(t) = H_h \Delta_2(t) E_h$$

where a priori knowledge about uncertainty is used for a direct derivation of known constant matrices H, H_c, E, E_b, H_d, E_d, H_h and E_h with appropriate dimensions. And $\Delta(t) \in \mathbb{R}^{\alpha_1 \times \alpha_2}$, $\Delta_1(t) \in \mathbb{R}^{\alpha_3 \times \alpha_4}$ and $\Delta_2(t) \in \mathbb{R}^{\alpha_5 \times \alpha_6}$ are unknown real time-varying matrices with Lebesgue measurable elements satisfying $\Delta(t)\Delta^T(t) \leq I$, $\Delta_1(t)\Delta_1^T(t) \leq I$ and $\Delta_2(t)\Delta_2^T(t) \leq I$ for all $t \in [0, t_f]$.

Let

$$p(t) \triangleq x(t - \tau) + x(t) \tag{7.3}$$

$$p(0) = g(-\tau) + g(0^-) \tag{7.4}$$

then the one-sided Laplace-domain representation of $p(t)$ is denoted by $\mathscr{P}(s) \triangleq \mathscr{L}\{p(t)\}$; e.g.,

$$\mathscr{P}(s) = \mathscr{L}\{x(t-\tau)\} + \mathscr{L}\{x(t)\} = \mathscr{X}(s)e^{-\tau s} + \mathscr{X}(s)$$

where $s \in \mathbb{C}$ is the Laplace transform variable and the state time-delay of τ is described by the Laplace-domain representation of $e^{-\tau s}$.

Naturally, the first important issue to be addressed is how to approximate $e^{-\tau s}$. Towards this end, it is convenient to recall its 1st-order Pade approximation

$$e^{-\tau s} \approx \frac{1 - \frac{1}{2}\tau s}{1 + \frac{1}{2}\tau s}$$

which leads to the result

$$\mathscr{P}(s) = \frac{2}{1 + \frac{1}{2}\tau s}\mathscr{X}(s). \tag{7.5}$$

According to the result (7.5), its time-domain equivalence can be obtained as follows

$$\dot{p}(t) = \frac{1}{\tau}(4x(t) - 2p(t)), \quad p(0). \tag{7.6}$$

Similarly, the focus is on the 1st-order Pade approximation of the input time-delay of η. Let $q(t) \triangleq u(t-\eta) + u(t)$ and $q(0) = h(-\eta) + h(0^-)$for all $t \in [0, t_f]$. Then, it is clearly that

$$\dot{q}(t) = \frac{1}{\eta}(4u(t) - 2q(t)), \quad q(0). \tag{7.7}$$

In the remainder of the section, for the sake of brevity, the attention will be restricted to a change of notations, namely

$$A_\Delta(t) \triangleq A + \Delta A(t); \quad A_{d\Delta}(t) \triangleq A_d + \Delta A_d(t); \quad C_\Delta(t) \triangleq C + \Delta C(t)$$
$$B_\Delta(t) \triangleq B + \Delta B(t); \quad B_{h\Delta}(t) \triangleq B_h + \Delta B_h(t); \quad F_\Delta(t) \triangleq F + \Delta F(t).$$

In this connection, it seems quite natural to rewrite the class of uncertain time-delay systems (7.1) as follows

$$\begin{aligned}\dot{x}(t) &= (A_\Delta(t) - A_{d\Delta}(t))x(t) + A_{d\Delta}(t)p(t) + B_{h\Delta}(t)q(t) \\ &\quad + (B_\Delta(t) - B_{h\Delta}(t))u(t) + D(t)w(t), \quad x(0) = x_0.\end{aligned} \tag{7.8}$$

As can be seen from (7.1) and (7.2), it is also worth noting that mutually uncorrelated non-zero mean stationary white noises $\{w(t) : t \in [0, t_f]\}$ and $\{v(t) : t \in [0, t_f]\}$ are idealizations of the incremental stationary Wiener processes with zero means and finite spectral bandwidths; e.g.,

$$dw(t) \triangleq \tilde{w}(t)dt = (w(t) - m_w(t))dt$$

$$dv(t) \triangleq \tilde{v}(t)dt = (v(t) - m_v(t))dt.$$

In this interpretation, what follows is an approximate system model that captures the 1st-order Pade approximations (7.6)–(7.7) of the state and input time-delays, uncertain dynamics (7.8), and the mechanisms for control decision updating based upon the received measurements (7.2)

$$dz(t) = (A_z(t)z(t) + B_z(t)u(t) + D_z(t)d(t))dt + G_z(t)dw(t) \tag{7.9}$$

$$dy(t) = (C_z(t)z(t) + F_z(t)u(t) + M_z(t)d(t))dt + M_z(t)d\xi(t) \tag{7.10}$$

where for each $t \in [0, t_f]$, the augmented system coefficients as shown in Fig. 7.2 are given by

$$A_z(t) \triangleq \begin{bmatrix} A_\Delta(t) - A_{d\Delta}(t) & A_{d\Delta}(t) & B_{h\Delta}(t) \\ \frac{4}{\tau} I_{n\times n} & -\frac{2}{\tau} I_{n\times n} & 0 \\ 0 & 0 & -\frac{2}{\eta} I_{m\times m} \end{bmatrix}; \quad B_z(t) \triangleq \begin{bmatrix} B_\Delta(t) - B_{h\Delta}(t) \\ 0 \\ \frac{4}{\eta} I_{m\times m} \end{bmatrix}$$

$$C_z(t) \triangleq \begin{bmatrix} C_\Delta(t) \; 0 \; 0 \end{bmatrix}; \quad D_z(t) \triangleq \begin{bmatrix} D(t) & 0 \\ 0 & 0 \\ 0 & 0 \end{bmatrix}; \quad G_z(t) \triangleq \begin{bmatrix} D(t) U_W \Lambda_W^{\frac{1}{2}} \\ 0 \\ 0 \end{bmatrix}$$

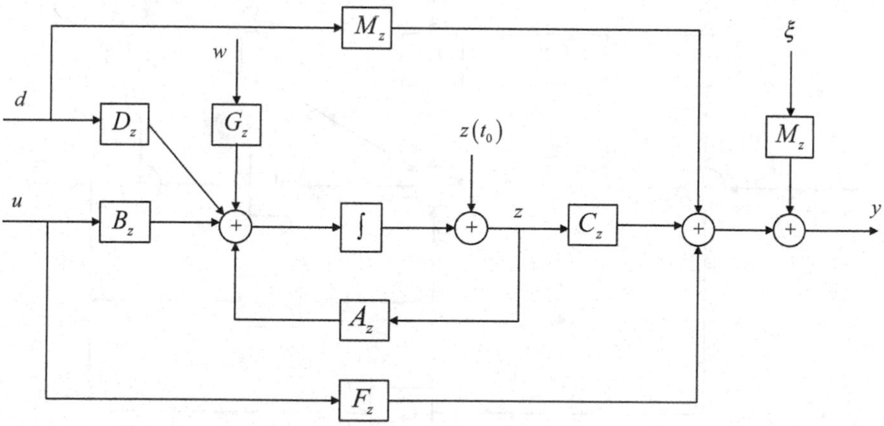

Fig. 7.2 Structure pertinent to aggregation of the approximate system model

$$F_z(t) \triangleq F_\Delta(t);\ M_z(t) \triangleq \left[R(t)U_W \Lambda_W^{\frac{1}{2}}\ T(t)U_V \Lambda_V^{\frac{1}{2}} \right];\ W \triangleq U_W \Lambda_W^{\frac{1}{2}} U_W^T;\ V \triangleq U_V \Lambda_V^{\frac{1}{2}} U_V^T$$

and the aggregate state variable $z(t)$, affine input $d(t)$, and mutually uncorrelated stationary Wiener process $\xi(t)$ with the correlation of independent increments $E\{[\xi(t_1) - \xi(t_2)][\xi(t_1) - \xi(t_2)]^T\} = \Xi|t_1 - t_2|$ for all $t_1,\ t_2 \in [0, t_f]$ are defined by

$$z(t) \triangleq \begin{bmatrix} x(t) \\ p(t) \\ q(t) \end{bmatrix};\ z(0) \triangleq \begin{bmatrix} x(0) \\ p(0) \\ q(0) \end{bmatrix};\ \xi(t) \triangleq \begin{bmatrix} w(t) \\ v(t) \end{bmatrix};\ d(t) \triangleq \begin{bmatrix} m_w(t) \\ m_v(t) \end{bmatrix};\ \Xi = \begin{bmatrix} I_{p\times p} & 0 \\ 0 & I_{q\times q} \end{bmatrix}$$

Further let the σ-algebras

$$\mathbb{F}_{0,t} \triangleq \sigma\{(w(\varsigma), v(\varsigma)) : 0 \leq \varsigma \leq t\}$$
$$\mathcal{G}_{0,t}^y \triangleq \sigma\{y(\varsigma) : 0 \leq \varsigma \leq t\}, \quad t \in [0, t_f]$$

then, the aim here is to investigate whether it is possible to devise a suitable state estimator whose framework not only preserves the inherent linear structure of (7.9)–(7.10) but also leverages the information available $\mathcal{G}_{t_f}^y \triangleq \{\mathcal{G}_{0,t}^y : t \in [0, t_f]\} \subset \{\mathbb{F}_{0,t} : t \in [0, t_f]\}$. To this end, the quest of filter estimates is met as depicted in Fig. 7.3; i.e.,

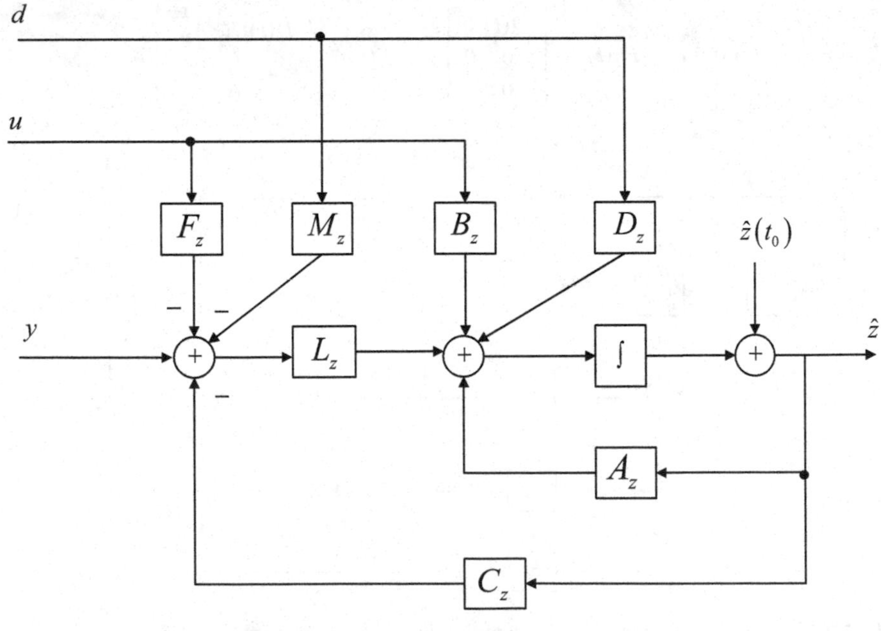

Fig. 7.3 State-space realization of the state estimator

$$d\hat{z}(t) = (A_z(t)\hat{z}(t) + B_z(t)u(t) + D_z(t)d(t))dt \\ +L_z(t)(dy(t) - (C_z(t)\hat{z}(t) + F_z(t)u(t) + M_z(t)d(t))dt), \quad \hat{z}(0) = z(0) \tag{7.11}$$

where the filtering gain $L_z(t)$ and state-estimate error covariance $\Sigma_z(t)$ are given by

$$L_z(t) = \Sigma_z(t)C_z^T(t)(M_z(t)\Xi M_z^T(t))^{-1} \tag{7.12}$$

$$\frac{d}{dt}\Sigma_z(t) = A_z(t)\Sigma_z(t) + \Sigma_z(t)A_z^T(t) + G_z(t)G_z^T(t) \\ - \Sigma_z(t)C_z^T(t)(M_z(t)\Xi M_z^T(t))^{-1}C_z(t)\Sigma_z(t), \quad \Sigma_z(0) = 0. \tag{7.13}$$

As well known, in this case, the estimate errors $\tilde{z}(t) \triangleq z(t) - \hat{z}(t)$ are satisfying the stochastic differential equation with the initial-value condition $\tilde{z}(0) = 0$

$$d\tilde{z}(t) = (A_z(t) - L_z(t)C_z(t))\tilde{z}(t)dt + G_z(t)dw(t) - L_z(t)M_z(t)d\xi(t). \tag{7.14}$$

The admissible set of feedback control laws is defined as follows

$$\mathbb{U}^y[0,t_f] \triangleq \{u \in \mathcal{L}^2_{\mathcal{G}^y_{t_f}}([0,t_f],\mathbb{R}^m), \mathbb{P} - a.s.\}$$

where $\mathbb{U}^y[0,t_f]$ is a closed convex subset of $\mathcal{L}^2_{\mathbb{F}_{t_f}}([0,t_f],\mathbb{R}^m)$.

Associated with an admissible initial condition $(z(\cdot),u(\cdot))$ is an integral-quadratic form cost $J : \mathbb{R}^{2n+m} \times \mathbb{U}^y[0,t_f] \mapsto \mathbb{R}^+$ on $[0,t_f]$

$$J(z_0,u(\cdot)) = z^T(t_f)Q_z^f z(t_f) + \int_0^{t_f} [z^T(\varsigma)Q_z(\varsigma)z(\varsigma) + u^T(\varsigma)R_z(\varsigma)u(\varsigma)]d\varsigma \tag{7.15}$$

where the design parameters $Q_z^f \in \mathbb{R}^{(2n+m)\times(2n+m)}$, $Q_z \in \mathscr{C}([t_0,t_f];\mathbb{R}^{(2n+m)\times(2n+m)})$ and invertible $R_z \in \mathscr{C}([t_0,t_f];\mathbb{R}^{m\times m})$ are deterministic, symmetric and positive semi-definite relative weighting of the terminal state, state trajectory, and control input.

In view of the linear structure of the system (7.9)–(7.10) and the quadratic nature of the performance measure (7.15), non-randomized control strategies are therefore, restricted to the linear mapping $\gamma : [0,t_f] \times \mathbb{R}^{2n+m} \mapsto \mathbb{U}^y[0,t_f]$ with the rule of actions

$$u(t) = \gamma(t,z(t)) \triangleq K_z(t)\hat{z}(t) + l_z(t) \tag{7.16}$$

where the deterministic matrix-valued functions $K_z \in \mathscr{C}([0,t_f];\mathbb{R}^{m\times(2n+m)})$ and vector-valued functions $l_z \in \mathscr{C}([0,t_f];\mathbb{R}^m)$ are the admissible feedback gain and feedforward input are yet to be defined.

Define the augmented vectors and matrices

$$\hbar(t) \triangleq \begin{bmatrix} \hat{z}(t) \\ \tilde{z}(t) \end{bmatrix}; \ \hbar(0) \triangleq \begin{bmatrix} z(0) \\ 0 \end{bmatrix}; \ \zeta(t) \triangleq \begin{bmatrix} w(t) \\ \xi(t) \end{bmatrix}; \quad G_\hbar(t) \triangleq \begin{bmatrix} 0 & L_z(t)M_z(t) \\ G_z(t) & -L_z(t)M_z(t) \end{bmatrix}$$

$$A_\hbar(t) \triangleq \begin{bmatrix} A_z(t)+B_z(t)K_z(t) & L_z(t)C_z(t) \\ 0 & A_z(t)-L_z(t)C_z(t) \end{bmatrix}; \ b_\hbar(t) \triangleq \begin{bmatrix} B_z(t)l_z(t)+D_z(t)d(t) \\ 0 \end{bmatrix}$$

$$\Xi_\hbar \triangleq \begin{bmatrix} I_{p\times p} & 0 \\ 0 & \Xi \end{bmatrix}; \ E\{[\zeta(t_1)-\zeta(t_2)][\zeta(t_1)-\zeta(t_2)]^T\} \triangleq \Xi_\hbar |t_1-t_2|, \forall t_1,t_2 \in [0,t_f]$$

Then, in the spirit of (7.16) and the decomposition aforementioned, the stochastic differential equation (7.9) is rewritten equivalently as follows

$$d\hbar(t) = (A_\hbar(t)\hbar(t) + b_\hbar(t))dt + G_\hbar(t)d\zeta(t), \quad \hbar(0) \tag{7.17}$$

together with the measure of performance (7.15)

$$J(h(0);K_z(\cdot),l_z(\cdot)) = \hbar^T(t_f)Q_\hbar^f\hbar(t_f) + \int_0^{t_f}[\hbar^T(\varsigma)Q_\hbar(\varsigma)\hbar(\varsigma) \\ +2\hbar^T(\varsigma)S_\hbar(\varsigma) + l_z^T(\varsigma)R_z(\varsigma)l_z(\varsigma)]d\varsigma \tag{7.18}$$

provided that

$$Q_\hbar^f \triangleq \begin{bmatrix} Q_z^f & Q_z^f \\ Q_z^f & Q_z^f \end{bmatrix}; \quad Q_\hbar(\varsigma) \triangleq \begin{bmatrix} Q_z(\varsigma)+K_z^T(\varsigma)R_z(\varsigma)K_z(\varsigma) & Q_z(\varsigma) \\ Q_z(\varsigma) & Q_z(\varsigma) \end{bmatrix}$$

$$S_\hbar(\varsigma) \triangleq \begin{bmatrix} K_z^T(\varsigma)R_z(\varsigma)l_z(\varsigma) \\ 0 \end{bmatrix}.$$

For further progress in an efficient methodology of computing all the statistical measures of performance risks, the following result derived from the cumulant-generating function reported herein is highly conducive to producing significant insights of performance distributions for the control analysis at hand.

Theorem 7.2.1 (Performance-Measure Statistics). *Suppose that (A_z, B_z) is uniformly stabilizable and (C_z, A_z) is uniformly detectable. The kth-order statistic of the chi-squared performance measure (7.18) is given by*

$$\kappa_k = \hbar^T(0)H_k(0)\hbar(0) + 2\hbar^T(0)\breve{D}_k(0) + D_k(0), \quad k \in \mathbb{Z}^+ \tag{7.19}$$

where the cumulant-generating components $H_k(\varsigma)$, $\breve{D}_k(\varsigma)$ and $D_k(\varsigma)$ evaluated at $\varsigma = 0$ satisfy the cumulant-generating equations (with the dependence of $H_k(\varsigma)$, $\breve{D}_k(\varsigma)$ and $D_k(\varsigma)$ upon K_z and l_z suppressed)

$$\frac{d}{d\varsigma}H_1(\varsigma) = -A_{\hbar}^T(\varsigma)H_1(\varsigma) - H_1(\varsigma)A_{\hbar}(\varsigma) - Q_{\hbar}(\varsigma), \quad H_1(t_f) = Q_{\hbar}^f \tag{7.20}$$

$$\frac{d}{d\varsigma}H_r(\varsigma) = -A_{\hbar}^T(\varsigma)H_r(\varsigma) - H_r(\varsigma)A_{\hbar}(\varsigma) - \sum_{\nu=1}^{r-1}\frac{2r!}{\nu!(r-\nu)!}H_\nu(\varsigma)G_{\hbar}(\varsigma)\Xi_{\hbar}G_{\hbar}^T(\varsigma)H_{r-\nu}(\varsigma),$$
$$H_r(t_f) = 0, \quad 2 \le r \le k \tag{7.21}$$

and

$$\frac{d}{d\varsigma}\breve{D}_1(\varsigma) = -A_{\hbar}^T(\varsigma)\breve{D}_1(\varsigma) - H_1(\varsigma)b_{\hbar}(\varsigma) - S_{\hbar}(\varsigma), \quad \breve{D}_1(t_f) = 0 \tag{7.22}$$

$$\frac{d}{d\varsigma}\breve{D}_r(\varsigma) = -A_{\hbar}^T(\varsigma)\breve{D}_r(\varsigma) - H_r(\varsigma)b_{\hbar}(\varsigma), \quad \breve{D}_r(t_f) = 0, \quad 2 \le r \le k \tag{7.23}$$

and, finally

$$\frac{d}{d\varsigma}D_1(\varsigma) = -\mathrm{Tr}\{H_1(\varsigma)G_{\hbar}(\varsigma)\Xi_{\hbar}G_{\hbar}^T(\varsigma)\} - 2\breve{D}_1^T(\varsigma)b_{\hbar}(\varsigma) - l_z^T(\varsigma)R_z(\varsigma)l_z(\varsigma) \tag{7.24}$$

$$\frac{d}{d\varsigma}D_r(\varsigma) = -\mathrm{Tr}\{H_r(\varsigma)G_{\hbar}(\varsigma)\Xi_{\hbar}G_{\hbar}^T(\varsigma)\} - 2\breve{D}_r^T(\varsigma)b_{\hbar}(\varsigma), \quad 2 \le r \le 2 \tag{7.25}$$

with the terminal-value conditions $D_r(t_f) = 0$ for $1 \le r \le k$.

Intuitively speaking, the recent result here permits the incorporation of classes of linear-feedback strategies in the statistical optimal control problems. Moreover, these performance-measure statistics (7.19) are further rewritten in terms of the system description of (7.9)–(7.10)

$$\kappa_k = z_0^T H_{11}^k(0)z_0 + 2z_0^T\breve{D}_{11}^k(0) + D^k(0), \quad k \in \mathbb{Z}^+ \tag{7.26}$$

according to the matrix and vector partitions

$$H_r(\varsigma) \triangleq \begin{bmatrix} H_{11}^r(\varsigma) & H_{12}^r(\varsigma) \\ (H_{12}^r)^T(\varsigma) & H_{22}^r(\varsigma) \end{bmatrix}; \quad \breve{D}_r(\varsigma) \triangleq \begin{bmatrix} \breve{D}_{11}^r(\varsigma) \\ \breve{D}_{21}^r(\varsigma) \end{bmatrix}; \quad D^k(\varsigma) \triangleq D_k(\varsigma) \tag{7.27}$$

from which the sub-matrix and vector components depend on K_z and l_z and satisfy the backward-in-time matrix-valued differential equations

$$\frac{d}{d\varsigma}H_{11}^1(\varsigma) = -(A_z(\varsigma) + B_z(\varsigma)K_z(\varsigma))^T H_{11}^1(\varsigma) - H_{11}^1(\varsigma)(A_z(\varsigma) + B_z(\varsigma)K_z(\varsigma)) - K_z^T(\varsigma)R_z(\varsigma)K_z(\varsigma) - Q_z(\varsigma), \quad H_{11}^1(t_f) = Q_z^f \tag{7.28}$$

$$\frac{d}{d\varsigma}H_{11}^r(\varsigma) = -(A_z(\varsigma) + B_z(\varsigma)K_z(\varsigma))^T H_{11}^r(\varsigma) - H_{11}^r(\varsigma)(A_z(\varsigma) + B_z(\varsigma)K_z(\varsigma)) - \sum_{\nu=1}^{r-1}\frac{2r!}{\nu!(r-\nu)!}(H_{11}^\nu(\varsigma)\Pi_1(\varsigma) + H_{12}^\nu(\varsigma)\Pi_3(\varsigma))H_{11}^{r-\nu}(\varsigma) - \sum_{\nu=1}^{r-1}\frac{2r!}{\nu!(r-\nu)!}(H_{11}^\nu(\varsigma)\Pi_2(\varsigma) + H_{12}^\nu(\varsigma)\Pi_4(\varsigma))H_{21}^{r-\nu}(\varsigma), \; H_{11}^r(t_f) = 0 \tag{7.29}$$

$$\frac{d}{d\varsigma}H_{12}^1(\varsigma) = -(A_z(\varsigma) + B_z(\varsigma)K_z(\varsigma))^T H_{12}^1(\varsigma) - H_{11}^1(\varsigma)L_z(\varsigma)C_z(\varsigma) - H_{12}^1(\varsigma)(A_z(\varsigma) - L_z(\varsigma)C_z(\varsigma)) - Q_z(\varsigma), \quad H_{12}^1(t_f) = Q_z^f \tag{7.30}$$

$$\frac{d}{d\varsigma}H_{12}^r(\varsigma) = -(A_z(\varsigma) + B_z(\varsigma)K_z(\varsigma))^T H_{12}^r(\varsigma) - H_{12}^r(\varsigma)(A_z(\varsigma) - L_z(\varsigma)C_z(\varsigma)) - H_{11}^r(\varsigma)L_z(\varsigma)C_z(\varsigma) - \sum_{\nu=1}^{r-1}\frac{2r!}{\nu!(r-\nu)!}(H_{11}^\nu(\varsigma)\Pi_1(\varsigma) + H_{12}^\nu(\varsigma)\Pi_3(\varsigma))H_{12}^{r-\nu}(\varsigma) - \sum_{\nu=1}^{r-1}\frac{2r!}{\nu!(r-\nu)!}(H_{11}^\nu(\varsigma)\Pi_2(\varsigma) + H_{12}^\nu(\varsigma)\Pi_4(\varsigma))H_{22}^{r-\nu}(\varsigma), \quad H_{12}^r(t_f) = 0 \tag{7.31}$$

$$\frac{d}{d\varsigma}H_{21}^1(\varsigma) = -H_{21}^1(\varsigma)(A_z(\varsigma) + B_z(\varsigma)K_z(\varsigma)) - (L_z(\varsigma)C_z(\varsigma))^T H_{11}^1(\varsigma) - (A_z(\varsigma) - L_z(\varsigma)C_z(\varsigma))^T H_{21}^1(\varsigma) - Q_z(\varsigma), \quad H_{21}^1(t_f) = Q_z^f \tag{7.32}$$

$$\frac{d}{d\varsigma}H_{21}^r(\varsigma) = -H_{21}^r(\varsigma)(A_z(\varsigma) + B_z(\varsigma)K_z(\varsigma)) - (A_z(\varsigma) - L_z(\varsigma)C_z(\varsigma))^T H_{21}^r(\varsigma) - (L_z(\varsigma)C_z(\varsigma))^T H_{11}^r(\varsigma) - \sum_{\nu=1}^{r-1}\frac{2r!}{\nu!(r-\nu)!}(H_{21}^\nu(\varsigma)\Pi_1(\varsigma) + H_{22}^\nu(\varsigma)\Pi_3(\varsigma))H_{11}^{r-\nu}(\varsigma) - \sum_{\nu=1}^{r-1}\frac{2r!}{\nu!(r-\nu)!}(H_{21}^\nu(\varsigma)\Pi_2(\varsigma) + H_{22}^\nu(\varsigma)\Pi_4(\varsigma))H_{21}^{r-\nu}(\varsigma), \quad H_{21}^r(t_f) = 0 \tag{7.33}$$

$$\frac{d}{d\varsigma}H_{22}^{1}(\varsigma) = -H_{22}^{1}(\varsigma)(A_z(\varsigma) - L_z(\varsigma)C_z(\varsigma)) - H_{21}^{1}(\varsigma)(L_z(\varsigma)C_z(\varsigma)) - Q_z(\varsigma)$$
$$- (L_z(\varsigma)C_z(\varsigma))^T H_{12}^{1}(\varsigma) - (A_z(\varsigma) - L_z(\varsigma)C_z(\varsigma))^T H_{22}^{1}(\varsigma), \quad H_{22}^{1}(t_f) = Q_z^f \tag{7.34}$$

$$\frac{d}{d\varsigma}H_{22}^{r}(\varsigma) = - H_{22}^{r}(\varsigma)(A_z(\varsigma) - L_z(\varsigma)C_z(\varsigma)) - H_{21}^{r}(\varsigma)(L_z(\varsigma)C_z(\varsigma))$$
$$- (L_z(\varsigma)C_z(\varsigma))^T H_{12}^{r}(\varsigma) - (A_z(\varsigma) - L_z(\varsigma)C_z(\varsigma))^T H_{22}^{r}(\varsigma)$$
$$- \sum_{\nu=1}^{r-1} \frac{2r!}{\nu!(r-\nu)!}(H_{21}^{\nu}(\varsigma)\Pi_1(\varsigma) + H_{22}^{\nu}(\varsigma)\Pi_3(\varsigma))H_{12}^{r-\nu}(\varsigma)$$
$$- \sum_{\nu=1}^{r-1} \frac{2r!}{\nu!(r-\nu)!}(H_{21}^{\nu}(\varsigma)\Pi_2(\varsigma)+H_{22}^{\nu}(\varsigma)\Pi_4(\varsigma))H_{22}^{r-\nu}(\varsigma), H_{22}^{r}(t_f)=0 \tag{7.35}$$

$$\frac{d}{d\varsigma}\breve{D}_{11}^{1}(\varsigma) = - (A_z(\varsigma)+B_z(\varsigma)K_z(\varsigma))^T \breve{D}_{11}^{1}(\varsigma)-H_{11}^{1}(\varsigma)(B_z(\varsigma)l_z(\varsigma)+D_z(\varsigma)d(\varsigma))$$
$$- K_z^T(\varsigma)R_z(\varsigma)l_z(\varsigma), \quad \breve{D}_{11}^{1}(t_f) = 0 \tag{7.36}$$

$$\frac{d}{d\varsigma}\breve{D}_{11}^{r}(\varsigma) = - (A_z(\varsigma) + B_z(\varsigma)K_z(\varsigma))^T \breve{D}_{11}^{r}(\varsigma)$$
$$- H_{11}^{r}(\varsigma)(B_z(\varsigma)l_z(\varsigma) + D_z(\varsigma)d(\varsigma)), \quad \breve{D}_{11}^{r}(t_f) = 0, \quad 2 \le r \le k \tag{7.37}$$

$$\frac{d}{d\varsigma}\breve{D}_{21}^{r}(\varsigma) = - (A_z(\varsigma)-L_z(\varsigma)C_z(\varsigma))^T \breve{D}_{21}^{r}(\varsigma)-H_{21}^{r}(\varsigma)(B_z(\varsigma)l_z(\varsigma)+D_z(\varsigma)d(\varsigma))$$
$$- (L_z(\varsigma)C_z(\varsigma))^T \breve{D}_{11}^{r}(\varsigma), \quad \breve{D}_{21}^{r}(t_f) = 0, \quad 1 \le r \le k \tag{7.38}$$

$$\frac{d}{d\varsigma}D_1(\varsigma) = - \mathrm{Tr}\{H_{11}^{1}(\varsigma)\Pi_1(\varsigma) + H_{12}^{1}(\varsigma)\Pi_3(\varsigma)\}$$
$$- \mathrm{Tr}\{H_{21}^{1}(\varsigma)\Pi_2(\varsigma) + H_{22}^{1}(\varsigma)\Pi_4(\varsigma)\} - l_z^T(\varsigma)R_z(\varsigma)l_z(\varsigma)$$
$$- 2(\breve{D}_{11}^{1})^T(\varsigma)(B_z(\varsigma)l_z(\varsigma) + D_z(\varsigma)d(\varsigma)), \quad D_1(t_f) = 0 \tag{7.39}$$

$$\frac{d}{d\varsigma}D_r(\varsigma) = - \mathrm{Tr}\{H_{11}^{r}(\varsigma)\Pi_1(\varsigma) + H_{12}^{r}(\varsigma)\Pi_3(\varsigma)\}$$
$$- \mathrm{Tr}\{H_{21}^{r}(\varsigma)\Pi_2(\varsigma) + H_{22}^{r}(\varsigma)\Pi_4(\varsigma)\}$$
$$- 2(\breve{D}_{11}^{r})^T(\varsigma)(B_z(\varsigma)l_z(\varsigma)+D_z(\varsigma)d(\varsigma)), \; D_r(t_f)=0, \; 2 \le r \le k \tag{7.40}$$

where $\Pi_1(\varsigma) \triangleq L_z(\varsigma)R(\varsigma)R^T(\varsigma)L_z^T(\varsigma) + L_z(\varsigma)T(\varsigma)T^T(\varsigma)L_z^T(\varsigma)$; $\Pi_2(\varsigma) = \Pi_3(\varsigma) \triangleq -\Pi_1(\varsigma)$ and $\Pi_4(\varsigma) \triangleq D(\varsigma)WD^T(\varsigma) + L_z(\varsigma)R(\varsigma)R^T(\varsigma)L_z^T(\varsigma) + L_z(\varsigma)T(\varsigma)T^T(\varsigma)L_z^T(\varsigma)$.

7.3 Participation in the System Control Problem

The results in Sect. 7.2 show where performance uncertainty is important. The mathematical statistics (7.26) associated with (7.15) or (7.18) to which the performance distribution is sensitive have been identified and considered as crucial state variables. Determining how these statistical measures of risks in the crucial state variables influence the feedback control decisions is the concern of the control optimization phase of the feedback design synthesis. The next step is to mitigate the uncertainty in performance measure (7.15) for each alternative (K_z, l_z) implied by the relationship of performance risk aversion to the time-backward trajectories (7.28)–(7.40) of the corresponding state variables, e.g., $H_{11}^r(\varsigma)$, $H_{12}^r(\varsigma)$, $H_{21}^r(\varsigma)$, $H_{22}^r(\tau)$, $\breve{D}_{11}^r(\varsigma)$, $\breve{D}_{21}^r(\varsigma)$ and $D^r(\varsigma)$.

For notational simplicity, the convenient mappings to denote the right members of (7.28)–(7.40) are put into use

$$\begin{aligned}
&\mathscr{F}_{11}^1(\varsigma, \mathscr{H}_{11}, \mathscr{H}_{12}, \mathscr{H}_{21}, K_z) \triangleq -(A_z(\varsigma) + B_z(\varsigma)K_z(\varsigma))^T \mathscr{H}_{11}^1(\varsigma) \\
&\quad - \mathscr{H}_{11}^1(\varsigma)(A_z(\varsigma) + B_z(\varsigma)K_z(\varsigma)) - K_z^T(\varsigma)R_z(\varsigma)K_z(\varsigma) - Q_z(\varsigma) \\
&\mathscr{F}_{11}^r(\varsigma, \mathscr{H}_{11}, \mathscr{H}_{12}, \mathscr{H}_{21}, K_z) \triangleq -(A_z(\varsigma) + B_z(\varsigma)K_z(\varsigma))^T \mathscr{H}_{11}^r(\varsigma) \\
&\quad - \mathscr{H}_{11}^r(\varsigma)(A_z(\varsigma) + B_z(\varsigma)K_z(\varsigma)) \\
&\quad - \sum_{\nu=1}^{r-1} \frac{2r!}{\nu!(r-\nu)!}(\mathscr{H}_{11}^\nu(\varsigma)\Pi_1(\varsigma) + \mathscr{H}_{12}^\nu(\varsigma)\Pi_3(\varsigma))\mathscr{H}_{11}^{r-\nu}(\varsigma) \\
&\quad - \sum_{\nu=1}^{r-1} \frac{2r!}{\nu!(r-\nu)!}(\mathscr{H}_{11}^\nu(\varsigma)\Pi_2(\varsigma) + \mathscr{H}_{12}^\nu(\varsigma)\Pi_4(\varsigma))\mathscr{H}_{21}^{r-\nu}(\varsigma), \quad 2 \le r \le k
\end{aligned}$$

$$\begin{aligned}
&\mathscr{F}_{12}^1(\varsigma, \mathscr{H}_{11}, \mathscr{H}_{12}, \mathscr{H}_{22}, K_z) \triangleq -(A_z(\varsigma) + B_z(\varsigma)K_z(\varsigma))^T \mathscr{H}_{12}^1(\varsigma) - \mathscr{H}_{11}^1(\varsigma)L_z(\varsigma)C_z(\varsigma) \\
&\quad - \mathscr{H}_{12}^1(\varsigma)(A_z(\varsigma) - L_z(\varsigma)C_z(\varsigma)) - Q_z(\varsigma) \\
&\mathscr{F}_{12}^r(\varsigma, \mathscr{H}_{11}, \mathscr{H}_{12}, \mathscr{H}_{22}, K_z) \triangleq -(A_z(\varsigma) + B_z(\varsigma)K_z(\varsigma))^T \mathscr{H}_{12}^r(\varsigma) - \mathscr{H}_{11}^r(\varsigma)L_z(\varsigma)C_z(\varsigma) \\
&\quad - \mathscr{H}_{12}^r(\varsigma)(A_z(\varsigma) - L_z(\varsigma)C_z(\varsigma)) \\
&\quad - \sum_{\nu=1}^{r-1} \frac{2r!}{\nu!(r-\nu)!}(\mathscr{H}_{11}^\nu(\varsigma)\Pi_1(\varsigma) + \mathscr{H}_{12}^\nu(\varsigma)\Pi_3(\varsigma))\mathscr{H}_{12}^{r-\nu}(\varsigma) \\
&\quad - \sum_{\nu=1}^{r-1} \frac{2r!}{\nu!(r-\nu)!}(\mathscr{H}_{11}^\nu(\varsigma)\Pi_2(\varsigma) + \mathscr{H}_{12}^\nu(\varsigma)\Pi_4(\varsigma))\mathscr{H}_{22}^{r-\nu}(\varsigma), \quad 2 \le r \le k
\end{aligned}$$

$$\mathscr{F}^1_{21}(\varsigma, \mathscr{H}_{11}, \mathscr{H}_{21}, \mathscr{H}_{22}, K_z) \triangleq -\mathscr{H}^1_{21}(\varsigma)(A_z(\varsigma) + B_z(\varsigma)K_z(\varsigma))$$
$$- (L_z(\varsigma)C_z(\varsigma))^T \mathscr{H}^1_{11}(\varsigma) - (A_z(\varsigma) - L_z(\varsigma)C_z(\varsigma))^T \mathscr{H}^1_{21}(\varsigma) - Q_z(\varsigma)$$
$$\mathscr{F}^r_{21}(\varsigma, \mathscr{H}_{11}, \mathscr{H}_{21}, \mathscr{H}_{22}, K_z) \triangleq -\mathscr{H}^r_{21}(\varsigma)(A_z(\varsigma) + B_z(\varsigma)K_z(\varsigma))$$
$$- (A_z(\varsigma) - L_z(\varsigma)C_z(\varsigma))^T \mathscr{H}^r_{21}(\varsigma) - (L_z(\varsigma)C_z(\varsigma))^T \mathscr{H}^r_{11}(\varsigma)$$
$$-\sum_{\nu=1}^{r-1} \frac{2r!}{\nu!(r-\nu)!}(\mathscr{H}^\nu_{21}(\varsigma)\Pi_1(\varsigma) + \mathscr{H}^\nu_{22}(\varsigma)\Pi_3(\varsigma))\mathscr{H}^{r-\nu}_{11}(\varsigma)$$
$$-\sum_{\nu=1}^{r-1} \frac{2r!}{\nu!(r-\nu)!}(\mathscr{H}^\nu_{21}(\varsigma)\Pi_2(\varsigma) + \mathscr{H}^\nu_{22}(\varsigma)\Pi_4(\varsigma))\mathscr{H}^{r-\nu}_{21}(\varsigma), \quad 2 \le r \le k$$

$$\mathscr{F}^1_{22}(\varsigma, \mathscr{H}_{22}, \mathscr{H}_{12}, \mathscr{H}_{21}) \triangleq -\mathscr{H}^1_{22}(\varsigma)(A_z(\varsigma) - L_z(\varsigma)C_z(\varsigma)) - \mathscr{H}^1_{21}(\varsigma)(L_z(\varsigma)C_z(\varsigma))$$
$$- (L_z(\varsigma)C_z(\varsigma))^T \mathscr{H}^1_{12}(\varsigma) - (A_z(\varsigma) - L_z(\varsigma)C_z(\varsigma))^T \mathscr{H}^1_{22}(\varsigma) - Q_z(\varsigma)$$
$$\mathscr{F}^r_{22}(\varsigma, \mathscr{H}_{22}, \mathscr{H}_{12}, \mathscr{H}_{21}) \triangleq -\mathscr{H}^r_{22}(\varsigma)(A_z(\varsigma) - L_z(\varsigma)C_z(\varsigma)) - \mathscr{H}^r_{21}(\varsigma)(L_z(\varsigma)C_z(\varsigma))$$
$$- (L_z(\varsigma)C_z(\varsigma))^T \mathscr{H}^r_{12}(\varsigma) - (A_z(\varsigma) - L_z(\varsigma)C_z(\varsigma))^T \mathscr{H}^r_{22}(\varsigma)$$
$$-\sum_{\nu=1}^{r-1} \frac{2r!}{\nu!(r-\nu)!}(\mathscr{H}^\nu_{21}(\varsigma)\Pi_1(\varsigma) + \mathscr{H}^\nu_{22}(\varsigma)\Pi_3(\varsigma))\mathscr{H}^{r-\nu}_{12}(\varsigma)$$
$$-\sum_{\nu=1}^{r-1} \frac{2r!}{\nu!(r-\nu)!}(\mathscr{H}^\nu_{21}(\varsigma)\Pi_2(\varsigma) + \mathscr{H}^\nu_{22}(\varsigma)\Pi_4(\varsigma))\mathscr{H}^{r-\nu}_{22}(\varsigma), \quad 2 \le r \le k$$

$$\breve{\mathscr{G}}^1_{11}(\varsigma, \breve{\mathscr{D}}_{11}, \mathscr{H}_{11}, K_z, l_z) \triangleq -(A_z(\varsigma) + B_z(\varsigma)K_z(\varsigma))^T \breve{\mathscr{D}}^1_{11}(\varsigma) - K_z^T(\varsigma)R_z(\varsigma)l_z(\varsigma)$$
$$- \mathscr{H}^1_{11}(\varsigma)(B_z(\varsigma)l_z(\varsigma) + D_z(\varsigma)d(\varsigma))$$
$$\breve{\mathscr{G}}^r_{11}(\varsigma, \breve{\mathscr{D}}_{11}, \mathscr{H}_{11}, K_z, l_z) \triangleq -(A_z(\varsigma) + B_z(\varsigma)K_z(\varsigma))^T \breve{\mathscr{D}}^r_{11}(\varsigma)$$
$$- \mathscr{H}^r_{11}(\varsigma)(B_z(\varsigma)l_z(\varsigma) + D_z(\varsigma)d(\varsigma)), \quad 2 \le r \le k$$

$$\breve{\mathscr{G}}^r_{21}(\varsigma, \breve{\mathscr{D}}_{11}, \breve{\mathscr{D}}_{21}, \mathscr{H}_{21}, l_z) \triangleq -(A_z(\varsigma) - L_z(\varsigma)C_z(\varsigma))^T \breve{\mathscr{D}}^r_{21}(\varsigma)$$
$$- \mathscr{H}^r_{21}(\varsigma)(B_z(\varsigma)l_z(\varsigma) + D_z(\varsigma)d(\varsigma)) - (L_z(\varsigma)C_z(\varsigma))^T \breve{\mathscr{D}}^r_{11}(\varsigma), \quad 1 \le r \le k$$

$$\mathscr{G}_1(\varsigma, \mathscr{H}_{11}, \mathscr{H}_{12}, \mathscr{H}_{21}, \mathscr{H}_{22}, \breve{\mathscr{D}}_{11}, l_z) \triangleq$$
$$- \mathrm{Tr}\{\mathscr{H}^1_{11}(\varsigma)\Pi_1(\varsigma) + \mathscr{H}^1_{12}(\varsigma)\Pi_3(\varsigma)\} - \mathrm{Tr}\{\mathscr{H}^1_{21}(\varsigma)\Pi_2(\varsigma) + \mathscr{H}^1_{22}(\varsigma)\Pi_4(\varsigma)\}$$

$$- 2(\breve{\mathscr{D}}_{11}^1)^T(\varsigma)(B_z(\varsigma)l_z(\varsigma) + D_z(\varsigma)d(\varsigma)) - l_z^T(\varsigma)R_z(\varsigma)l_z(\varsigma)$$

$$\mathscr{G}_r(\varsigma, \mathscr{H}_{11}, \mathscr{H}_{12}, \mathscr{H}_{21}, \mathscr{H}_{22}, \breve{\mathscr{D}}_{11}, l_z) \triangleq$$
$$- \mathrm{Tr}\{\mathscr{H}_{11}^r(\varsigma)\Pi_1(\varsigma) + \mathscr{H}_{12}^r(\varsigma)\Pi_3(\varsigma)\} - \mathrm{Tr}\{\mathscr{H}_{21}^r(\varsigma)\Pi_2(\varsigma) + \mathscr{H}_{22}^r(\varsigma)\Pi_4(\varsigma)\}$$
$$- 2(\breve{\mathscr{D}}_{11}^r)^T(\varsigma)(B_z(\varsigma)l_z(\varsigma) + D_z(\varsigma)d(\varsigma)), \quad 2 \le r \le k$$

where the k-tuple variables $\mathscr{H}_{11}$, $\mathscr{H}_{12}$, $\mathscr{H}_{21}$, $\mathscr{H}_{22}$, $\breve{\mathscr{D}}_{11}$, $\breve{\mathscr{D}}_{21}$ and $\mathscr{D}$ defined by

$$\mathscr{H}_{11}(\cdot) \triangleq (\mathscr{H}_{11}^1(\cdot), \ldots, \mathscr{H}_{11}^k(\cdot)), \qquad \mathscr{H}_{12}(\cdot) \triangleq (\mathscr{H}_{12}^1(\cdot), \ldots, \mathscr{H}_{12}^k(\cdot))$$
$$\mathscr{H}_{21}(\cdot) \triangleq (\mathscr{H}_{21}^1(\cdot), \ldots, \mathscr{H}_{21}^k(\cdot)), \qquad \mathscr{H}_{22}(\cdot) \triangleq (\mathscr{H}_{22}^1(\cdot), \ldots, \mathscr{H}_{22}^k(\cdot))$$
$$\breve{\mathscr{D}}_{11}(\cdot) \triangleq (\breve{\mathscr{D}}_{11}^1(\cdot), \ldots, \breve{\mathscr{D}}_{11}^k(\cdot)), \qquad \breve{\mathscr{D}}_{21}(\cdot) \triangleq (\breve{\mathscr{D}}_{21}^1(\cdot), \ldots, \breve{\mathscr{D}}_{21}^k(\cdot))$$
$$\mathscr{D}(\cdot) \triangleq (\mathscr{D}^1(\cdot), \ldots, \mathscr{D}^k(\cdot))$$

within which the continuous-time, differentiable and matrix-valued $\mathscr{H}_{11}^r$, $\mathscr{H}_{12}^r$, $\mathscr{H}_{21}^r$, $\mathscr{H}_{22}^r \in \mathscr{C}^1([0, t_f]; \mathbb{R}^{(2n+m)\times(2n+m)})$; vector-valued $\breve{\mathscr{D}}_{11}^r$, $\breve{\mathscr{D}}_{21}^r \in \mathscr{C}^1([0, t_f]; \mathbb{R}^{2n+m})$ and scalar-valued $\mathscr{D}^r \in \mathscr{C}^1([0, t_f]; \mathbb{R})$ have the representations $\mathscr{H}_{11}^r(\cdot) \triangleq H_{11}^r(\cdot)$, $\mathscr{H}_{12}^r(\cdot) \triangleq H_{12}^r(\cdot)$, $\mathscr{H}_{21}^r(\cdot) \triangleq H_{21}^r(\cdot)$, $\mathscr{H}_{22}^r(\cdot) \triangleq H_{22}^r(\cdot)$, $\breve{\mathscr{D}}_{11}^r(\cdot) \triangleq \breve{D}_{11}^r(\cdot)$, $\breve{\mathscr{D}}_{21}^r(\cdot) \triangleq \breve{D}_{21}^r(\cdot)$, and $\mathscr{D}^r(\cdot) \triangleq D^r(\cdot)$.

Indeed, the essence of the Cartesian products of the crucial state variables as being illustrated above is the construction of a structural model of feedback control parameters K_z and l_z in a form suitable for control optimization and manipulation; the realization of the product mappings which are clearly bounded and Lipschitz continuous on $[0, t_f]$ of the dynamical equations (7.28)–(7.40)

$$\mathscr{F}_{11} : [0, t_f] \times (\mathbb{R}^{(2n+m)\times(2n+m)})^{3k} \times \mathbb{R}^{m\times(2n+m)} \mapsto (\mathbb{R}^{(2n+m)\times(2n+m)})^k$$
$$\mathscr{F}_{12} : [0, t_f] \times (\mathbb{R}^{(2n+m)\times(2n+m)})^{3k} \times \mathbb{R}^{m\times(2n+m)} \mapsto (\mathbb{R}^{(2n+m)\times(2n+m)})^k$$
$$\mathscr{F}_{21} : [0, t_f] \times (\mathbb{R}^{(2n+m)\times(2n+m)})^{3k} \times \mathbb{R}^{m\times(2n+m)} \mapsto (\mathbb{R}^{(2n+m)\times(2n+m)})^k$$
$$\mathscr{F}_{22} : [0, t_f] \times (\mathbb{R}^{(2n+m)\times(2n+m)})^{3k} \mapsto (\mathbb{R}^{(2n+m)\times(2n+m)})^k$$
$$\breve{\mathscr{G}}_{11} : [0, t_f] \times (\mathbb{R}^{(2n+m)\times(2n+m)})^k \times (\mathbb{R}^{2n+m})^k \times \mathbb{R}^{m\times(2n+m)} \times \mathbb{R}^m \mapsto (\mathbb{R}^{2n+m})^k$$
$$\breve{\mathscr{G}}_{21} : [0, t_f] \times (\mathbb{R}^{(2n+m)\times(2n+m)})^k \times (\mathbb{R}^{2n+m})^{2k} \times \mathbb{R}^m \mapsto (\mathbb{R}^{2n+m})^k$$
$$\mathscr{G} : [0, t_f] \times (\mathbb{R}^{(2n+m)\times(2n+m)})^{4k} \times (\mathbb{R}^{2n+m})^k \times \mathbb{R}^m \mapsto \mathbb{R}^k$$

in the statistical optimal control for the class of uncertain time-delayed systems, have the rules of action given by

$$\frac{d}{d\varsigma}\mathscr{H}_{11}(\varsigma) = \mathscr{F}_{11}(\varsigma, \mathscr{H}_{11}(\varsigma), \mathscr{H}_{12}(\varsigma), \mathscr{H}_{21}(\varsigma), K_z(\varsigma)), \quad \mathscr{H}_{11}(t_f) \tag{7.41}$$

$$\frac{d}{d\varsigma}\mathscr{H}_{12}(\varsigma) = \mathscr{F}_{12}(\varsigma, \mathscr{H}_{11}(\varsigma), \mathscr{H}_{12}(\varsigma), \mathscr{H}_{22}(\varsigma), K_z(\varsigma)), \quad \mathscr{H}_{12}(t_f) \tag{7.42}$$

$$\frac{d}{d\varsigma}\mathscr{H}_{21}(\varsigma) = \mathscr{F}_{21}(\varsigma, \mathscr{H}_{11}(\varsigma), \mathscr{H}_{12}(\varsigma), \mathscr{H}_{22}(\varsigma), K_z(\varsigma)), \quad \mathscr{H}_{21}(t_f) \tag{7.43}$$

$$\frac{d}{d\varsigma}\mathscr{H}_{22}(\varsigma) = \mathscr{F}_{22}(\varsigma, \mathscr{H}_{12}(\varsigma), \mathscr{H}_{21}(\varsigma), \mathscr{H}_{22}(\varsigma)), \quad \mathscr{H}_{22}(t_f) \tag{7.44}$$

$$\frac{d}{d\varsigma}\breve{\mathscr{D}}_{11}(\varsigma) = \breve{\mathscr{G}}_{11}(\varsigma, \breve{\mathscr{D}}_{11}, \mathscr{H}_{11}(\varsigma), K_z(\varsigma), l_z(\varsigma)), \quad \breve{\mathscr{D}}_{11}(t_f) \tag{7.45}$$

$$\frac{d}{d\varsigma}\breve{\mathscr{D}}_{21}(\varsigma) = \breve{\mathscr{G}}_{21}(\varsigma, \breve{\mathscr{D}}_{11}, \breve{\mathscr{D}}_{21}, \mathscr{H}_{21}(\varsigma), l_z(\varsigma)), \quad \breve{\mathscr{D}}_{21}(t_f) \tag{7.46}$$

$$\frac{d}{d\varsigma}\mathscr{D}(\varsigma) = \mathscr{G}(\varsigma, \mathscr{H}_{11}(\varsigma), \mathscr{H}_{12}(\varsigma), \mathscr{H}_{21}(\varsigma), \mathscr{H}_{22}(\varsigma), \breve{\mathscr{D}}_{11}(\varsigma), l_z(\varsigma)) \tag{7.47}$$

under the following definitions

$$\mathscr{F}_{11} \triangleq \mathscr{F}_{11}^1 \times \cdots \times \mathscr{F}_{11}^k, \quad \mathscr{F}_{12} \triangleq \mathscr{F}_{12}^1 \times \cdots \times \mathscr{F}_{12}^k, \quad \mathscr{F}_{21} \triangleq \mathscr{F}_{21}^1 \times \cdots \times \mathscr{F}_{21}^k$$
$$\mathscr{F}_{22} \triangleq \mathscr{F}_{22}^1 \times \cdots \times \mathscr{F}_{22}^k, \quad \breve{\mathscr{G}}_{11} \triangleq \breve{\mathscr{G}}_{11}^1 \times \cdots \times \breve{\mathscr{G}}_{11}^k, \quad \breve{\mathscr{G}}_{21} \triangleq \breve{\mathscr{G}}_{21}^1 \times \cdots \times \breve{\mathscr{G}}_{21}^k$$
$$\mathscr{G} \triangleq \mathscr{G}^1 \times \cdots \times \mathscr{G}^k$$

and the terminal-value conditions

$$\mathscr{H}_{11}(t_f) = \mathscr{H}_{12}(t_f) = \mathscr{H}_{21}(t_f) = \mathscr{H}_{22}(t_f) \triangleq (Q_z^f, 0, \ldots, 0)$$
$$\breve{\mathscr{D}}_{11}(t_f) = \breve{\mathscr{D}}_{21}(t_f) \triangleq (0, \ldots, 0), \quad \mathscr{D}(t_f) \triangleq (0, \ldots, 0).$$

Consideration is next given to the fact that the product system (7.41)–(7.47) uniquely determines $\mathscr{H}_{11}$, $\mathscr{H}_{12}$, $\mathscr{H}_{21}$, $\mathscr{H}_{22}$, $\breve{\mathscr{D}}_{11}$, $\breve{\mathscr{D}}_{21}$ and $\mathscr{D}$ once admissible feedback parameters K_z and l_z are specified. Therefore, it further implies that $\mathscr{H}_{11} \equiv \mathscr{H}_{11}(\cdot, K_z)$, $\mathscr{H}_{12} \equiv \mathscr{H}_{12}(\cdot, K_z)$, $\mathscr{H}_{21} \equiv \mathscr{H}_{21}(\cdot, K_z)$, $\mathscr{H}_{22} \equiv \mathscr{H}_{22}(\cdot, K_z)$, $\breve{\mathscr{D}}_{11} \equiv \breve{\mathscr{D}}_{11}(\cdot, K_z, l_z)$, $\breve{\mathscr{D}}_{21} \equiv \breve{\mathscr{D}}_{21}(\cdot, K_z, l_z)$, and $\mathscr{D} \equiv \mathscr{D}(\cdot, K_z, l_z)$.

Now all must move on to the inclusion of performance uncertainty, to the establishment of control decision goals that are reflected in risk aversion and expected value preferences. One of the most widely used measures for performance reliability is expected valued approaches to summarize the underlying performance variations. However, other aspects of performance distributions, that do not appear in most of the existing progress, are variance, skewness, flatness, etc. Specifically, to be consistent in selecting the best control decisions, the performance index that follows is proposed to show the explicit need for expected value and risk aversion for selection of the best alternative by means of the feedback control parameters K_z and l_z.

Definition 7.3.1 (Mean-Risk Aware Performance Index). Fix $k \in \mathbb{Z}^+$ and the sequence $\mu = \{\mu_i \geq 0\}_{i=1}^k$ with $\mu_1 > 0$. Then, the performance index with risk consequences for the uncertain time-delay stochastic system (7.9)–(7.10) is given by

$$\phi_0 : \{0\} \times (\mathbb{R}^{(2n+m)\times(2n+m)})^k \times (\mathbb{R}^{2n+m})^k \times \times \mathbb{R}^k \mapsto \mathbb{R}^+$$

with the rule of action

$$\phi_0\left(0, \mathscr{H}_{11}(0), \breve{\mathscr{D}}_{11}(0), \mathscr{D}(0)\right) \triangleq \underbrace{\mu_1\kappa_1}_{\text{Mean Measure}} + \underbrace{\mu_2\kappa_2 + \cdots + \mu_k\kappa_k}_{\text{Risk Measures}}$$

$$= \sum_{r=1}^{k} \mu_r \left[z_0^T \mathscr{H}_{11}^r(t_0)z_0 + 2z_0^T \breve{\mathscr{D}}_{11}^r(0) + \mathscr{D}^r(0)\right] \tag{7.48}$$

where additional parametric design of freedom μ_r, chosen by risk-averse controller designers, represent different levels of robustness prioritization according to the importance of the resulting performance-measure statistics to the probabilistic performance distribution. And the unique solutions $\{\mathscr{H}_{11}^r(\varsigma)\}_{r=1}^k$, $\left\{\breve{\mathscr{D}}_{11}^r(\varsigma)\right\}_{r=1}^k$, and $\{\mathscr{D}^r(\varsigma)\}_{r=1}^k$ evaluated at $\varsigma = 0$ satisfy the dynamical equations (7.41)–(7.47).

Of note, the performance index (7.48) has described its comprehensiveness and vitality as a mode of the control decision and thus, its ability to place alternative evaluation on performance uncertainty. Thereafter, it is becoming clear to control designs that dealing with probabilistic phenomena (i.e., rather than expected value ones) requires more formal approaches to the problem of selecting feedback parameters K_z and l_z that bears on an admissible feedback law.

Definition 7.3.2 (Admissible Feedback Parameters). For the given terminal data $(t_f, \mathscr{H}_{11}^f, \mathscr{H}_{12}^f, \mathscr{H}_{21}^f, \mathscr{H}_{22}^f, \breve{\mathscr{D}}_{11}^f, \mathscr{D}^f)$, the classes of admissible feedback gains and feedforward inputs are defined as follows. Let compact subsets $\overline{K}_z \subset \mathbb{R}^{m\times(2n+m)}$ and $\overline{L}_z \subset \mathbb{R}^{2n+m}$ be the sets of allowable gain and input values. With $k \in \mathbb{Z}^+$ and the sequence $\mu = \{\mu_r \geq 0\}_{r=1}^k$ and $\mu_1 > 0$ given, the sets of admissible $\mathscr{K}^z_{t_f,\mathscr{H}_{11}^f,\mathscr{H}_{12}^f,\mathscr{H}_{21}^f,\mathscr{H}_{22}^f,\breve{\mathscr{D}}_{11}^f,\mathscr{D}^f;\mu}$ and $\mathscr{L}^z_{t_f,\mathscr{H}_{11}^f,\mathscr{H}_{12}^f,\mathscr{H}_{21}^f,\mathscr{H}_{22}^f,\breve{\mathscr{D}}_{11}^f,\mathscr{D}^f;\mu}$ are the classes of time-continuous matrices $\mathscr{C}([0, t_f]; \mathbb{R}^{m\times(2n+m)})$ and vectors $\mathscr{C}([0, t_f]; \mathbb{R}^{2n+m})$ with values $K_z(\cdot) \in \overline{K}_z$ and $l_z(\cdot) \in \overline{L}_z$ for which the solutions to the dynamical equations (7.41)–(7.47) exist on the finite interval of optimization $[0, t_f]$.

What follows are the traditional end-point problem and the corresponding use of dynamic programming as a prerequisite for more appropriate modifications in the sequence of results [8] and the introduction of the terminology associated with the statistical optimal control.

Definition 7.3.3 (Optimization Problem of Mayer Type). Suppose $k \in \mathbb{Z}^+$ and the sequence $\mu = \{\mu_r \geq 0\}_{r=1}^k$ with $\mu_1 > 0$ are fixed. Then, the optimization problem over $[0, t_f]$ is defined as the minimization of (7.48) with respect to (K_z, l_z) in $\mathscr{K}^z_{t_f, \mathscr{H}^f_{11}, \mathscr{H}^f_{12}, \mathscr{H}^f_{21}, \mathscr{H}^f_{22}, \breve{\mathscr{D}}^f_{11}, \mathscr{D}^f; \mu} \times \mathscr{L}^z_{t_f, \mathscr{H}^f_{11}, \mathscr{H}^f_{12}, \mathscr{H}^f_{21}, \mathscr{H}^f_{22}, \breve{\mathscr{D}}^f_{11}, \mathscr{D}^f; \mu}$ and subject to the dynamical equations (7.41)–(7.47).

Matters concerning the aforementioned optimization are dealt with next. In particular, the terminal time and states $(t_f, \mathscr{H}^f_{11}, \mathscr{H}^f_{12}, \mathscr{H}^f_{21}, \mathscr{H}^f_{22}, \breve{\mathscr{D}}^f_{11}, \mathscr{D}^f)$ are parameterized as $(\varepsilon, \mathscr{Y}_{11}, \mathscr{Y}_{12}, \mathscr{Y}_{21}, \mathscr{Y}_{22}, \breve{\mathscr{Z}}_{11}, \mathscr{Z})$ through the help of a reachable set.

Definition 7.3.4 (Reachable Set). Let a reachable set

$$\mathscr{Q} \triangleq \Big\{(\varepsilon, \mathscr{Y}_{11}, \mathscr{Y}_{12}, \mathscr{Y}_{21}, \mathscr{Y}_{22}, \breve{\mathscr{Z}}_{11}, \mathscr{Z}) \in [0, t_f] \times (\mathbb{R}^{(2n+m)\times(2n+m)})^{4k} \times (\mathbb{R}^{2n+m})^k \times \mathbb{R}^k$$

$$\text{such that } \mathscr{K}^z_{t_f, \mathscr{H}^f_{11}, \mathscr{H}^f_{12}, \mathscr{H}^f_{21}, \mathscr{H}^f_{22}, \breve{\mathscr{D}}^f_{11}, \mathscr{D}^f; \mu} \times \mathscr{L}^z_{t_f, \mathscr{H}^f_{11}, \mathscr{H}^f_{12}, \mathscr{H}^f_{21}, \mathscr{H}^f_{22}, \breve{\mathscr{D}}^f_{11}, \mathscr{D}^f; \mu} \neq \emptyset\Big\}.$$

Therefore, the value function for this optimization problem is now depending on parameterizations of the terminal-value conditions.

Definition 7.3.5 (Value Function). Suppose that $(\varepsilon, \mathscr{Y}_{11}, \mathscr{Y}_{12}, \mathscr{Y}_{21}, \mathscr{Y}_{22}, \breve{\mathscr{Z}}_{11}, \mathscr{Z}) \in \mathscr{Q}$ be given. Then, the value function $\mathscr{V}(\varepsilon, \mathscr{Y}_{11}, \breve{\mathscr{Z}}_{11}, \mathscr{Z})$ is defined by

$$\mathscr{V}(\varepsilon, \mathscr{Y}_{11}, \breve{\mathscr{Z}}_{11}, \mathscr{Z}) = \inf_{K_z \in \overline{K}_z, l_z \in L_z} \phi_0(0, \mathscr{H}_{11}(0), \breve{\mathscr{D}}_{11}(0), \mathscr{D}(0)).$$

By convention, $\mathscr{V}(\varepsilon, \mathscr{Y}_{11}, \breve{\mathscr{Z}}_{11}, \mathscr{Z}) = +\infty$ when the Cartesian product that follows $\mathscr{K}^z_{t_f, \mathscr{H}^f_{11}, \mathscr{H}^f_{12}, \mathscr{H}^f_{21}, \mathscr{H}^f_{22}, \breve{\mathscr{D}}^f_{11}, \mathscr{D}^f; \mu} \times \mathscr{L}^z_{t_f, \mathscr{H}^f_{11}, \mathscr{H}^f_{12}, \mathscr{H}^f_{21}, \mathscr{H}^f_{22}, \breve{\mathscr{D}}^f_{11}, \mathscr{D}^f; \mu}$ is empty.

A brief historical background of [8] is followed by presentation of the initial-cost problem and adaptation of the corespondent terminologies in statistical optimal control and these activities are analyzed in terms of the Hamilton-Jacobi-Bellman (HJB) equation and verification theorem.

Theorem 7.3.1 (HJB Equation for Mayer Problem). *Let* $(\varepsilon, \mathscr{Y}_{11}, \mathscr{Y}_{12}, \mathscr{Y}_{21}, \mathscr{Y}_{22}, \breve{\mathscr{Z}}_{11}, \mathscr{Z})$ *be any interior point of* $\mathscr{Q}$ *at which the value function* $\mathscr{V}(\varepsilon, \mathscr{Y}_{11}, \breve{\mathscr{Z}}_{11}, \mathscr{Z})$ *is differentiable. If there exist optimal feedback gain* $K_z^* \in \mathscr{K}^z_{t_f, \mathscr{H}^f_{11}, \mathscr{H}^f_{12}, \mathscr{H}^f_{21}, \mathscr{H}^f_{22}, \breve{\mathscr{D}}^f_{11}, \mathscr{D}^f; \mu}$ *and affine input* $l_z^* \in \mathscr{L}^z_{t_f, \mathscr{H}^f_{11}, \mathscr{H}^f_{12}, \mathscr{H}^f_{21}, \mathscr{H}^f_{22}, \breve{\mathscr{D}}^f_{11}, \mathscr{D}^f; \mu}$*, then the partial differential equation of dynamic programming*

$$\begin{aligned} 0 = \min_{K_z \in \overline{K}_z, l_z \in \overline{L}_z} \Big\{ & \frac{\partial}{\partial \varepsilon} \mathscr{V}\left(\varepsilon, \mathscr{Y}_{11}, \breve{\mathscr{Z}}_{11}, \mathscr{Z}\right) \\ & + \frac{\partial}{\partial \operatorname{vec}(\mathscr{Y}_{11})} \mathscr{V}\left(\varepsilon, \mathscr{Y}_{11}, \breve{\mathscr{Z}}_{11}, \mathscr{Z}\right) \operatorname{vec}\left(\mathscr{F}_{11}\left(\varepsilon, \mathscr{Y}_{11}, \mathscr{Y}_{12}, \mathscr{Y}_{21}, K_z\right)\right) \\ & + \frac{\partial}{\partial \operatorname{vec}(\breve{\mathscr{Z}}_{11})} \mathscr{V}\left(\varepsilon, \mathscr{Y}_{11}, \breve{\mathscr{Z}}_{11}, \mathscr{Z}\right) \operatorname{vec}(\breve{\mathscr{G}}_{11}(\varepsilon, \breve{\mathscr{Z}}_{11}, \mathscr{Y}_{11}, K_z, l_z)) \\ & + \frac{\partial}{\partial \operatorname{vec}(\mathscr{Z})} \mathscr{V}\left(\varepsilon, \mathscr{Y}_{11}, \breve{\mathscr{Z}}_{11}, \mathscr{Z}\right) \operatorname{vec}(\mathscr{G}(\varepsilon, \mathscr{Y}_{11}, \mathscr{Y}_{12}, \mathscr{Y}_{21}, \mathscr{Y}_{22}, \breve{\mathscr{Z}}_{11}, l_z))\Big\} \end{aligned} \quad (7.49)$$

is satisfied. The boundary condition of (7.49) is given by

$$\mathscr{V}\left(0, \mathscr{H}_{11}(0), \breve{\mathscr{D}}_{11}(0), \mathscr{D}(0)\right) = \phi_0\left(0, \mathscr{H}_{11}(0), \breve{\mathscr{D}}_{11}(0), \mathscr{D}(0)\right).$$

Proof. A rigorous proof of the necessary condition herein can be found in [9].

As is expected, the next result examines the development and provision of the sufficient condition for optimality.

Theorem 7.3.2 (Verification Theorem). *Fix $k \in \mathbb{Z}^+$ and let $\mathscr{W}(\varepsilon, \mathscr{Y}_{11}, \breve{\mathscr{Z}}_{11}, \mathscr{Z})$ be a continuously differentiable solution of (7.49) which satisfies the boundary*

$$\mathscr{W}(0, \mathscr{H}_{11}(0), \breve{\mathscr{D}}_{11}(0), \mathscr{D}(0)) = \phi_0\left(0, \mathscr{H}_{11}(0), \breve{\mathscr{D}}_{11}(0), \mathscr{D}(0)\right).$$

Let the terminal-value condition $(t_f, \mathscr{H}_{11}^f, \mathscr{H}_{12}^f, \mathscr{H}_{21}^f, \mathscr{H}_{22}^f, \breve{\mathscr{D}}_{11}^f, \mathscr{D}^f)$ be in $\mathscr{Q}$; the 2-tuple feedback (K_z, l_z) in $\mathscr{K}^z_{t_f,\mathscr{H}_{11}^f,\mathscr{H}_{12}^f,\mathscr{H}_{21}^f,\mathscr{H}_{22}^f,\breve{\mathscr{D}}_{11}^f,\mathscr{D}^f;\mu} \times \mathscr{L}^z_{t_f,\mathscr{H}_{11}^f,\mathscr{H}_{12}^f,\mathscr{H}_{21}^f,\mathscr{H}_{22}^f,\breve{\mathscr{D}}_{11}^f,\mathscr{D}^f;\mu}$; the trajectory solutions $\mathscr{H}_{11}$, $\breve{\mathscr{D}}_{11}$ and $\mathscr{D}$ of the dynamical equations (7.41)–(7.47). Then, $\mathscr{W}(\varepsilon, \mathscr{H}_{11}(\varepsilon), \breve{\mathscr{D}}_{11}(\varepsilon), \mathscr{D}(\varepsilon))$ is a time-backward increasing function of ε. If (K_z^*, l_z^*) is in $\mathscr{K}^z_{t_f,\mathscr{H}_{11}^f,\mathscr{H}_{12}^f,\mathscr{H}_{21}^f,\mathscr{H}_{22}^f,\breve{\mathscr{D}}_{11}^f,\mathscr{D}^f;\mu} \times \mathscr{L}^z_{t_f,\mathscr{H}_{11}^f,\mathscr{H}_{12}^f,\mathscr{H}_{21}^f,\mathscr{H}_{22}^f,\breve{\mathscr{D}}_{11}^f,\mathscr{D}^f;\mu}$ defined on $[0, t_f]$ with the corresponding solutions $\mathscr{H}_{11}^*$, $\breve{\mathscr{D}}_{11}^*$ and $\mathscr{D}^*$ of the dynamical equations (7.41)–(7.47) such that, for $\tau \in [0, t_f]$

$$\begin{aligned}
0 &= \frac{\partial}{\partial \varepsilon}\mathscr{W}(\tau, \mathscr{H}_{11}^*(\tau), \breve{\mathscr{D}}_{11}^*(\tau), \mathscr{D}^*(\tau)) + \frac{\partial}{\partial\, \mathrm{vec}(\mathscr{Y}_{11})}\mathscr{W}(\tau, \mathscr{H}_{11}^*(\tau), \breve{\mathscr{D}}_{11}^*(\tau), \mathscr{D}^*(\tau)) \\
&\cdot \mathrm{vec}(\mathscr{F}_{11}(\tau, \mathscr{H}_{11}^*(\tau), \mathscr{H}_{12}^*(\tau), \mathscr{H}_{21}^*(\tau), K_z^*(\tau))) \\
&+ \frac{\partial}{\partial\, \mathrm{vec}(\breve{\mathscr{Z}}_{11})}\mathscr{W}(\tau, \mathscr{H}_{11}^*(\tau), \breve{\mathscr{D}}_{11}^*(\tau), \mathscr{D}^*(\tau)) \\
&\cdot \mathrm{vec}(\breve{\mathscr{G}}_{11}(\tau, \breve{\mathscr{D}}_{11}^*(\tau), \mathscr{H}_{11}(\tau), K_z^*(\tau), l_z^*(\tau))) \\
&+ \frac{\partial}{\partial\, \mathrm{vec}(\mathscr{Z})}\mathscr{W}(\tau, \mathscr{H}_{11}^*(\tau), \breve{\mathscr{D}}_{11}^*(\tau), \mathscr{D}^*(\tau)) \\
&\cdot \mathrm{vec}(\mathscr{G}(\tau, \mathscr{H}_{11}^*(\tau), \mathscr{H}_{12}^*(\tau), \mathscr{H}_{21}^*(\tau), \mathscr{H}_{22}^*(\tau), l_z^*(\tau)))
\end{aligned} \tag{7.50}$$

then K_z^* and l_z^* are optimal feedback and feedforward parameters. Moreover

$$\mathscr{W}(\varepsilon, \mathscr{Y}_{11}, \breve{\mathscr{Z}}_{11}, \mathscr{Z}) = \mathscr{V}(\varepsilon, \mathscr{Y}_{11}, \breve{\mathscr{Z}}_{11}, \mathscr{Z}) \tag{7.51}$$

where $\mathscr{V}(\varepsilon, \mathscr{Y}_{11}, \breve{\mathscr{Z}}_{11}, \mathscr{Z})$ is the value function.

Proof. It is already contained in the rigorous proof of the sufficiency with the essential conditions aforementioned from [9].

7.4 Framing Resilient Control Solutions

Notwithstanding that the dynamic optimization under investigation here is in "Mayer form" and can therefore be solved by an adaptation of the Mayer-form verification theorem of dynamic programming given in [8]. In conformity of the dynamic programming framework, it is required to denote the terminal time and states of a family of optimization problems by $(\varepsilon, \mathscr{Y}_{11}, \mathscr{Y}_{12}, \mathscr{Y}_{21}, \mathscr{Y}_{22}, \breve{\mathscr{Z}}_{11} \mathscr{Z})$ rather than $(t_f, \mathscr{H}_{11}^f, \mathscr{H}_{12}^f, \breve{\mathscr{H}}_{21}^f, \mathscr{H}_{22}^f, \breve{\mathscr{D}}_{11}^f, \mathscr{D}^f)$. Subsequently, the value of the optimization problem depends on the terminal conditions. For instance, for any $\varepsilon \in [0, t_f]$, the states of the equations (7.41)–(7.47) are denoted by $\mathscr{H}_{11}(\varepsilon) = \mathscr{Y}_{11}$, $\mathscr{H}_{12}(\varepsilon) = \mathscr{Y}_{12}$, $\mathscr{H}_{21}(\varepsilon) = \mathscr{Y}_{21}$, $\mathscr{H}_{22}(\varepsilon) = \mathscr{Y}_{22}$, $\breve{\mathscr{D}}_{11}(\varepsilon) = \breve{\mathscr{Z}}_{11}$, and $\mathscr{D}(\varepsilon) = \mathscr{Z}$. Then, the quadratic-affine nature of (7.48) implies that a solution to the HJB equation (7.49) is of the form as follows.

Fix $k \in \mathbb{Z}^+$ and let $\left(\varepsilon, \mathscr{Y}_{11}, \mathscr{Y}_{12}, \mathscr{Y}_{21}, \mathscr{Y}_{22}, \breve{\mathscr{Z}}_{11}, \mathscr{Z}\right)$ be any interior point of the reachable set $\mathscr{Q}$ at which the real-valued function $\mathscr{W}\left(\varepsilon, \mathscr{Y}_{11}, \breve{\mathscr{Z}}_{11}, \mathscr{Z}\right)$ described by

$$\mathscr{W}(\varepsilon, \mathscr{Y}_{11}, \breve{\mathscr{Z}}_{11}, \mathscr{Z}) = z_0^T \sum_{r=1}^{k} \mu_r (\mathscr{Y}_{11}^r + \mathscr{E}_{11}^r(\varepsilon)) z_0$$

$$+2 z_0^T \sum_{r=1}^{k} \mu_r (\breve{\mathscr{Z}}_{11}^r + \breve{\mathscr{T}}_{11}^r(\varepsilon)) + \sum_{r=1}^{k} \mu_r (\mathscr{Z}^r + \mathscr{T}^r(\varepsilon)) \quad (7.52)$$

is differentiable. The time-parametric functions $\mathscr{E}_{11}^r \in \mathscr{C}^1([0, t_f]; \mathbb{R}^{(2n+m)\times(2n+m)})$, $\breve{\mathscr{T}}_{11}^r \in \mathscr{C}^1([0, t_f]; \mathbb{R}^{2n+m})$, and $\mathscr{T}^r \in \mathscr{C}^1([0, t_f]; \mathbb{R})$ are yet to be determined. Next the derivative of $\mathscr{W}\left(\varepsilon, \mathscr{Y}_{11}, \breve{\mathscr{Z}}_{11}, \mathscr{Z}\right)$ with respect to ε is obtained as illustrated in [9]

$$\frac{d}{d\varepsilon}\mathscr{W}(\varepsilon, \mathscr{Y}_{11}, \mathscr{Y}_{12}, \mathscr{Y}_{22}, \mathscr{Z}) = z_0^T \sum_{r=1}^{k} \mu_r (\mathscr{F}_{11}^r(\varepsilon, \mathscr{Y}_{11}, \mathscr{Y}_{12}, \mathscr{Y}_{21}, K_z) + \frac{d}{d\varepsilon}\mathscr{E}_{11}^r(\varepsilon)) z_0$$

$$+ 2 z_0^T \sum_{r=1}^{k} \mu_r (\breve{\mathscr{G}}_{11}^r \left(\varepsilon, \mathscr{Y}_{11}, \breve{\mathscr{Z}}_{11}, K_z, l_z\right) + \frac{d}{d\varepsilon}\breve{\mathscr{T}}_{11}^r(\varepsilon))$$

$$+ \sum_{r=1}^{k} \mu_r (\mathscr{G}^r \left(\varepsilon, \mathscr{Y}_{11}, \mathscr{Y}_{12}, \mathscr{Y}_{21}, \mathscr{Y}_{22}, \breve{\mathscr{Z}}_{11}, l_z\right) + \frac{d}{d\varepsilon}\mathscr{T}^r(\varepsilon))$$

$$(7.53)$$

provided that the admissible 2-tuple $(K_z, l_z) \in \overline{K}_z \times \overline{L}_z$.

Replacing the candidate solution of (7.52) and the result (7.53) into the HJB equation (7.49), one can deduce that

$$\min_{(K_z,l_z)\in\overline{K}_z\times\overline{L}_z}\Big\{z_0^T\sum_{r=1}^{k}\mu_r(\mathscr{F}_{11}^r(\varepsilon,\mathscr{Y}_{11},\mathscr{Y}_{12},\mathscr{Y}_{21},K_z)+\frac{d}{d\varepsilon}\mathscr{E}_{11}^r(\varepsilon))z_0$$
$$+2z_0^T\sum_{r=1}^{k}\mu_r(\breve{\mathscr{G}}_{11}^r(\varepsilon,\mathscr{Y}_{11},\breve{\mathscr{Z}}_{11},K_z,l_z)+\frac{d}{d\varepsilon}\breve{\mathscr{T}}_{11}^r(\varepsilon))$$
$$+\sum_{r=1}^{k}\mu_r(\mathscr{G}^r(\varepsilon,\mathscr{Y}_{11},\mathscr{Y}_{12},\mathscr{Y}_{21},\mathscr{Y}_{22},\breve{\mathscr{Z}}_{11},l_z)+\frac{d}{d\varepsilon}\mathscr{T}^r(\varepsilon))\Big\}\equiv 0. \quad (7.54)$$

As it turns out, the differentiation of the expression within the bracket (7.54) with respect to K_z and l_z yields the necessary conditions for an interior extremum of the performance index with risk consequences (7.48) on $[0,t_f]$. In other words, the extremizing K_z and l_z are given by

$$K_z(\varepsilon,\mathscr{Y}_{11})=-R_z^{-1}(\varepsilon)B_z^T(\varepsilon)\sum_{r=1}^{k}\hat{\mu}_r\mathscr{Y}_{11}^r \quad (7.55)$$

$$l_z(\varepsilon,\breve{\mathscr{Z}}_{11})=-R_z^{-1}(\varepsilon)B_z^T(\varepsilon)\sum_{r=1}^{k}\hat{\mu}_r\breve{\mathscr{Z}}_{11}^r, \qquad \hat{\mu}_r=\mu_r/\mu_1. \quad (7.56)$$

In view of (7.55) and (7.56), the zero value of the expression inside of the bracket of (7.54) that is pursued here for any $\varepsilon\in[0,t_f]$ when $\mathscr{Y}_{11}^r$, $\breve{\mathscr{Z}}_{11}^r$ and $\mathscr{Z}^r$ evaluated at the time-backward differential equations (7.41)–(7.47) requires

$$\frac{d}{d\varepsilon}\mathscr{E}_{11}^1(\varepsilon)=(A_z(\varepsilon)+B_z(\varepsilon)K_z(\varepsilon))^T\mathscr{H}_{11}^1(\varepsilon)+\mathscr{H}_{11}^1(\varepsilon)(A_z(\varepsilon)+B_z(\varepsilon)K_z(\varepsilon))$$
$$+K_z^T(\varepsilon)R_z(\varepsilon)K_z(\varepsilon)+Q_z(\varepsilon) \quad (7.57)$$

$$\frac{d}{d\varepsilon}\mathscr{E}_{11}^r(\varepsilon)=(A_z(\varepsilon)+B_z(\varepsilon)K_z(\varepsilon))^T\mathscr{H}_{11}^r(\varepsilon)+\mathscr{H}_{11}^r(\varepsilon)(A_z(\varepsilon)+B_z(\varepsilon)K_z(\varepsilon))$$
$$+\sum_{\nu=1}^{r-1}\frac{2r!}{\nu!(r-\nu)!}(\mathscr{H}_{11}^\nu(\varepsilon)\Pi_1(\varepsilon)+\mathscr{H}_{12}^\nu(\varepsilon)\Pi_3(\varepsilon))\mathscr{H}_{11}^{r-\nu}(\varepsilon)$$
$$+\sum_{\nu=1}^{r-1}\frac{2r!}{\nu!(r-\nu)!}(\mathscr{H}_{11}^\nu(\varepsilon)\Pi_2(\varepsilon)+\mathscr{H}_{12}^\nu(\varepsilon)\Pi_4(\varepsilon))\mathscr{H}_{21}^{r-\nu}(\varepsilon) \quad (7.58)$$

$$\frac{d}{d\varepsilon}\breve{\mathscr{T}}_{11}^1(\varepsilon)=(A_z(\varepsilon)+B_z(\varepsilon)K_z(\varepsilon))^T\breve{\mathscr{D}}_{11}^1(\varepsilon)+\mathscr{H}_{11}^1(\varepsilon)(B_z(\varepsilon)l_z(\varepsilon)+D_z(\varepsilon)d(\varepsilon))$$
$$+K_z^T(\varepsilon)R_z(\varepsilon)l_z(\varepsilon) \quad (7.59)$$

$$\frac{d}{d\varepsilon}\breve{\mathscr{T}}^r_{11}(\varepsilon) = (A_z(\varepsilon) + B_z(\varepsilon)K_z(\varepsilon))^T \breve{\mathscr{D}}^r_{11}(\varepsilon)$$
$$+ \mathscr{H}^r_{11}(\varepsilon)(B_z(\varepsilon)l_z(\varepsilon) + D_z(\varepsilon)d(\varepsilon)), \quad 2 \le r \le k \tag{7.60}$$

$$\frac{d}{d\epsilon}\mathscr{T}_1(\varepsilon) = \mathrm{Tr}\{\mathscr{H}^1_{11}(\varepsilon)\Pi_1(\varepsilon) + \mathscr{H}^1_{12}(\varepsilon)\Pi_3(\varepsilon)\}$$
$$+ \mathrm{Tr}\{\mathscr{H}^1_{21}(\varepsilon)\Pi_2(\varepsilon) + \mathscr{H}^1_{22}(\varepsilon)\Pi_4(\varepsilon)\} + l_z^T(\varepsilon)R_z(\varepsilon)l_z(\varepsilon)$$
$$+ 2(\breve{\mathscr{D}}^1_{11})^T(\varepsilon)(B_z(\varepsilon)l_z(\varepsilon) + D_z(\varepsilon)d(\varepsilon)) \tag{7.61}$$

$$\frac{d}{d\varepsilon}\mathscr{T}_r(\varepsilon) = \mathrm{Tr}\{\mathscr{H}^r_{11}(\varepsilon)\Pi_1(\varepsilon) + \mathscr{H}^r_{12}(\varepsilon)\Pi_3(\varepsilon)\}$$
$$+ \mathrm{Tr}\{\mathscr{H}^r_{21}(\varepsilon)\Pi_2(\varepsilon) + \mathscr{H}^r_{22}(\varepsilon)\Pi_4(\varepsilon)\}$$
$$+ 2(\breve{\mathscr{D}}^r_{11})^T(\varepsilon)(B_z(\varepsilon)l_z(\varepsilon) + D_z(\varepsilon)d(\varepsilon)), \quad 2 \le r \le k \tag{7.62}$$

The boundary condition of $\mathscr{W}(\varepsilon, \mathscr{Y}_{11}, \breve{\mathscr{Z}}_{11}, \mathscr{Z})$ implies that the initial-value conditions $\mathscr{E}^r_{11}(0) = 0$, $\breve{\mathscr{T}}^r_{11}(0) = 0$, and $\mathscr{T}^r(0) = 0$ for the forward-in-time differential equations (7.57)–(7.62) and yields a value function

$$\mathscr{W}(\varepsilon, \mathscr{Y}_{11}, \breve{\mathscr{Z}}_{11}, \mathscr{Z}) = \mathscr{V}(\varepsilon, \mathscr{Y}_{11}, \breve{\mathscr{Z}}_{11}, \mathscr{Z})$$
$$= z_0^T \sum_{r=1}^{k} \mu_r \mathscr{H}^r_{11}(0) z_0 + 2 z_0^T \sum_{r=1}^{k} \mu_r \breve{\mathscr{D}}^r_{11}(0) + \sum_{r=1}^{k} \mu_r \mathscr{D}^r(0)$$

for which the sufficient condition (7.50) of the verification theorem is satisfied so that the feedback and feedforward parameters (7.55)–(7.56) become optimal

$$K_z^*(\varepsilon) = -R_z^{-1}(\varepsilon)B_z^T(\varepsilon)\sum_{r=1}^{k}\hat{\mu}_r \mathscr{H}^{*r}_{11}(\varepsilon)$$

$$l_z^*(\varepsilon) = -R_z^{-1}(\varepsilon)B_z^T(\varepsilon)\sum_{r=1}^{k}\hat{\mu}_r \breve{\mathscr{D}}^{*r}_{11}(\varepsilon).$$

Brief summary in Fig. 7.4 that follows is of interest both to uncertainty quantification and to anyone concerned with development of resilient controls with risk consequences for the linear-quadratic class of uncertain time-delay stochastic systems.

Theorem 7.4.1 (Risk-Averse Control for Uncertain Time-Delay Stochastic Systems). *Under the assumptions of (A_z, B_z) uniformly stabilizable and (C_z, A_z) uniformly detectable, the uncertain time-delayed dynamics governed by (7.9)–(7.10) is subject to the chi-squared measure of performance (7.15). Suppose $k \in \mathbb{Z}^+$ and the sequence $\mu = \{\mu_r \ge 0\}_{r=1}^k$ with $\mu_1 > 0$ are fixed. Then, the statistical optimal control solution over $[0, t_f]$ is a two-degrees-of-freedom controller*

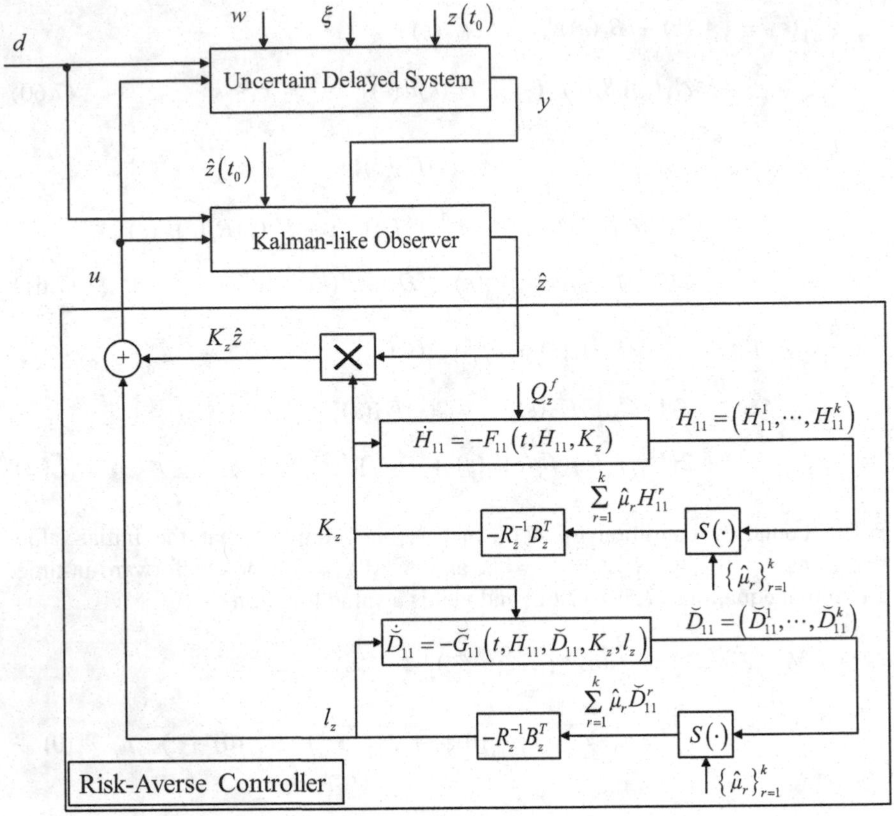

Fig. 7.4 Implementation of risk-averse control for uncertain time-delay stochastic systems

$$u^*(t) = K_z^*(t)\hat{z}^*(t) + l_z^*(t), \quad t = t_f - \tau \tag{7.63}$$

$$K_z^*(\tau) = -R_z^{-1}(\tau)B_z^T(\tau)\sum_{r=1}^{k}\hat{\mu}_r \mathscr{H}_{11}^{r*}(\tau) \tag{7.64}$$

$$l_z^*(\tau) = -R_z^{-1}(\tau)B_z^T(\tau)\sum_{r=1}^{k}\hat{\mu}_r \breve{\mathscr{D}}_{11}^{r*}(\tau) \tag{7.65}$$

where $\hat{\mu}_r = \mu_r/\mu_1$ represent different levels of influence as they deem important to the performance distribution. Finally, $\{\mathscr{H}_{11}^{r*}(\tau)\}_{r=1}^{k}$ and $\left\{\breve{\mathscr{D}}_{11}^{r*}(\tau)\right\}_{r=1}^{k}$ are the optimal solutions of the backward-in-time differential equations

$$\begin{aligned}\frac{d}{d\tau}\mathscr{H}_{11}^{1*}(\tau) = &-(A_z(\tau)+B_z(\tau)K_z^*(\tau))^T\mathscr{H}_{11}^{1*}(\tau)-\mathscr{H}_{11}^{1*}(\tau)(A_z(\tau)+B_z(\tau)K_z^*(\tau))\\ &-K_z^{*T}(\tau)R_z(\tau)K_z^*(\tau)-Q_z(\tau)\end{aligned} \tag{7.66}$$

$$\frac{d}{d\tau}\mathscr{H}_{11}^{r*}(\tau) = -(A_z(\tau)+B_z(\tau)K_z^*(\tau))^T\mathscr{H}_{11}^{r*}(\tau)-\mathscr{H}_{11}^{r*}(\tau)(A_z(\tau)+B_z(\tau)K_z^*(\tau))$$
$$-\sum_{\nu=1}^{r-1}\frac{2r!}{\nu!(r-\nu)!}(\mathscr{H}_{11}^{\nu*}(\tau)\Pi_1(\tau)+\mathscr{H}_{12}^{\nu*}(\tau)\Pi_3(\tau))\mathscr{H}_{11}^{r-\nu*}(\tau)$$
$$-\sum_{\nu=1}^{r-1}\frac{2r!}{\nu!(r-\nu)!}(\mathscr{H}_{11}^{\nu*}(\tau)\Pi_2(\tau)+\mathscr{H}_{12}^{\nu*}(\tau)\Pi_4(\tau))\mathscr{H}_{21}^{r-\nu*}(\tau) \tag{7.67}$$

and

$$\frac{d}{d\tau}\breve{\mathscr{D}}_{11}^{1*}(\tau) = -(A_z(\tau)+B_z(\tau)K_z^*(\tau))^T\breve{\mathscr{D}}_{11}^{1*}(\tau)-\mathscr{H}_{11}^{1*}(\tau)(B_z(\tau)l_z^*(\tau)+D_z(\tau)d(\tau))$$
$$-K_z^{*T}(\tau)R_z(\tau)l_z^*(\tau) \tag{7.68}$$

$$\frac{d}{d\tau}\breve{\mathscr{D}}_{11}^{r*}(\tau) = -(A_z(\tau)+B_z(\tau)K_z^*(\tau))^T\breve{\mathscr{D}}_{11}^{r*}(\tau)$$
$$-\mathscr{H}_{11}^{r*}(\tau)(B_z(\tau)l_z^*(\tau)+D_z(\tau)d(\tau)),\quad 2\le r\le k \tag{7.69}$$

whereby the terminal-value conditions $\mathscr{H}_{11}^{1*}(t_f) = Q_z^f$, $\mathscr{H}_{11}^{r*}(t_f) = 0$ *for* $2 \le r \le k$ *as well as* $\breve{\mathscr{D}}_{11}^{r*}(t_f) = 0$ *for* $1 \le r \le k$.

7.5 Chapter Summary

Being able to show the impact of uncertainties on performance measures of the class of uncertain time-delay stochastic linear systems is one of the most important features of resilient control analysis. It leads directly to the next step which is finding the most effective information gathering scheme. In fact, the results herein demonstrated that the mathematical statistics associated with the performance measures with the chi-squared random nature is found to be beneficial for gathering insights about performance uncertainty. Once the basic structure of statistical measures of performance risks is efficiently captured, it is now in an excellent position to incorporate such information in the form of a utility function; e.g., mean-risk aware performance index for expected value and risk aversion preferences. Thereafter, as the result of the performance uncertainty analysis the risk-averse feedback controller with two-degrees-of-freedom is subsequently obtained as the best control strategy in dealing with probabilistic and time-delay phenomena.

References

1. Campbell, S.L., Nikoukhah, R.: Auxiliary Signal Design for Failure Detection. Princeton Series in Applied Mathematics. Princeton University Press, Princeton (2004)
2. Chen, J., Patton, R.J.: Robust Model-Based Fault Diagnosis for Dynamic Systems. Kluwer Academic, Norwell (1999)
3. Patton, R.J.: Fault-tolerant control: the 1997 situation survey. In: Proceedings of the IFAC Symposium on Fault Detection, Supervision and Safety for Technical Processes: SAFEPROCESS'97, Pergamon 1998, pp. 1029–1052. University of Hull (1997)
4. Zhou, D.H., Frank, P.M.: Fault diagnosis and fault tolerant control. IEEE Trans. Aerosp. Electron. Syst. **34**(2), 420–427 (1998)
5. Mahmoud, M.S.: Resilient Control of Uncertain Dynamical Systems. Lecture Notes in Control and Information Sciences, vol. 303. Springer, Berlin/New York (2004)
6. Pham, K.D.: Linear-quadratic controls in risk-averse decision making: performance-measure statistics and control decision optimization. Springer Briefs in Optimization, ISBN 978-1-4614-5078-8 (2012)
7. Pham, K.D.: Risk-averse feedback for stochastic fault-tolerant control systems with actuator failure accommodation. In: Proceedings of the 20th Mediterranean Conference on Control and Automation, Barcelona, pp. 504–511 (2012)
8. Fleming, W.H., Rishel, R.W.: Deterministic and Stochastic Optimal Control. Springer, New York (1975)
9. Pham, K.D.: Performance-reliability-aided decision-making in multiperson quadratic decision games against jamming and estimation confrontations. In: Giannessi F. (ed.) J. Optim. Theory Appl. **149**(1), 599–629 (2011)

Chapter 8
Networked Control with Time Delay Measurements and Communications Channel Constraints

8.1 Introduction

The compensation of network time delays, communication channel constraints, etc. is an important problem in networked control. Typical motivating applications of networked control as well as decentralized control with communications between controllers and controlled systems arise, for instance, in control of large interconnected power distribution systems which have time-delay measurements; and firms consisting of many divisions, etc. which are widely distributed in space and subject to communications constraints. In such networks, there are local controllers at the nodes of the network, each having local information about the states of the network but no global information. Of note, there are many publications proposing various solutions. The aspect of the problem that asks for control under communications was considered in [1]. Other research have been carried out by [2, 3], and [4].

For the purpose of stabilization analysis and feedback control synthesis, many existing methods have been developed utilizing the knowledge of quadratic costs or the assumption of normal distribution functions of costs. It has been known for some time [5] that the traditional expected value approach and its variants are of limited generality and not realistic as they rule out asymmetry or skewness in the probability distributions of the non-normal random costs. Therefore, a control objective based on mean or expected value alone is indeed not justifiable on theoretical grounds but is merely an approximate workaround for the main purpose of mathematical tractability. Perhaps so motivated, this chapter will investigate the problem of network control of linear stochastic systems with network measurements and control actuation subject to network time delay and communication channel constraints, respectively. The admissible control for ordering uncertain prospects will leverage the complete knowledge of the entire chi-squared distribution of the finite-horizon integral quadratic form cost; not just means and/or variances as often can be seen from the existing literature.

K.D. Pham, *Resilient Controls for Ordering Uncertain Prospects*, Springer Optimization and Its Applications 98, DOI 10.1007/978-3-319-08705-4_8

The structure of Chap. 8 is outlined as follows. Section 8.2 attempts to understand the rationales for performance uncertainty. The two driving attributes of Sect. 8.3 are that the control problem statements are seen as central to the development of an admissible set of control laws with delayed feedback and constraint communications and that the recently advocated mean-risk aware performance index involves naturally two parameters, one of which is always the mean but the other parameter is the appropriate measures of performance dispersion. It is certainly progress when Sect. 8.4 summarizes the algorithmic development to synthesize such a feasible control policy that can order uncertain prospects of the chi-squared random cost. And so the conclusion of Sect. 8.5 is not surprising: The results herein provide a strong impetus for further research in the development of resilient controls for performance robustness while subject to networking effects.

8.2 Pursuing Performance Robustness

The development hereafter assumes a fixed probability space $(\Omega, \mathbb{F}, \{\mathbb{F}_{t_0,t} : t \in [t_0, t_f]\}, \mathbb{P})$ with filtration satisfying the usual conditions. Throughout the chapter, all filtrations are right continuous and complete and $\mathbb{F}_{t_f} \triangleq \{\mathbb{F}_{t_0,t} : t \in [t_0, t_f]\}$. In addition, let $\mathcal{L}^2_{\mathbb{F}_{t_f}}([t_0, t_f]; \mathbb{R}^n)$ denote the space of $\mathbb{F}_{t_f}$-adapted random processes $\{\hbar(t) : t \in [t_0, t_f]$ such that $E\{\int_{t_0}^{t_f} ||\hbar(t)||^2 \, dt\} < \infty\}$.

On recognizing the restrictiveness of the problem of stochastic control with time delay observations and communications constraints, the controlled system, as shown in Fig. 8.1 is described by a finite dimensional stochastic differential equation with the known initial state $x(t_0)$

$$\dot{x}(t) = A(t)x(t) + B(t)u(t) + G(t)w(t), \quad x(t_0) \tag{8.1}$$

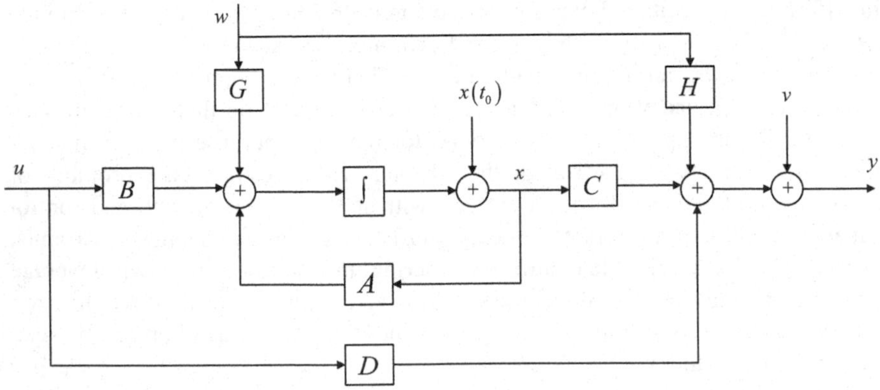

Fig. 8.1 Structure of stochastic control

where $x(t)$ is the controlled state process valued in $\mathbb{R}^n$; $u(t)$ is the control process valued in a closed convex subset of an appropriate metric space; and $w(t)$ is an exogenous state noise taking the form of a mutually uncorrelated stationary $\mathbb{F}_{t_f}$-adapted Gaussian process with its mean $E\{w(t)\} = m_w(t)$ and covariance $cov\{w(t_1), w(t_2)\} = W\delta(t_1 - t_2)$ for all t_1, $t_2 \in [t_0, t_f]$ and $W > 0$.

In addition, there is a partial information structure available to the controller which is generated by noisy observations

$$y(t) = C(t)x(t) + D(t)u(t) + H(t)w(t) + v(t) \tag{8.2}$$

whereas the exogenous measurement noise $v(t)$ is another mutually uncorrelated stationary $\mathbb{F}_{t_f}$-adapted Gaussian process with its mean $E\{v(t)\} = m_v(t)$ and covariance $cov\{v(t_1), v(t_2)\} = V\delta(t_1 - t_2)$ for all t_1, $t_2 \in [t_0, t_f]$ and $V > 0$.

In view of (8.1) and (8.2), both white noises $\{w(t) : t \in [t_0, t_f]\}$ and $\{v(t) : t \in [t_0, t_f]\}$ can be interpreted mathematically rigorous as the corresponding Brownian motions adapted to $\mathbb{F}_{t_f}$; in fact, they are now rewritten as mutually uncorrelated stationary Wiener random processes

$$dw(t) \triangleq \tilde{w}(t)dt = (w(t) - m_w(t))dt$$

$$dv(t) \triangleq \tilde{v}(t)dt = (v(t) - m_v(t))dt$$

together with the singular-value decompositions of intensity covariances W and V

$$W \triangleq U_W \Lambda_W^{1/2} U_W^T, \quad V \triangleq U_V \Lambda_V^{1/2} U_V^T.$$

In this Ito interpretation, the controlled stochastic system dynamics (8.1)–(8.2) is equivalently described by

$$dx(t) = (A(t)x(t) + B(t)u(t) + G(t)m_w(t))dt + G(t)U_W \Lambda_W^{1/2} dw(t), \quad x_0 \tag{8.3}$$

$$dy(t) = (C(t)x(t) + D(t)u(t) + H(t)m_w(t) + m_v(t))dt + H(t)dw(t) + dv(t). \tag{8.4}$$

To help account for the network delay effects, partial observations of the system (8.1)–(8.2) are subject to a priori time delay, τ_{pg} seconds whose the Laplace-domain representation is governed by $e^{-s\tau_{pg}}$ and s the Laplace transformation variable. As noted earlier and depicted in Fig. 8.2, the 1st-order Pade approximation of $e^{-\tau_{pg}s}$ is given by

$$\dot{x}_D(t) = A_D(\tau_{pg})x_D(t) + B_D(\tau_{pg})u_D(t), \quad x_D(t_0) \tag{8.5}$$

$$y_D(t) = C_D x_D(t) + D_D u_D(t) \tag{8.6}$$

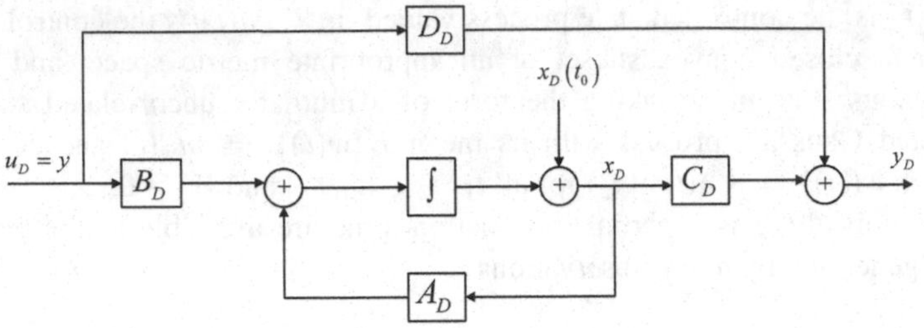

Fig. 8.2 Pade approximation structure for delayed observations

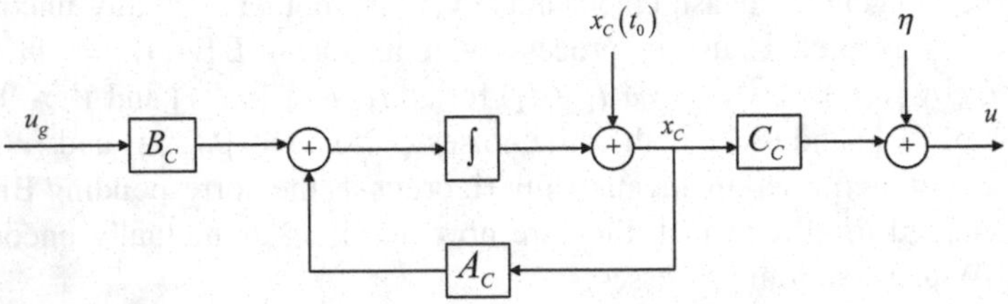

Fig. 8.3 Realization of bandlimited communications channel

where $x_D(t) \in \mathbb{R}^{m_u}$, $u_D(t) \in \mathbb{R}^{m_u}$ and $y_D(t) \in \mathbb{R}^{m_u}$ are the state, input and output vectors. Moreover, the system coefficients of the Pade approximation (8.5)–(8.6) are defined by $A_D(\tau_{pg}) \triangleq diag\{a_1, \ldots, a_{m_u}\}$, $B_D(\tau_{pg}) \triangleq diag\{b_1, \ldots, b_{m_u}\}$, $C_D \triangleq diag\{c_1, \ldots, c_{m_u}\}$ and $D_D \triangleq diag\{d_1, \ldots, d_{m_u}\}$ whereas $a_i = -\frac{2}{\tau_{pg}}$, $b_i = \frac{2}{\tau_{pg}}$, $c_i = 2$, $d_i = -1$ and $i = 1, \ldots, m_u$.

Recall that the controller uses its local measurements which are subject to the network time delay governed by (8.5)–(8.6) to ensure that its control objective is met as well as possible; e.g.,

$$dx_D(t) = (A_D(\tau_{pg})x_D(t) + B_D(\tau_{pg})C(t)x(t) + B_D(\tau_{pg})D(t)u(t) + B_D(\tau_{pg})H(t)m_w(t) + B_D(\tau_{pg})m_v(t))dt + B_D(\tau_{pg})H(t)dw(t) + B_D(\tau_{pg})dv(t) \tag{8.7}$$

$$dy_D(t) = (C_D x_D(t) + D_D C(t)x(t) + D_D D(t)u(t) + D_D H(t)m_w(t) + D_D m_v(t))dt + D_D H(t)dw(t) + D_D dv(t). \tag{8.8}$$

Perhaps so motivated, a further constraint on the communication channels between the controller and the controlled system is essential to the problem of networked controls. The proposed approach, as illustrated in Fig. 8.3, to communications channel constraints is best focused on a realization of the low-pass filter, which characterizes the bandwidth of the channel

$$dx_c(t) = (A_c(t)x_c(t) + B_c(t)u_g(t))dt, \quad x_c(t_0) \tag{8.9}$$

$$u(t)dt = C_c(t)x_c(t)dt + d\eta(t) \tag{8.10}$$

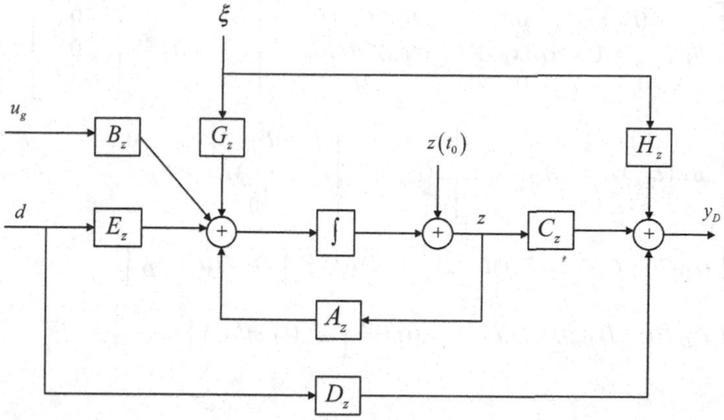

Fig. 8.4 Implementation by controlled system, Pade approximation, and bandlimited channel

where the controller generates its intended control input $u_g(t)$ through the channel dynamics (8.9); however, $u(t)dt$ is now considered as the actual communicated signal entering the system input of (8.3)–(8.4) and is realistically determined via (8.10), up to an exogenous measurement noise, by the uncorrelated stationary Wiener process with the independent increments of $E\{[\eta(t_1) - \eta(t_2)][\eta(t_1) - \eta(t_2)]^T\} = N|t_1 - t_2|$ for all t_1, $t_2 \in [t_0, t_f]$.

As is to be theoretically expected and after some algebraic simplifications, the integrated model as shown in Fig. 8.4 for communications and control over the network that follows, will convey the understanding of the inherent issues of time-delay observations and communications channel constraints between the controlled system and the controller

$$dz(t) = (A_z(t)z(t) + B_z(t)u_g(t) + E_z(t)d(t))dt + G_z(t)d\xi(t), \quad z(t_0) \tag{8.11}$$

$$dy_D(t) = (C_z(t)z(t) + D_z(t)d(t))dt + H_z(t)d\xi(t) \tag{8.12}$$

where for each $t \in [t_0, t_f]$, the augmented state variables $z(t)$ and the exogenous process/measurement noises $\xi(t)$ are given by

$$z(t) \triangleq \begin{bmatrix} x(t) \\ x_D(t) \\ x_c(t) \end{bmatrix}, \quad z(t_0) \triangleq \begin{bmatrix} x(t_0) \\ x_D(t_0) \\ x_c(t_0) \end{bmatrix}, \quad \xi(t) \triangleq \begin{bmatrix} w(t) \\ v(t) \\ \eta(t) \end{bmatrix}, \quad \Xi \triangleq \begin{bmatrix} I_{p\times p} & 0 & 0 \\ 0 & V & 0 \\ 0 & 0 & N \end{bmatrix}$$

and the correlations of independent increments $E\left\{[\xi(t_1) - \xi(t_2)][\xi(t_1) - \xi(t_2)]^T\right\} = \Xi|t_1 - t_2|$ for all t_1, $t_2 \in [t_0, t_f]$. In addition, the aggregate model coefficients and inputs are defined by

$$A_z(t) \triangleq \begin{bmatrix} A(t) & 0 & B(t)C_c(t) \\ B_D(\tau_{pg})C(t) & A_D(\tau_{pg}) & B_D(\tau_{pg})D(t)C_c(t) \\ 0 & 0 & A_c(t) \end{bmatrix}; \quad B_z(t) \triangleq \begin{bmatrix} 0 \\ 0 \\ B_c(t) \end{bmatrix}$$

$$E_z(t) \triangleq \begin{bmatrix} G(t) & 0 \\ B_D(\tau_{pg})H(t) & B_D(\tau_{pg}) \\ 0 & 0 \end{bmatrix}; G_z(t) \triangleq \begin{bmatrix} G(t)U_W \Lambda_W^{1/2} & 0 & B(t) \\ B_D(\tau_{pg})H(t) & B_D(\tau_{pg}) & B_D(\tau_{pg})D(t) \\ 0 & 0 & 0 \end{bmatrix}$$

$$C_z(t) \triangleq \begin{bmatrix} D_D C(t) & C_D & D_D D(t) C_c(t) \end{bmatrix}; \quad D_z(t) \triangleq \begin{bmatrix} D_D H(t) & D_D \end{bmatrix}$$

$$H_z(t) \triangleq \begin{bmatrix} D_D H(t) & D_D & D_D D(t) \end{bmatrix}; \quad d(t) \triangleq \begin{bmatrix} m_w^T(t) & m_v^T(t) \end{bmatrix}^T.$$

Given uncertainties of the fact by the σ-algebra

$$\mathbb{F}_{t_0,t} \triangleq \sigma\{(w(\tau), v(\tau), \eta(\tau)) : t_0 \le \tau \le t\},$$

the assessment of $\mathcal{G}_{t_0,t}^{y_D} \triangleq \sigma\{y_D(\tau) : t_0 \le \tau \le t\}$ for all $t \in [t_0, t_f]$ is necessarily contingent. More likely, a reasonable state estimator $\hat{z}(t) \triangleq E\{z(t)|\mathcal{G}_{t_f}^{y_D}\}$, as shown in Fig. 8.5, is possible with two emphases: the first on the framework not only preserving the inherent linear Gaussian structure of (8.11)–(8.12), and the second on the mechanism but also leveraging the information available $\mathcal{G}_{t_f}^{y_D} \triangleq \{\mathcal{G}_{t_0,t}^{y_D} : t \in [t_0, t_f]\} \subset \{\mathbb{F}_{t_0,t} : t \in [t_0, t_f]\}$

$$\begin{aligned} d\hat{z}(t) &= (A_z(t)\hat{z}(t) + B_z(t)u_g(t) + E_z(t)d(t))dt \\ &\quad + L_z(t)(dy_D(t) - (C_z(t)\hat{z}(t) + D_z(t)d(t))dt), \quad \hat{z}(t_0) = z(t_0) \end{aligned} \tag{8.13}$$

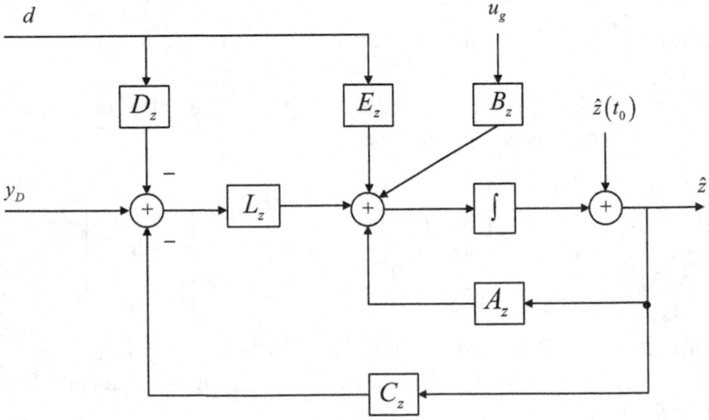

Fig. 8.5 Structure illustrating the system state estimation

where the filtering gain, $L_z(t)$ and state-estimate error covariance, $\Sigma_z(t) \triangleq E\{[z(t)-\hat{z}(t)][z(t)-\hat{z}(t)]^T|\mathcal{G}_{t_f}^{y_D}\}$ are given by

$$L_z(t) = \Sigma_z(t)C_z^T(t)(H_z(t)\Xi H_z^T(t))^{-1} \tag{8.14}$$

$$\frac{d}{dt}\Sigma_z(t) = A_z(t)\Sigma_z(t) + \Sigma_z(t)A_z^T(t) + G_z(t)\Xi G_z^T(t) \\ - \Sigma_z(t)C_z^T(t)(H_z(t)\Xi H_z^T(t))^{-1}C_z(t)\Sigma_z(t), \quad \Sigma_z(t_0) = 0. \tag{8.15}$$

As well known, in this case, the state-estimate errors $\tilde{z}(t) \triangleq z(t) - \hat{z}(t)$ are satisfying the stochastic differential equation with the initial-value condition $\tilde{z}(t_0) = 0$

$$d\tilde{z}(t) = (A_z(t) - L_z(t)C_z(t))\tilde{z}(t)dt + (G_z(t) - L_z(t)H_z(t))d\xi(t). \tag{8.16}$$

As noted earlier, the information structure $\mathcal{G}_{t_f}^{y_D}$ which is defined by a communication channel with memory feedback via the stochastic differential equation (8.12). Admissible sets of feedback control laws therefore reduce to

$$\mathbb{U}^{y_D}[t_0,t_f] \triangleq \{u_g \in \mathcal{L}^2_{\mathcal{G}_{t_f}^{y_D}}([t_0,t_f],\mathbb{R}^m), \mathbb{P}-a.s.\}$$

where $\mathbb{U}^{y_D}[t_0,t_f]$ is a closed convex subset of $\mathcal{L}^2_{\mathbb{F}_{t_f}}([t_0,t_f],\mathbb{R}^m)$.

Resilient controls under uncertainty may be viewed as control strategies between probabilistic performance alternatives, and the control designer chooses between them in accordance to a consistent set of expected value and risk preferences. Of note, the restrictive class of performance measures that is consistent with observed tradeoffs between state regulation and control constraint has spawned a considerable body in practical applications; e.g., $J : \mathbb{U}^{y_D}[t_0,t_f] \mapsto \mathbb{R}^+$ with the rule of action

$$J(u) = z^T(t_f)Q_z^f(t_f) \\ + \int_{t_0}^{t_f} [z^T(\tau)Q_z(\tau)z(\tau) + (u_g(\tau)-u_r(\tau))^T R_z(\tau)(u_g(\tau)-u_r(\tau))]d\tau \tag{8.17}$$

where the terminal penalty weighting $Q_z^f \in \mathbb{R}^{(n+m+c)\times(n+m+c)}$, the state weighting $Q_z(t) \in \mathbb{R}^{(n+m+c)\times(n+m+c)}$, and control weighting $R_z(t) \in \mathbb{R}^{m\times m}$ are the continuous-time matrix functions with the properties of symmetry and positive semi-definiteness. In addition, $R_z(t)$ is invertible. And $u_r(t) \in \mathbb{R}^m$ is a soft control constraint.

For the case of normal distribution, the conditional probability density $p(z(t)|\mathcal{G}_{t_f}^{y_D})$ which is the probability density of $z(t)$ conditioned on $\mathcal{G}_{t_f}^{y_D}$ represents the sufficient statistics for describing the conditional stochastic effects of future family of admissible feedback control laws. Quite remarkably, the complete knowledge of the entire class of distribution function $p(z(t)|\mathcal{G}_{t_f}^{y_D})$ involves naturally only two parameters, i.e., $\hat{z}(t)$ and $\Sigma_z(t)$. With certain restrictions on the class of linear stochastic systems (8.11)–(8.12) and quadratic performance measure (8.17), a reasonable approximation to optimal feedback control laws, $u_g(t) = \gamma_u(t, \hat{z}(t))$ is further reduced to a simple manageable form; e.g., a linear time-varying feedback law generated from the accessible states $\hat{z}(t)$ by

$$u_g(t) = K_z(t)\hat{z}(t) + l_z(t) + u_r(t), \quad \forall t \in [t_0, t_f] \tag{8.18}$$

with the admissible gains $K_z \in \mathcal{C}([t_0, t_f]; \mathbb{R}^{m\times(n+m+c)})$ and $l_z \in \mathcal{C}([t_0, t_f]; \mathbb{R}^m)$ to which further defining properties will be defined subsequently.

Under the assumption of Lipschitz and linear growth conditions, uniformly in $t \in [t_0, t_f]$, the controlled system (8.11)–(8.12) has a unique $\mathbb{F}_{t_f}$-adapted continuous solution. In view of the control policy (8.18), the controlled system (8.11)–(8.12) is further rewritten as

$$dq(t) = (A_q(t)q(t) + b_q(t))dt + G_q(t)d\xi(t), \quad q(t_0) \tag{8.19}$$

where for each $t \in [t_0, t_f]$, the augmented state variables and system coefficients are defined by

$$q(t) \triangleq \begin{bmatrix} \hat{z}(t) \\ \tilde{z}(t) \end{bmatrix}, q(t_0) \triangleq \begin{bmatrix} z(t_0) \\ 0 \end{bmatrix}, A_q(t) \triangleq \begin{bmatrix} A_z(t) + B_z(t)K_z(t) & L_z(t)C_z(t) \\ 0 & A_z(t) - L_z(t)C_z(t) \end{bmatrix}$$

$$b_q(t) \triangleq \begin{bmatrix} B_z(t)l_z(t) + B_z(t)u_r(t) + E_z(t)d(t) \\ 0 \end{bmatrix}, \quad G_q(t) \triangleq \begin{bmatrix} L_z(t)H_z(t) \\ G_z(t) - L_z(t)H_z(t) \end{bmatrix}.$$

Similarly, the performance measure (8.17) which also incorporates new state variables, $q(t)$ is the quadratic utility function rewritten by

$$J = q^T(t_f)Q_q^f q(t_f) + \int_{t_0}^{t_f} [q^T(\tau)Q_q(\tau)q(\tau) + 2q^T(\tau)S_q(\tau) + l_z^T(\tau)R_z(\tau)l_z(\tau)]d\tau \tag{8.20}$$

where the continuous-time weightings are given by

$$Q_q(\tau) \triangleq \begin{bmatrix} Q_z(\tau) + K_z^T(\tau)R_z(\tau)K_z(\tau) & Q_z(\tau) \\ Q_z(\tau) & Q_z(\tau) \end{bmatrix}, \quad S_q(\tau) \triangleq \begin{bmatrix} K_z^T(\tau)R_z(\tau)l_z(\tau) \\ 0 \end{bmatrix}$$

$$Q_q^f \triangleq \begin{bmatrix} Q_z^f & Q_z^f \\ Q_z^f & Q_z^f \end{bmatrix}.$$

Conversely, it is evident that the closed-loop performance (8.20) pertaining to the problem of networked control with communication constraints here is a random variable of the generalized chi-squared type. This insight provides for explicit consideration of asymmetry or skewness in the probability distributions. Thereafter, it is to be preferred to control objectives based on mean, variance, skewness, etc. of the closed-loop distribution function of the chi-squared type. Indeed, on theoretical grounds, the approach to modeling and management of all statistical measures of performance risks can be easily shown by the following result.

Theorem 8.2.1 (Cumulant-Generating Function). *Let $q(\cdot)$ be a state variable of the networked control system (8.19) with the initial-value condition $q(\tau) \equiv q_\tau$ and $\tau \in [t_0, t_f]$. Further let the moment-generating function be defined by*

$$\varphi(\tau, q_\tau, \theta) \triangleq \varrho(\tau, \theta) \exp\{q_\tau^T \Upsilon(\tau, \theta) q_\tau + 2q_\tau^T \eta(\tau, \theta)\} \tag{8.21}$$

$$\upsilon(\tau, \theta) \triangleq \ln\{\varrho(\tau, \theta)\}, \qquad \theta \in \mathbb{R}^+. \tag{8.22}$$

Then, the cumulant-generating function has the form of quadratic affine

$$\psi(\tau, q_\tau, \theta) = q_\tau^T \Upsilon(\tau, \theta) q_\tau + 2q_\tau^T \eta(\tau, \theta) + \upsilon(\tau, \theta) \tag{8.23}$$

where the scalar solution $\upsilon(\tau, \theta)$ solves the backward-in-time differential equation

$$\begin{aligned}\frac{d}{d\tau}\upsilon(\tau, \theta) = &- \mathrm{Tr}\left\{\Upsilon(\tau, \theta) G_q(\tau)\, \Xi\, G_q^T(\tau)\right\} - 2\eta^T(\tau, \theta) b_q(\tau) \\ &- \theta l_z^T(\tau) R_z(\tau) l_z(\tau), \quad \upsilon\left(t_f, \theta\right) = 0,\end{aligned} \tag{8.24}$$

the matrix solution $\Upsilon(\tau, \theta)$ satisfies the backward-in-time differential equation

$$\begin{aligned}\frac{d}{d\tau}\Upsilon(\tau, \theta) = &- A_q^T(\tau)\Upsilon(\tau, \theta) - \Upsilon(\tau, \theta) A_q(\tau) - 2\Upsilon(\tau, \theta) G_q(\tau)\Xi G_q^T(\tau)\Upsilon(\tau, \theta) \\ &- \theta Q_q(\tau), \quad \Upsilon(t_f, \theta) = \theta Q_q^f,\end{aligned} \tag{8.25}$$

and the vector solution $\eta(\tau, \theta)$ satisfies the backward-in-time differential equation

$$\frac{d}{d\tau}\eta(\tau, \theta) = - A_q^T(\tau)\eta(\tau, \theta) - \Upsilon(\tau, \theta) b_q(\tau) - \theta S_q(\tau), \quad \eta(t_f, \theta) = 0. \tag{8.26}$$

Also, the scalar solution $\varrho(\tau, \theta)$ satisfies the time-backward differential equation

$$\begin{aligned}\frac{d}{d\tau}\varrho(\tau, \theta) = &- \varrho(\tau, \theta) [\mathrm{Tr}\left\{\Upsilon(\tau, \theta) G_q(\tau)\, \Xi\, G_q^T(\tau)\right\} + 2\eta^T(\tau, \theta) b_q(\tau) \\ &+ \theta l_z^T(\tau) R_z(\tau) l_z(\tau)], \quad \varrho\left(t_f, \theta\right) = 1.\end{aligned} \tag{8.27}$$

Proof. In the course of the proof, it is convenient to have $\varpi(\tau, q_\tau, \theta) \triangleq e^{\{\theta J(\tau, q_\tau)\}}$ in which the performance measure (8.20) is rewritten as the cost-to-go function from an arbitrary state q_τ at a running time $\tau \in [t_0, t_f]$, that is,

$$
\begin{aligned}
J(\tau, q_\tau) &= q^T(t_f) Q_q^f q(t_f) \\
&\quad + \int_\tau^{t_f} [q^T(t) Q_q(t) q(t) + 2q^T(t) S_q(t) + l_z^T(t) R_z(t) l_z(t)] dt
\end{aligned} \tag{8.28}
$$

subject to

$$
dq(t) = (A_q(t)q(t) + b_q(t))dt + G_q(t)d\xi(t), \quad q(\tau) = q_\tau. \tag{8.29}
$$

By definition, the moment-generating function is

$$
\varphi(\tau, q_\tau, \theta) \triangleq E\{\varpi(\tau, q_\tau, \theta)\}.
$$

Henceforth, the total time derivative of $\varphi(\tau, q_\tau, \theta)$ is obtained as

$$
\frac{d}{d\tau}\varphi(\tau, q_\tau, \theta) = -\theta[q_\tau^T Q_q(\tau) q_\tau + 2q_\tau^T S_q(\tau) + l_z^T(\tau) R_z(\tau) l_z(\tau)]\varphi(\tau, q_\tau, \theta).
$$

An application of the standard formula of Ito calculus results in

$$
\begin{aligned}
d\varphi(\tau, q_\tau, \theta) &= E\{d\varpi(\tau, q_\tau, \theta)\} \\
&= E\Big\{\varpi_\tau(\tau, q_\tau, \theta)\, d\tau + \varpi_{q_\tau}(\tau, q_\tau, \theta)\, dq_\tau \\
&\quad + \frac{1}{2}\mathrm{Tr}\left\{\varpi_{q_\tau q_\tau}(\tau, q_\tau, \theta) G_q(\tau) \Xi G_q^T(\tau)\right\} d\tau\Big\} \\
&= \varphi_\tau(\tau, q_\tau; \theta) d\tau + \varphi_{q_\tau}(\tau, q_\tau, \theta)(A_q(\tau) q_\tau + b_q(\tau)) d\tau \\
&\quad + \frac{1}{2}\mathrm{Tr}\left\{\varphi_{q_\tau q_\tau}(\tau, q_\tau, \theta) G_q(\tau) \Xi G_q^T(\tau)\right\} d\tau
\end{aligned}
$$

which under the definition $\varphi(\tau, q_\tau, \theta) = \varrho(\tau, \theta) \exp\left\{q_\tau^T \Upsilon(\tau, \theta) q_\tau + 2q_\tau^T \eta(\tau, \theta)\right\}$ and its partial derivatives leads to the result

$$
\begin{aligned}
&-\theta[q_\tau^T Q_q(\tau) q_\tau + 2q_\tau^T S_q(\tau) + l_z^T(\tau) R_z(\tau) l_z(\tau)]\varphi(\tau, q_\tau, \theta) \\
&= \Bigg\{\frac{\frac{d}{d\tau}\varrho(\tau, \theta)}{\varrho(\tau, \theta)} + 2q_\tau^T[\frac{d}{d\tau}\eta(\tau, \theta) + A_q^T(\tau)\eta(\tau, \theta) + \Upsilon(\tau, \theta) b_q(\tau)] \\
&\quad + \mathrm{Tr}\left\{\Upsilon(\tau, \theta) G_q(\tau) \Xi G_q^T(\tau)\right\} + 2\eta^T(\tau, \theta) b_q(\tau)
\end{aligned}
$$

$$+ q_\tau^T [\frac{d}{d\tau}\Upsilon(\tau,\theta) + A_q^T(\tau)\Upsilon(\tau,\theta) + \Upsilon(\tau,\theta)A_q(\tau)$$
$$+ 2\Upsilon(\tau,\theta)G_q(\tau)\Xi G_q^T(\tau)\Upsilon(\tau,\theta)]q_\tau\Big\}\varphi(\tau,q_\tau,\theta).$$

In view of the common acceptance among the constant, linear and quadratic terms being independent of arbitrary initial values q_τ, one can easily obtain the results (8.24)–(8.27); and hereafter the proof is now complete.

For ordering uncertain prospects of (8.20), it is easily identifiable by looking at the appropriate parameters that characterize the dispersion of the distribution of (8.20). In the statistics literature, statistical measures of risks are considered as the shape parameters which can be next determined by means of a Maclaurin series expansion of (8.23)

$$\psi(\tau,q_\tau,\theta) = \sum_{r=1}^{\infty} \frac{\partial^{(r)}}{\partial\theta^{(r)}}\psi(\tau,q_\tau,\theta)\bigg|_{\theta=0} \frac{\theta^r}{r!} \tag{8.30}$$

in which all $\kappa_r \triangleq \frac{\partial^{(r)}}{\partial\theta^{(r)}}\psi(\tau,q_\tau,\theta)\Big|_{\theta=0}$ are called rth-order performance-measure statistics. Moreover, the series expansion coefficients are computed by using the cumulant-generating function (8.23)

$$\frac{\partial^{(r)}}{\partial\theta^{(r)}}\psi(\tau,q_\tau,\theta)\bigg|_{\theta=0} = q_\tau^T \frac{\partial^{(r)}}{\partial\theta^{(r)}}\Upsilon(\tau,\theta)\bigg|_{\theta=0} q_\tau$$
$$+ 2q_\tau^T \frac{\partial^{(r)}}{\partial\theta^{(r)}}\eta(\tau,\theta)\bigg|_{\theta=0} + \frac{\partial^{(r)}}{\partial\theta^{(r)}}\upsilon(\tau,\theta)\bigg|_{\theta=0}. \tag{8.31}$$

In view of the definition (8.30), the rth performance-measure statistic becomes

$$\kappa_r = q_\tau^T \frac{\partial^{(r)}}{\partial\theta^{(r)}}\Upsilon(\tau,\theta)\bigg|_{\theta=0} q_\tau + 2q_\tau^T \frac{\partial^{(r)}}{\partial\theta^{(r)}}\eta(\tau,\theta)\bigg|_{\theta=0} + \frac{\partial^{(r)}}{\partial\theta^{(r)}}\upsilon(\tau,\theta)\bigg|_{\theta=0} \tag{8.32}$$

for any finite $1 \le r < \infty$. For notational convenience, there is, in turn, a need of changing notations as illustrated below

$$H_r(\tau) \triangleq \frac{\partial^{(r)}\Upsilon(\tau,\theta)}{\partial\theta^{(r)}}\bigg|_{\theta=0}, \quad \breve{D}_r(\tau) \triangleq \frac{\partial^{(r)}\eta(\tau,\theta)}{\partial\theta^{(r)}}\bigg|_{\theta=0} \tag{8.33}$$

$$D_r(\tau) \triangleq \frac{\partial^{(r)}\upsilon(\tau,\theta)}{\partial\theta^{(r)}}\bigg|_{\theta=0}. \tag{8.34}$$

It thus leads to the next result which effectively forecast the higher-order characteristics associated with (8.20). With appropriate choices of K_z and l_z, one can reasonably reshape all the mathematical statistics of (8.20) and hence even characterize the closed-loop performance distributions in accordance to a priori criteria for stochastic dominance.

Theorem 8.2.2 (Performance-Measure Statistics). *Suppose that (A_q, B_q) is uniformly stabilizable and (C_q, A_q) is uniformly detectable. The kth-order statistic of the chi-squared performance measure (8.20) is given by*

$$\kappa_k = q^T(t_0)H_k(t_0)q(t_0) + 2q^T(t_0)\breve{D}_k(t_0) + D_k(t_0), \quad k \in \mathbb{Z}^+ \tag{8.35}$$

where the cumulant-generating components $H_k(\varsigma)$, $\breve{D}_k(\varsigma)$ and $D_k(\varsigma)$ evaluated at $\varsigma = 0$ satisfy the cumulant-generating equations (with the dependence of $H_k(\varsigma)$, $\breve{D}_k(\varsigma)$ and $D_k(\varsigma)$ upon K_z and l_z suppressed)

$$\frac{d}{d\varsigma}H_1(\varsigma) = -A_q^T(\varsigma)H_1(\varsigma) - H_1(\varsigma)A_q(\varsigma) - Q_q(\varsigma), \quad H_1(t_f) = Q_q^f \tag{8.36}$$

$$\frac{d}{d\varsigma}H_r(\varsigma) = -A_q^T(\varsigma)H_r(\varsigma) - H_r(\varsigma)A_q(\varsigma) - \sum_{\nu=1}^{r-1}\frac{2r!}{\nu!(r-\nu)!}H_\nu(\varsigma)G_q(\varsigma)\Xi G_q^T(\varsigma)H_{r-\nu}(\varsigma), \quad H_r(t_f) = 0, \quad 2 \le r \le k \tag{8.37}$$

and

$$\frac{d}{d\varsigma}\breve{D}_1(\varsigma) = -A_q^T(\varsigma)\breve{D}_1(\varsigma) - H_1(\varsigma)b_q(\varsigma) - S_q(\varsigma), \quad \breve{D}_1(t_f) = 0 \tag{8.38}$$

$$\frac{d}{d\varsigma}\breve{D}_r(\varsigma) = -A_q^T(\varsigma)\breve{D}_r(\varsigma) - H_r(\varsigma)b_q(\varsigma), \quad \breve{D}_r(t_f) = 0, \quad 2 \le r \le k \tag{8.39}$$

and, finally

$$\frac{d}{d\varsigma}D_1(\varsigma) = -\operatorname{Tr}\{H_1(\varsigma)G_q(\varsigma)\Xi G_q^T(\varsigma)\} - 2\breve{D}_1^T(\varsigma)b_q(\varsigma) - l_z^T(\varsigma)R_z(\varsigma)l_z(\varsigma) \tag{8.40}$$

$$\frac{d}{d\varsigma}D_r(\varsigma) = -\operatorname{Tr}\{H_r(\varsigma)G_q(\varsigma)\Xi_q G_q^T(\varsigma)\} - 2\breve{D}_r^T(\varsigma)b_q(\varsigma), \quad 2 \le r \le 2 \tag{8.41}$$

with the terminal-value conditions $D_r(t_f) = 0$ for $1 \le r \le k$.

Again, it is important to note that the use of mathematical statistics of (8.20) as measures of risk for non-symmetric distributions has been the focal of the present chapter. All the statistical measures of risks are tractable mathematically as can been judged herein. A procedure for obtaining these performance-measure statistics is further described in terms of the system description of (8.11)–(8.12)

$$\kappa_k = z_0^T H_{11}^k(t_0) z_0 + 2 z_0^T \breve{D}_{11}^k(t_0) + D^k(t_0), \quad k \in \mathbb{Z}^+ \tag{8.42}$$

according to the matrix and vector partitions

$$H_r(\varsigma) \triangleq \begin{bmatrix} H_{11}^r(\varsigma) & H_{12}^r(\varsigma) \\ (H_{12}^r)^T(\varsigma) & H_{22}^r(\varsigma) \end{bmatrix}; \quad \breve{D}_r(\varsigma) \triangleq \begin{bmatrix} \breve{D}_{11}^r(\varsigma) \\ \breve{D}_{21}^r(\varsigma) \end{bmatrix}; \quad D^k(\varsigma) \triangleq D_k(\varsigma) \tag{8.43}$$

from which the sub-matrix and vector components depend on K_z and l_z and satisfy the backward-in-time matrix-valued differential equations

$$\frac{d}{d\varsigma} H_{11}^1(\varsigma) = -(A_z(\varsigma) + B_z(\varsigma)K_z(\varsigma))^T H_{11}^1(\varsigma) - H_{11}^1(\varsigma)(A_z(\varsigma) + B_z(\varsigma)K_z(\varsigma)) - K_z^T(\varsigma)R_z(\varsigma)K_z(\varsigma) - Q_z(\varsigma), \quad H_{11}^1(t_f) = Q_z^f \tag{8.44}$$

$$\frac{d}{d\varsigma} H_{11}^r(\varsigma) = -(A_z(\varsigma) + B_z(\varsigma)K_z(\varsigma))^T H_{11}^r(\varsigma) - H_{11}^r(\varsigma)(A_z(\varsigma) + B_z(\varsigma)K_z(\varsigma)) - \sum_{\nu=1}^{r-1} \frac{2r!}{\nu!(r-\nu)!}(H_{11}^\nu(\varsigma)\Pi_1(\varsigma) + H_{12}^\nu(\varsigma)\Pi_3(\varsigma))H_{11}^{r-\nu}(\varsigma) - \sum_{\nu=1}^{r-1} \frac{2r!}{\nu!(r-\nu)!}(H_{11}^\nu(\varsigma)\Pi_2(\varsigma)+H_{12}^\nu(\varsigma)\Pi_4(\varsigma))H_{21}^{r-\nu}(\varsigma), \quad H_{11}^r(t_f) = 0 \tag{8.45}$$

$$\frac{d}{d\varsigma} H_{12}^1(\varsigma) = -(A_z(\varsigma) + B_z(\varsigma)K_z(\varsigma))^T H_{12}^1(\varsigma) - H_{11}^1(\varsigma)L_z(\varsigma)C_z(\varsigma) - H_{12}^1(\varsigma)(A_z(\varsigma) - L_z(\varsigma)C_z(\varsigma)) - Q_z(\varsigma), \quad H_{12}^1(t_f) = Q_z^f \tag{8.46}$$

$$\frac{d}{d\varsigma} H_{12}^r(\varsigma) = -(A_z(\varsigma) + B_z(\varsigma)K_z(\varsigma))^T H_{12}^r(\varsigma) - H_{12}^r(\varsigma)(A_z(\varsigma) - L_z(\varsigma)C_z(\varsigma)) - H_{11}^r(\varsigma)L_z(\varsigma)C_z(\varsigma) - \sum_{\nu=1}^{r-1} \frac{2r!}{\nu!(r-\nu)!}(H_{11}^\nu(\varsigma)\Pi_1(\varsigma) + H_{12}^\nu(\varsigma)\Pi_3(\varsigma))H_{12}^{r-\nu}(\varsigma) - \sum_{\nu=1}^{r-1} \frac{2r!}{\nu!(r-\nu)!}(H_{11}^\nu(\varsigma)\Pi_2(\varsigma) + H_{12}^\nu(\varsigma)\Pi_4(\varsigma))H_{22}^{r-\nu}(\varsigma), \quad H_{12}^r(t_f) = 0 \tag{8.47}$$

$$\frac{d}{d\varsigma}H_{21}^{1}(\varsigma) = -H_{21}^{1}(\varsigma)(A_z(\varsigma)+B_z(\varsigma)K_z(\varsigma)) - (L_z(\varsigma)C_z(\varsigma))^T H_{11}^{1}(\varsigma) \\ -(A_z(\varsigma)-L_z(\varsigma)C_z(\varsigma))^T H_{21}^{1}(\varsigma) - Q_z(\varsigma), \quad H_{21}^{1}(t_f) = Q_z^f \tag{8.48}$$

$$\frac{d}{d\varsigma}H_{21}^{r}(\varsigma) = -H_{21}^{r}(\varsigma)(A_z(\varsigma)+B_z(\varsigma)K_z(\varsigma)) - (A_z(\varsigma)-L_z(\varsigma)C_z(\varsigma))^T H_{21}^{r}(\varsigma) \\ -(L_z(\varsigma)C_z(\varsigma))^T H_{11}^{r}(\varsigma) - \sum_{\nu=1}^{r-1}\frac{2r!}{\nu!(r-\nu)!}(H_{21}^{\nu}(\varsigma)\Pi_1(\varsigma) \\ + H_{22}^{\nu}(\varsigma)\Pi_3(\varsigma))H_{11}^{r-\nu}(\varsigma) - \sum_{\nu=1}^{r-1}\frac{2r!}{\nu!(r-\nu)!}(H_{21}^{\nu}(\varsigma)\Pi_2(\varsigma) \\ + H_{22}^{\nu}(\varsigma)\Pi_4(\varsigma))H_{21}^{r-\nu}(\varsigma), \quad H_{21}^{r}(t_f) = 0 \tag{8.49}$$

$$\frac{d}{d\varsigma}H_{22}^{1}(\varsigma) = -H_{22}^{1}(\varsigma)(A_z(\varsigma)-L_z(\varsigma)C_z(\varsigma)) - H_{21}^{1}(\varsigma)(L_z(\varsigma)C_z(\varsigma)) - Q_z(\varsigma) \\ -(L_z(\varsigma)C_z(\varsigma))^T H_{12}^{1}(\varsigma) - (A_z(\varsigma)-L_z(\varsigma)C_z(\varsigma))^T H_{22}^{1}(\varsigma), \quad H_{22}^{1}(t_f) = Q_z^f \tag{8.50}$$

$$\frac{d}{d\varsigma}H_{22}^{r}(\varsigma) = -H_{22}^{r}(\varsigma)(A_z(\varsigma)-L_z(\varsigma)C_z(\varsigma)) - H_{21}^{r}(\varsigma)(L_z(\varsigma)C_z(\varsigma)) \\ -(L_z(\varsigma)C_z(\varsigma))^T H_{12}^{r}(\varsigma) - (A_z(\varsigma)-L_z(\varsigma)C_z(\varsigma))^T H_{22}^{r}(\varsigma) \\ -\sum_{\nu=1}^{r-1}\frac{2r!}{\nu!(r-\nu)!}(H_{21}^{\nu}(\varsigma)\Pi_1(\varsigma) + H_{22}^{\nu}(\varsigma)\Pi_3(\varsigma))H_{12}^{r-\nu}(\varsigma) \\ -\sum_{\nu=1}^{r-1}\frac{2r!}{\nu!(r-\nu)!}(H_{21}^{\nu}(\varsigma)\Pi_2(\varsigma) + H_{22}^{\nu}(\varsigma)\Pi_4(\varsigma))H_{22}^{r-\nu}(\varsigma), \quad H_{22}^{r}(t_f) = 0 \tag{8.51}$$

$$\frac{d}{d\varsigma}\breve{D}_{11}^{1}(\varsigma) = -(A_z(\varsigma)+B_z(\varsigma)K_z(\varsigma))^T \breve{D}_{11}^{1}(\varsigma) - K_z^T(\varsigma)R_z(\varsigma)l_z(\varsigma) \\ -H_{11}^{1}(\varsigma)(B_z(\varsigma)l_z(\varsigma) + B_z(\varsigma)u_r(\varsigma) + E_z(\varsigma)d(\varsigma)), \quad \breve{D}_{11}^{1}(t_f) = 0 \tag{8.52}$$

$$\frac{d}{d\varsigma}\breve{D}_{11}^{r}(\varsigma) = -H_{11}^{r}(\varsigma)(B_z(\varsigma)l_z(\varsigma) + B_z(\varsigma)u_r(\varsigma) + E_z(\varsigma)d(\varsigma)) \\ -(A_z(\varsigma)+B_z(\varsigma)K_z(\varsigma))^T \breve{D}_{11}^{r}(\varsigma), \quad \breve{D}_{11}^{r}(t_f) = 0,, \quad 2 \le r \le k \tag{8.53}$$

$$\frac{d}{d\varsigma}\breve{D}_{21}^r(\varsigma) = -(A_z(\varsigma)-L_z(\varsigma)C_z(\varsigma))^T\breve{D}_{21}^r(\varsigma) - (L_z(\varsigma)C_z(\varsigma))^T\breve{D}_{11}^r(\varsigma)$$
$$- H_{21}^r(\varsigma)(B_z(\varsigma)l_z(\varsigma)+B_z(\varsigma)u_r(\varsigma)+E_z(\varsigma)d(\varsigma)), \quad \breve{D}_{21}^r(t_f)=0, \quad r\geq 1 \tag{8.54}$$

$$\frac{d}{d\varsigma}D_1(\varsigma) = -\operatorname{Tr}\{H_{11}^1(\varsigma)\Pi_1(\varsigma)+H_{12}^1(\varsigma)\Pi_3(\varsigma)\}$$
$$-\operatorname{Tr}\{H_{21}^1(\varsigma)\Pi_2(\varsigma)+H_{22}^1(\varsigma)\Pi_4(\varsigma)\}-l_z^T(\varsigma)R_z(\varsigma)l_z(\varsigma)$$
$$-2(\breve{D}_{11}^1)^T(\varsigma)(B_z(\varsigma)l_z(\varsigma)+B_z(\varsigma)u_r(\varsigma)+E_z(\varsigma)d(\varsigma)), \quad D_1(t_f)=0 \tag{8.55}$$

$$\frac{d}{d\varsigma}D_r(\varsigma) = -\operatorname{Tr}\{H_{11}^r(\varsigma)\Pi_1(\varsigma) + H_{12}^r(\varsigma)\Pi_3(\varsigma)\}$$
$$-2(\breve{D}_{11}^r)^T(\varsigma)(B_z(\varsigma)l_z(\varsigma) + B_z(\tau)u_r(\varsigma) + E_z(\varsigma)d(\varsigma))$$
$$-\operatorname{Tr}\{H_{21}^r(\varsigma)\Pi_2(\varsigma) + H_{22}^r(\varsigma)\Pi_4(\varsigma)\}, \quad D_r(t_f)=0, \quad 2\leq r\leq k \tag{8.56}$$

provided that

$$\Pi_1(\varsigma) \triangleq L_z(\varsigma)H_z(\varsigma)\Xi(L_z(\varsigma)H_z(\varsigma))^T$$
$$\Pi_2(\varsigma) \triangleq L_z(\varsigma)H_z(\varsigma)\Xi(G_z(\varsigma)-L_z(\varsigma)H_z(\varsigma))^T = \Pi_3^T(\varsigma)$$
$$\Pi_4(\varsigma) \triangleq (G_z(\varsigma)-L_z(\varsigma)H_z(\varsigma))\Xi(G_z(\varsigma)-L_z(\varsigma)H_z(\varsigma))^T.$$

8.3 Asserting Problem Statements

When the system model (8.19)–(8.20) has dispersion measures (8.42) for the class of continuous distributions of (8.20), a risk-averse control considered addressing performance uncertainty then faces the problem of designing feedback control alternatives (K_z, l_z). At this stage the only framework addressing uncertain prospects of (8.20) for the linear-quadratic class of stochastic systems is the one reported in [5]. In fact, there are certain technicalities that must be addressed at the information structures as governed by (8.44)–(8.56) together with the crucial state variables $H_{11}^r(\varsigma)$, $H_{12}^r(\varsigma)$, $H_{21}^r(\varsigma)$, $H_{22}^r(\tau)$, $\breve{D}_{11}^r(\varsigma)$, $\breve{D}_{21}^r(\varsigma)$ and $D^r(\varsigma)$. In order to have notational simplicity, it is required to denote the right members of (8.44)–(8.56)

$$\mathscr{F}_{11}^1(\varsigma, \mathscr{H}_{11}, \mathscr{H}_{12}, \mathscr{H}_{21}, K_z) \triangleq -(A_z(\varsigma) + B_z(\varsigma)K_z(\varsigma))^T \mathscr{H}_{11}^1(\varsigma)$$
$$- \mathscr{H}_{11}^1(\varsigma)(A_z(\varsigma) + B_z(\varsigma)K_z(\varsigma)) - K_z^T(\varsigma)R_z(\varsigma)K_z(\varsigma) - Q_z(\varsigma)$$
$$\mathscr{F}_{11}^r(\varsigma, \mathscr{H}_{11}, \mathscr{H}_{12}, \mathscr{H}_{21}, K_z) \triangleq -(A_z(\varsigma) + B_z(\varsigma)K_z(\varsigma))^T \mathscr{H}_{11}^r(\varsigma)$$
$$- \mathscr{H}_{11}^r(\varsigma)(A_z(\varsigma) + B_z(\varsigma)K_z(\varsigma)) - \sum_{\nu=1}^{r-1} \frac{2r!}{\nu!(r-\nu)!}(\mathscr{H}_{11}^\nu(\varsigma)\Pi_1(\varsigma)$$
$$- \sum_{\nu=1}^{r-1} \frac{2r!}{\nu!(r-\nu)!}(\mathscr{H}_{11}^\nu(\varsigma)\Pi_2(\varsigma) + \mathscr{H}_{12}^\nu(\varsigma)\Pi_3(\varsigma))\mathscr{H}_{11}^{r-\nu}(\varsigma)$$
$$+ \mathscr{H}_{12}^\nu(\varsigma)\Pi_4(\varsigma))\mathscr{H}_{21}^{r-\nu}(\varsigma), \quad 2 \le r \le k$$

$$\mathscr{F}_{12}^1(\varsigma, \mathscr{H}_{11}, \mathscr{H}_{12}, \mathscr{H}_{22}, K_z) \triangleq -(A_z(\varsigma)+B_z(\varsigma)K_z(\varsigma))^T \mathscr{H}_{12}^1(\varsigma) - \mathscr{H}_{11}^1(\varsigma)L_z(\varsigma)C_z(\varsigma)$$
$$- \mathscr{H}_{12}^1(\varsigma)(A_z(\varsigma) - L_z(\varsigma)C_z(\varsigma)) - Q_z(\varsigma)$$
$$\mathscr{F}_{12}^r(\varsigma, \mathscr{H}_{11}, \mathscr{H}_{12}, \mathscr{H}_{22}, K_z) \triangleq -(A_z(\varsigma) + B_z(\varsigma)K_z(\varsigma))^T \mathscr{H}_{12}^r(\varsigma) - \mathscr{H}_{11}^r(\varsigma)L_z(\varsigma)C_z(\varsigma)$$
$$- \mathscr{H}_{12}^r(\varsigma)(A_z(\varsigma) - L_z(\varsigma)C_z(\varsigma)) - \sum_{\nu=1}^{r-1} \frac{2r!}{\nu!(r-\nu)!}(\mathscr{H}_{11}^\nu(\varsigma)\Pi_1(\varsigma)$$
$$+ \mathscr{H}_{12}^\nu(\varsigma)\Pi_3(\varsigma))\mathscr{H}_{12}^{r-\nu}(\varsigma) - \sum_{\nu=1}^{r-1} \frac{2r!}{\nu!(r-\nu)!}(\mathscr{H}_{11}^\nu(\varsigma)\Pi_2(\varsigma)$$
$$+ \mathscr{H}_{12}^\nu(\varsigma)\Pi_4(\varsigma))\mathscr{H}_{22}^{r-\nu}(\varsigma), \quad 2 \le r \le k$$

$$\mathscr{F}_{21}^1(\varsigma, \mathscr{H}_{11}, \mathscr{H}_{21}, \mathscr{H}_{22}, K_z) \triangleq -\mathscr{H}_{21}^1(\varsigma)(A_z(\varsigma) + B_z(\varsigma)K_z(\varsigma))$$
$$- (L_z(\varsigma)C_z(\varsigma))^T \mathscr{H}_{11}^1(\varsigma) - (A_z(\varsigma) - L_z(\varsigma)C_z(\varsigma))^T \mathscr{H}_{21}^1(\varsigma) - Q_z(\varsigma)$$
$$\mathscr{F}_{21}^r(\varsigma, \mathscr{H}_{11}, \mathscr{H}_{21}, \mathscr{H}_{22}, K_z) \triangleq -\mathscr{H}_{21}^r(\varsigma)(A_z(\varsigma) + B_z(\varsigma)K_z(\varsigma))$$
$$- (A_z(\varsigma) - L_z(\varsigma)C_z(\varsigma))^T \mathscr{H}_{21}^r(\varsigma) - (L_z(\varsigma)C_z(\varsigma))^T \mathscr{H}_{11}^r(\varsigma)$$
$$- \sum_{\nu=1}^{r-1} \frac{2r!}{\nu!(r-\nu)!}(\mathscr{H}_{21}^\nu(\varsigma)\Pi_1(\varsigma) + \mathscr{H}_{22}^\nu(\varsigma)\Pi_3(\varsigma))\mathscr{H}_{11}^{r-\nu}(\varsigma)$$
$$- \sum_{\nu=1}^{r-1} \frac{2r!}{\nu!(r-\nu)!}(\mathscr{H}_{21}^\nu(\varsigma)\Pi_2(\varsigma) + \mathscr{H}_{22}^\nu(\varsigma)\Pi_4(\varsigma))\mathscr{H}_{21}^{r-\nu}(\varsigma), \quad 2 \le r \le k$$

$$\mathscr{F}_{22}^1(\varsigma, \mathscr{H}_{22}, \mathscr{H}_{12}, \mathscr{H}_{21}) \triangleq -\mathscr{H}_{22}^1(\varsigma)(A_z(\varsigma) - L_z(\varsigma)C_z(\varsigma)) - \mathscr{H}_{21}^1(\varsigma)(L_z(\varsigma)C_z(\varsigma))$$
$$- (L_z(\varsigma)C_z(\varsigma))^T \mathscr{H}_{12}^1(\varsigma) - (A_z(\varsigma) - L_z(\varsigma)C_z(\varsigma))^T \mathscr{H}_{22}^1(\varsigma) - Q_z(\varsigma)$$

$$
\begin{aligned}
&\mathscr{F}^r_{22}(\varsigma,\mathscr{H}_{22},\mathscr{H}_{12},\mathscr{H}_{21}) \triangleq -\mathscr{H}^r_{22}(\varsigma)(A_z(\varsigma)-L_z(\varsigma)C_z(\varsigma)) - \mathscr{H}^r_{21}(\varsigma)(L_z(\varsigma)C_z(\varsigma))\\
&\quad -(L_z(\varsigma)C_z(\varsigma))^T\mathscr{H}^r_{12}(\varsigma) - (A_z(\varsigma)-L_z(\varsigma)C_z(\varsigma))^T\mathscr{H}^r_{22}(\varsigma)\\
&\quad -\sum_{\nu=1}^{r-1}\frac{2r!}{\nu!(r-\nu)!}(\mathscr{H}^\nu_{21}(\varsigma)\Pi_1(\varsigma)+\mathscr{H}^\nu_{22}(\varsigma)\Pi_3(\varsigma))\mathscr{H}^{r-\nu}_{12}(\varsigma)\\
&\quad -\sum_{\nu=1}^{r-1}\frac{2r!}{\nu!(r-\nu)!}(\mathscr{H}^\nu_{21}(\varsigma)\Pi_2(\varsigma)+\mathscr{H}^\nu_{22}(\varsigma)\Pi_4(\varsigma))\mathscr{H}^{r-\nu}_{22}(\varsigma), \quad 2\le r\le k
\end{aligned}
$$

$$
\begin{aligned}
&\breve{\mathscr{G}}^1_{11}(\varsigma,\breve{\mathscr{D}}_{11},\mathscr{H}_{11},K_z,l_z) \triangleq -(A_z(\varsigma)+B_z(\varsigma)K_z(\varsigma))^T\breve{\mathscr{D}}^1_{11}(\varsigma) - K_z^T(\varsigma)R_z(\varsigma)l_z(\varsigma)\\
&\quad -\mathscr{H}^1_{11}(\varsigma)(B_z(\varsigma)l_z(\varsigma)+B_z(\varsigma)u_r(\varsigma)+E_z(\varsigma)d(\varsigma))\\
&\breve{\mathscr{G}}^r_{11}(\varsigma,\breve{\mathscr{D}}_{11},\mathscr{H}_{11},K_z,l_z) \triangleq -(A_z(\varsigma)+B_z(\varsigma)K_z(\varsigma))^T\breve{\mathscr{D}}^r_{11}(\varsigma)\\
&\quad -\mathscr{H}^r_{11}(\varsigma)(B_z(\varsigma)l_z(\varsigma)+B_z(\varsigma)u_r(\varsigma)+E_z(\varsigma)d(\varsigma)), \quad 2\le r\le k
\end{aligned}
$$

$$
\begin{aligned}
&\breve{\mathscr{G}}^r_{21}(\varsigma,\breve{\mathscr{D}}_{11},\breve{\mathscr{D}}_{21},\mathscr{H}_{21},l_z) \triangleq (A_z(\varsigma)-L_z(\varsigma)C_z(\varsigma))^T\breve{\mathscr{D}}^r_{21}(\varsigma) - (L_z(\varsigma)C_z(\varsigma))^T\breve{\mathscr{D}}^r_{11}(\varsigma)\\
&\quad -\mathscr{H}^r_{21}(\varsigma)(B_z(\varsigma)l_z(\varsigma)+B_z(\varsigma)u_r(\varsigma)+E_z(\varsigma)d(\varsigma)), \quad 1\le r\le k
\end{aligned}
$$

$$
\begin{aligned}
&\mathscr{G}_1(\varsigma,\mathscr{H}_{11},\mathscr{H}_{12},\mathscr{H}_{21},\mathscr{H}_{22},\breve{\mathscr{D}}_{11},l_z) \triangleq\\
&\quad -\mathrm{Tr}\{\mathscr{H}^1_{11}(\varsigma)\Pi_1(\varsigma)+\mathscr{H}^1_{12}(\varsigma)\Pi_3(\varsigma)\} - \mathrm{Tr}\{\mathscr{H}^1_{21}(\varsigma)\Pi_2(\varsigma)+\mathscr{H}^1_{22}(\varsigma)\Pi_4(\varsigma)\}\\
&\quad -2(\breve{\mathscr{D}}^1_{11})^T(\varsigma)(B_z(\varsigma)l_z(\varsigma)+B_z(\varsigma)u_r(\varsigma)+E_z(\varsigma)d(\varsigma)) - l_z^T(\varsigma)R_z(\varsigma)l_z(\varsigma)\\
&\mathscr{G}_r(\varsigma,\mathscr{H}_{11},\mathscr{H}_{12},\mathscr{H}_{21},\mathscr{H}_{22},\breve{\mathscr{D}}_{11},l_z) \triangleq\\
&\quad -\mathrm{Tr}\{\mathscr{H}^r_{11}(\varsigma)\Pi_1(\varsigma)+\mathscr{H}^r_{12}(\varsigma)\Pi_3(\varsigma)\} - \mathrm{Tr}\{\mathscr{H}^r_{21}(\varsigma)\Pi_2(\varsigma)+\mathscr{H}^r_{22}(\varsigma)\Pi_4(\varsigma)\}\\
&\quad -2(\breve{\mathscr{D}}^r_{11})^T(\varsigma)(B_z(\varsigma)l_z(\varsigma)+B_z(\tau)u_r(\varsigma)+E_z(\varsigma)d(\varsigma)), \quad 2\le r\le k
\end{aligned}
$$

where the k-tuple variables $\mathscr{H}_{11}$, $\mathscr{H}_{12}$, $\mathscr{H}_{21}$, $\mathscr{H}_{22}$, $\breve{\mathscr{D}}_{11}$, $\breve{\mathscr{D}}_{21}$ and $\mathscr{D}$ defined by

$$
\begin{aligned}
&\mathscr{H}_{11}(\cdot) \triangleq (\mathscr{H}^1_{11}(\cdot),\ldots,\mathscr{H}^k_{11}(\cdot)), && \mathscr{H}_{12}(\cdot) \triangleq (\mathscr{H}^1_{12}(\cdot),\ldots,\mathscr{H}^k_{12}(\cdot))\\
&\mathscr{H}_{21}(\cdot) \triangleq (\mathscr{H}^1_{21}(\cdot),\ldots,\mathscr{H}^k_{21}(\cdot)), && \mathscr{H}_{22}(\cdot) \triangleq (\mathscr{H}^1_{22}(\cdot),\ldots,\mathscr{H}^k_{22}(\cdot))\\
&\breve{\mathscr{D}}_{11}(\cdot) \triangleq (\breve{\mathscr{D}}^1_{11}(\cdot),\ldots,\breve{\mathscr{D}}^k_{11}(\cdot)), && \breve{\mathscr{D}}_{21}(\cdot) \triangleq (\breve{\mathscr{D}}^1_{21}(\cdot),\ldots,\breve{\mathscr{D}}^k_{21}(\cdot))\\
&\mathscr{D}(\cdot) \triangleq (\mathscr{D}^1(\cdot),\ldots,\mathscr{D}^k(\cdot))
\end{aligned}
$$

whereas all the continuous-time and matrix-valued states $\mathscr{H}^r_{11}, \mathscr{H}^r_{12}, \mathscr{H}^r_{21}, \mathscr{H}^r_{22} \in \mathscr{C}^1([t_0,t_f];\mathbb{R}^{(n+m+c)\times(n+m+c)})$; vector-valued states $\breve{\mathscr{D}}^r_{11}, \breve{\mathscr{D}}^r_{21} \in \mathscr{C}^1([t_0,t_f];$

$\mathbb{R}^{n+m+c}$) and scalar-valued states $\mathscr{D}^r \in \mathscr{C}^1([t_0, t_f]; \mathbb{R})$ have the representations $\mathscr{H}_{11}^r(\cdot) \triangleq H_{11}^r(\cdot)$, $\mathscr{H}_{12}^r(\cdot) \triangleq H_{12}^r(\cdot)$, $\mathscr{H}_{21}^r(\cdot) \triangleq H_{21}^r(\cdot)$, $\mathscr{H}_{22}^r(\cdot) \triangleq H_{22}^r(\cdot)$, $\breve{\mathscr{D}}_{11}^r(\cdot) \triangleq \breve{D}_{11}^r(\cdot)$, $\breve{\mathscr{D}}_{21}^r(\cdot) \triangleq \breve{D}_{21}^r(\cdot)$, and $\mathscr{D}^r(\cdot) \triangleq D^r(\cdot)$.

Next it is essential to establish and maintain precise product mappings of the dynamical equations (8.44)–(8.56) which can be shown to be bounded and Lipschitz continuous on $[t_0, t_f]$; see [6]. Subsequently, the development that follows is beneficial from the Cartesian product mappings for the system control formulation

$$\mathscr{F}_{11} : [t_0, t_f] \times (\mathbb{R}^{(n+m+c)\times(n+m+c)})^{3k} \times \mathbb{R}^{m\times(n+m+c)} \mapsto (\mathbb{R}^{(n+m+c)\times(n+m+c)})^k$$

$$\mathscr{F}_{12} : [t_0, t_f] \times (\mathbb{R}^{(n+m+c)\times(n+m+c)})^{3k} \times \mathbb{R}^{m\times(n+m+c)} \mapsto (\mathbb{R}^{(n+m+c)\times(n+m+c)})^k$$

$$\mathscr{F}_{21} : [t_0, t_f] \times (\mathbb{R}^{(n+m+c)\times(n+m+c)})^{3k} \times \mathbb{R}^{m\times(n+m+c)} \mapsto (\mathbb{R}^{(n+m+c)\times(n+m+c)})^k$$

$$\mathscr{F}_{22} : [t_0, t_f] \times (\mathbb{R}^{(n+m+c)\times(n+m+c)})^{3k} \mapsto (\mathbb{R}^{(n+m+c)\times(n+m+c)})^k$$

$$\breve{\mathscr{G}}_{11} : [t_0, t_f] \times (\mathbb{R}^{(n+m+c)\times(n+m+c)})^k \times (\mathbb{R}^{n+m+c})^k \times \mathbb{R}^{m\times(n+m+c)} \times \mathbb{R}^m \mapsto (\mathbb{R}^{n+m+c})^k$$

$$\breve{\mathscr{G}}_{21} : [t_0, t_f] \times (\mathbb{R}^{(n+m+c)\times(n+m+c)})^k \times (\mathbb{R}^{n+m+c})^{2k} \times \mathbb{R}^m \mapsto (\mathbb{R}^{n+m+c})^k$$

$$\mathscr{G} : [t_0, t_f] \times (\mathbb{R}^{(n+m+c)\times(n+m+c)})^{4k} \times (\mathbb{R}^{n+m+c})^k \times \mathbb{R}^m \mapsto \mathbb{R}^k$$

whose the rules of action are governed by

$$\frac{d}{d\varsigma}\mathscr{H}_{11}(\varsigma) = \mathscr{F}_{11}(\varsigma, \mathscr{H}_{11}(\varsigma), \mathscr{H}_{12}(\varsigma), \mathscr{H}_{21}(\varsigma), K_z(\varsigma)), \quad \mathscr{H}_{11}(t_f) \tag{8.57}$$

$$\frac{d}{d\varsigma}\mathscr{H}_{12}(\varsigma) = \mathscr{F}_{12}(\varsigma, \mathscr{H}_{11}(\varsigma), \mathscr{H}_{12}(\varsigma), \mathscr{H}_{22}(\varsigma), K_z(\varsigma)), \quad \mathscr{H}_{12}(t_f) \tag{8.58}$$

$$\frac{d}{d\varsigma}\mathscr{H}_{21}(\varsigma) = \mathscr{F}_{21}(\varsigma, \mathscr{H}_{11}(\varsigma), \mathscr{H}_{12}(\varsigma), \mathscr{H}_{22}(\varsigma), K_z(\varsigma)), \quad \mathscr{H}_{21}(t_f) \tag{8.59}$$

$$\frac{d}{d\varsigma}\mathscr{H}_{22}(\varsigma) = \mathscr{F}_{22}(\varsigma, \mathscr{H}_{12}(\varsigma), \mathscr{H}_{21}(\varsigma), \mathscr{H}_{22}(\varsigma)), \quad \mathscr{H}_{22}(t_f) \tag{8.60}$$

$$\frac{d}{d\varsigma}\breve{\mathscr{D}}_{11}(\varsigma) = \breve{\mathscr{G}}_{11}(\varsigma, \breve{\mathscr{D}}_{11}, \mathscr{H}_{11}(\varsigma), K_z(\varsigma), l_z(\varsigma)), \quad \breve{\mathscr{D}}_{11}(t_f) \tag{8.61}$$

$$\frac{d}{d\varsigma}\breve{\mathscr{D}}_{21}(\varsigma) = \breve{\mathscr{G}}_{21}(\varsigma, \breve{\mathscr{D}}_{11}, \breve{\mathscr{D}}_{21}, \mathscr{H}_{21}(\varsigma), l_z(\varsigma)), \quad \breve{\mathscr{D}}_{21}(t_f) \tag{8.62}$$

$$\frac{d}{d\varsigma}\mathscr{D}(\varsigma) = \mathscr{G}(\varsigma, \mathscr{H}_{11}(\varsigma), \mathscr{H}_{12}(\varsigma), \mathscr{H}_{21}(\varsigma), \mathscr{H}_{22}(\varsigma), \breve{\mathscr{D}}_{11}(\varsigma), l_z(\varsigma)) \tag{8.63}$$

under the following definitions

$$\mathscr{F}_{11} \triangleq \mathscr{F}_{11}^1 \times \cdots \times \mathscr{F}_{11}^k, \quad \mathscr{F}_{12} \triangleq \mathscr{F}_{12}^1 \times \cdots \times \mathscr{F}_{12}^k, \quad \mathscr{F}_{21} \triangleq \mathscr{F}_{21}^1 \times \cdots \times \mathscr{F}_{21}^k$$

$$\mathscr{F}_{22} \triangleq \mathscr{F}_{22}^1 \times \cdots \times \mathscr{F}_{22}^k, \quad \breve{\mathscr{G}}_{11} \triangleq \breve{\mathscr{G}}_{11}^1 \times \cdots \times \breve{\mathscr{G}}_{11}^k, \quad \breve{\mathscr{G}}_{21} \triangleq \breve{\mathscr{G}}_{21}^1 \times \cdots \times \breve{\mathscr{G}}_{21}^k$$

$$\mathscr{G} \triangleq \mathscr{G}^1 \times \cdots \times \mathscr{G}^k$$

and the terminal-value conditions

$$\mathscr{H}_{11}(t_f) = \mathscr{H}_{12}(t_f) = \mathscr{H}_{21}(t_f) = \mathscr{H}_{22}(t_f) \triangleq \left(Q_z^f, 0, \ldots, 0\right)$$

$$\breve{\mathscr{D}}_{11}(t_f) = \breve{\mathscr{D}}_{21}(t_f) \triangleq (0, \ldots, 0), \quad \mathscr{D}(t_f) \triangleq (0, \ldots, 0).$$

Recall that the product system (8.57)–(8.63) uniquely determines $\mathscr{H}_{11}$, $\mathscr{H}_{12}$, $\mathscr{H}_{21}$, $\mathscr{H}_{22}$, $\breve{\mathscr{D}}_{11}$, $\breve{\mathscr{D}}_{21}$ and $\mathscr{D}$ once admissible feedback parameters K_z and l_z are specified. Thereby, $\mathscr{H}_{11} \equiv \mathscr{H}_{11}(\cdot, K_z)$, $\mathscr{H}_{12} \equiv \mathscr{H}_{12}(\cdot, K_z)$, $\mathscr{H}_{21} \equiv \mathscr{H}_{21}(\cdot, K_z)$, $\mathscr{H}_{22} \equiv \mathscr{H}_{22}(\cdot, K_z)$, $\breve{\mathscr{D}}_{11} \equiv \breve{\mathscr{D}}_{11}(\cdot, K_z, l_z)$, $\breve{\mathscr{D}}_{21} \equiv \breve{\mathscr{D}}_{21}(\cdot, K_z, l_z)$, and $\mathscr{D} \equiv \mathscr{D}(\cdot, K_z, l_z)$.

Of note, the use of expected values and variances as measures of risk for non-symmetric distributions of the random variable (8.20) has been questioned by financial theorists [7] and [8]. It has been shown that the subclass that can be ordered by the mean-variance rule is indeed small. This popular approach and its variants, thus appear to be of limited generality. Of practical interest in engineering and control applications, a finite linear combination of performance-measure statistics of (8.20) provides a strong rationale for using mean, variance, skewness, flatness, etc. to order uncertain prospects of the chi-squared random cost (8.20). As the result, this new paradigm is attempted to address both necessary and sufficient conditions as needed when ordering stochastic dominance for the class of probabilistic distributions with equal and unequal means herein. Subsequently, the feedback control designer is interested in optimizing the feedback control parameters K_z and l_z in accordance with this stochastic dominance rule of preference, which is now referred as the mean-risk aware performance index as specifically defined hereafter.

Definition 8.3.1 (Mean-Risk Aware Performance Index). Fix $k \in \mathbb{Z}^+$ and the sequence $\mu = \{\mu_i \geq 0\}_{i=1}^k$ with $\mu_1 > 0$. Then, the performance index with risk consequences for the stochastic system with network communication effects (8.19)–(8.20) is given by

$$\phi_0 : \{t_0\} \times (\mathbb{R}^{(n+m+c)\times(n+m+c)})^k \times (\mathbb{R}^{n+m+c})^k \times \mathbb{R}^k \mapsto \mathbb{R}^+$$

with the rule of action

$$\phi_0\left(t_0, \mathscr{H}_{11}(t_0), \breve{\mathscr{D}}_{11}(t_0), \mathscr{D}(t_0)\right) \triangleq \underbrace{\mu_1 \kappa_1}_{\text{Mean Measure}} + \underbrace{\mu_2 \kappa_2 + \cdots + \mu_k \kappa_k}_{\text{Risk Measures}}$$

$$= \sum_{r=1}^{k} \mu_r \left[z_0^T \mathscr{H}_{11}^r(t_0) z_0 + 2 z_0^T \breve{\mathscr{D}}_{11}^r(t_0) + \mathscr{D}^r(t_0) \right] \tag{8.64}$$

where additional parametric design of freedom μ_r, chosen by risk-averse controller designers, represent different levels of robustness prioritization according to the importance of the resulting performance-measure statistics to the probabilistic performance distribution. And the unique solutions $\{\mathscr{H}_{11}^r(\varsigma)\}_{r=1}^k$, $\left\{\breve{\mathscr{D}}_{11}^r(\varsigma)\right\}_{r=1}^k$, and $\{\mathscr{D}^r(\varsigma)\}_{r=1}^k$ evaluated at $\varsigma = t_0$ satisfy the dynamical equations (8.57)–(8.63).

For the performance-measure statistics (8.64) to underpin the natural measures of dispersion, it is necessary for appropriate choices of parameters K_z and l_z in the admissible set of feedback control laws to be further defined as follows.

Definition 8.3.2 (Admissible Feedback Parameters). For the given terminal data $(t_f, \mathscr{H}_{11}^f, \mathscr{H}_{12}^f, \mathscr{H}_{21}^f, \mathscr{H}_{22}^f, \breve{\mathscr{D}}_{11}^f, \mathscr{D}^f)$, the classes of admissible feedback gains and feedforward inputs are defined as follows. Let compact subsets $\overline{K}_z \subset \mathbb{R}^{m\times(n+m+c)}$ and $\overline{L}_z \subset \mathbb{R}^{n+m+c}$ be the sets of allowable gain and input values. With $k \in \mathbb{Z}^+$ and the sequence $\mu = \{\mu_r \geq 0\}_{r=1}^k$ and $\mu_1 > 0$ given, the sets of admissible $\mathscr{K}^z_{t_f,\mathscr{H}_{11}^f,\mathscr{H}_{12}^f,\mathscr{H}_{21}^f,\mathscr{H}_{22}^f,\breve{\mathscr{D}}_{11}^f,\mathscr{D}^f;\mu}$ and $\mathscr{L}^z_{t_f,\mathscr{H}_{11}^f,\mathscr{H}_{12}^f,\mathscr{H}_{21}^f,\mathscr{H}_{22}^f,\breve{\mathscr{D}}_{11}^f,\mathscr{D}^f;\mu}$ are the classes of time-continuous matrices $\mathscr{C}([t_0, t_f]; \mathbb{R}^{m\times(n+m+c)})$ and vectors $\mathscr{C}([t_0, t_f]; \mathbb{R}^{n+m+c})$ with values $K_z(\cdot) \in \overline{K}_z$ and $l_z(\cdot) \in \overline{L}_z$ for which the solutions to the dynamical equations (8.57)–(8.63) exist on the finite interval $[t_0, t_f]$.

The development in the sequel is motivated by the excellent treatment in [9] and is intended to follow it closely. Because the development therein embodies the traditional end-point problem and corresponding use of dynamic programming, it is necessary to make appropriate modifications in the sequence of results, as well as to introduce the terminology of statistical optimal control.

Definition 8.3.3 (Optimization Problem of Mayer Type). Suppose $k \in \mathbb{Z}^+$ and the sequence $\mu = \{\mu_r \geq 0\}_{r=1}^k$ with $\mu_1 > 0$ are fixed. Then, the optimization problem over $[t_0, t_f]$ is defined as the minimization of the performance index (8.64) with respect to (K_z, l_z) in $\mathscr{K}^z_{t_f,\mathscr{H}_{11}^f,\mathscr{H}_{12}^f,\mathscr{H}_{21}^f,\mathscr{H}_{22}^f,\breve{\mathscr{D}}_{11}^f,\mathscr{D}^f;\mu} \times \mathscr{L}^z_{t_f,\mathscr{H}_{11}^f,\mathscr{H}_{12}^f,\mathscr{H}_{21}^f,\mathscr{H}_{22}^f,\breve{\mathscr{D}}_{11}^f,\mathscr{D}^f;\mu}$ and subject to the dynamical equations (8.57)–(8.63).

Admittedly, the traditional approach to the aim of embedding the Mayer-type optimization into a larger optimal control problem is to parameterize the terminal time and states $(t_f, \mathscr{H}_{11}^f, \mathscr{H}_{12}^f, \mathscr{H}_{21}^f, \mathscr{H}_{22}^f, \breve{\mathscr{D}}_{11}^f, \mathscr{D}^f)$ as $(\varepsilon, \mathscr{Y}_{11}, \mathscr{Y}_{12}, \mathscr{Y}_{21}, \mathscr{Y}_{22}, \breve{\mathscr{Z}}_{11}, \mathscr{Z})$ instead.

Definition 8.3.4 (Reachable Set). $\mathscr{Q} \triangleq \Big\{(\varepsilon, \mathscr{Y}_{11}, \mathscr{Y}_{12}, \mathscr{Y}_{21}, \mathscr{Y}_{22}, \breve{\mathscr{Z}}_{11}, \mathscr{Z}) \in [t_0, t_f] \times (\mathbb{R}^{(n+m+c)\times(n+m+c)})^{4k} \times (\mathbb{R}^{n+m+c})^k \times \mathbb{R}^k$ such that $\mathscr{K}^z_{t_f,\mathscr{H}_{11}^f,\mathscr{H}_{12}^f,\mathscr{H}_{21}^f,\mathscr{H}_{22}^f,\breve{\mathscr{D}}_{11}^f,\mathscr{D}^f;\mu} \times \mathscr{L}^z_{t_f,\mathscr{H}_{11}^f,\mathscr{H}_{12}^f,\mathscr{H}_{21}^f,\mathscr{H}_{22}^f,\breve{\mathscr{D}}_{11}^f,\mathscr{D}^f;\mu} \neq \emptyset\Big\}$.

Therefore, the value function for this optimization problem is now depending on parameterizations of the terminal-value conditions.

Definition 8.3.5 (Value Function). Suppose that $(\varepsilon, \mathscr{Y}_{11}, \mathscr{Y}_{12}, \mathscr{Y}_{21}, \mathscr{Y}_{22}, \breve{\mathscr{Z}}_{11}, \mathscr{Z}) \in \mathscr{Q}$ be given. Then, the value function $\mathscr{V}(\varepsilon, \mathscr{Y}_{11}, \breve{\mathscr{Z}}_{11}, \mathscr{Z})$ is defined by

$$\mathscr{V}(\varepsilon, \mathscr{Y}_{11}, \breve{\mathscr{Z}}_{11}, \mathscr{Z}) = \inf_{K_z \in \overline{K}_z, l_z \in \overline{L}_z} \phi_0(t_0, \mathscr{H}_{11}(t_0), \breve{\mathscr{D}}_{11}(t_0), \mathscr{D}(t_0)).$$

By convention, $\mathscr{V}(\varepsilon, \mathscr{Y}_{11}, \breve{\mathscr{Z}}_{11}, \mathscr{Z}) = +\infty$ when the Cartesian product that follows $\mathscr{K}^z_{t_f, \mathscr{H}_{11}^f, \mathscr{H}_{12}^f, \mathscr{H}_{21}^f, \mathscr{H}_{22}^f, \breve{\mathscr{D}}_{11}^f, \mathscr{D}^f; \mu} \times \mathscr{L}^z_{t_f, \mathscr{H}_{11}^f, \mathscr{H}_{12}^f, \mathscr{H}_{21}^f, \mathscr{H}_{22}^f, \breve{\mathscr{D}}_{11}^f, \mathscr{D}^f; \mu}$ is empty.

Besides adapting to the initial-cost problem to solve the control optimization of Mayer type, the following results relating to the Hamilton-Jacobi-Bellman (HJB) equation and verification theorem also maintain a strong commitment to provide a comparative analysis to the existing literature [9] through various unique terminologies in statistical optimal control.

Theorem 8.3.1 (HJB Equation for Mayer Problem). *Let any interior point of* $\mathscr{Q}$ *be denoted as* $(\varepsilon, \mathscr{Y}_{11}, \mathscr{Y}_{12}, \mathscr{Y}_{21}, \mathscr{Y}_{22}, \breve{\mathscr{Z}}_{11}, \mathscr{Z})$ *at which* $\mathscr{V}(\varepsilon, \mathscr{Y}_{11}, \breve{\mathscr{Z}}_{11}, \mathscr{Z})$ *is differentiable. If there exist optimal feedback gain* $K_z^* \in \mathscr{K}^z_{t_f, \mathscr{H}_{11}^f, \mathscr{H}_{12}^f, \mathscr{H}_{21}^f, \mathscr{H}_{22}^f, \breve{\mathscr{D}}_{11}^f, \mathscr{D}^f; \mu}$ *and feedforward input* $l_z^* \in \mathscr{L}^z_{t_f, \mathscr{H}_{11}^f, \mathscr{H}_{12}^f, \mathscr{H}_{21}^f, \mathscr{H}_{22}^f, \breve{\mathscr{D}}_{11}^f, \mathscr{D}^f; \mu}$, *then the partial differential equation of dynamic programming*

$$
\begin{aligned}
0 = \min_{K_z \in \overline{K}_z, l_z \in \overline{L}_z} \Big\{ & \frac{\partial}{\partial \varepsilon} \mathscr{V}\left(\varepsilon, \mathscr{Y}_{11}, \breve{\mathscr{Z}}_{11}, \mathscr{Z}\right) \\
& + \frac{\partial}{\partial \operatorname{vec}(\mathscr{Y}_{11})} \mathscr{V}\left(\varepsilon, \mathscr{Y}_{11}, \breve{\mathscr{Z}}_{11}, \mathscr{Z}\right) \operatorname{vec}\left(\mathscr{F}_{11}\left(\varepsilon, \mathscr{Y}_{11}, \mathscr{Y}_{12}, \mathscr{Y}_{21}, K_z\right)\right) \\
& + \frac{\partial}{\partial \operatorname{vec}(\breve{\mathscr{Z}}_{11})} \mathscr{V}\left(\varepsilon, \mathscr{Y}_{11}, \breve{\mathscr{Z}}_{11}, \mathscr{Z}\right) \operatorname{vec}(\breve{\mathscr{G}}_{11}(\varepsilon, \breve{\mathscr{Z}}_{11}, \mathscr{Y}_{11}, K_z, l_z)) \\
& + \frac{\partial}{\partial \operatorname{vec}(\mathscr{Z})} \mathscr{V}\left(\varepsilon, \mathscr{Y}_{11}, \breve{\mathscr{Z}}_{11}, \mathscr{Z}\right) \operatorname{vec}(\mathscr{G}(\varepsilon, \mathscr{Y}_{11}, \mathscr{Y}_{12}, \mathscr{Y}_{21}, \mathscr{Y}_{22}, \breve{\mathscr{Z}}_{11}, l_z)) \Big\}
\end{aligned}
\tag{8.65}
$$

is satisfied. The boundary condition of (8.65) is given by

$$
\mathscr{V}\left(t_0, \mathscr{H}_{11}(t_0), \breve{\mathscr{D}}_{11}(t_0), \mathscr{D}(t_0)\right) = \phi_0\left(t_0, \mathscr{H}_{11}(t_0), \breve{\mathscr{D}}_{11}(t_0), \mathscr{D}(t_0)\right).
$$

Proof. A rigorous proof of the necessary condition herein can be found in [10].

At present, it is also important to acknowledge the contribution being made by the testing of the sufficient condition for optimality within the dynamic programming framework, particularly the verification theorem.

Theorem 8.3.2 (Verification Theorem). *Fix* $k \in \mathbb{Z}^+$ *and let* $\mathscr{W}(\varepsilon, \mathscr{Y}_{11}, \breve{\mathscr{Z}}_{11}, \mathscr{Z})$ *be a continuously differentiable solution of (8.65) and satisfies the boundary condition*

$$
\mathscr{W}(t_0, \mathscr{H}_{11}(t_0), \breve{\mathscr{D}}_{11}(t_0), \mathscr{D}(t_0)) = \phi_0\left(t_0, \mathscr{H}_{11}(t_0), \breve{\mathscr{D}}_{11}(t_0), \mathscr{D}(t_0)\right).
$$

Let the terminal-value condition $(t_f, \mathscr{H}_{11}^f, \mathscr{H}_{12}^f, \mathscr{H}_{21}^f, \mathscr{H}_{22}^f, \breve{\mathscr{D}}_{11}^f, \mathscr{D}^f)$ *be in* $\mathscr{Q}$; *the 2-tuple feedback* (K_z, l_z) *in* $\mathscr{K}^z_{t_f, \mathscr{H}_{11}^f, \mathscr{H}_{12}^f, \mathscr{H}_{21}^f, \mathscr{H}_{22}^f, \breve{\mathscr{D}}_{11}^f, \mathscr{D}^f; \mu} \times \mathscr{L}_{t_f, \mathscr{H}_{11}^f, \mathscr{H}_{12}^f, \mathscr{H}_{21}^f, \mathscr{H}_{22}^f, \breve{\mathscr{D}}_{11}^f,}$

$\mathscr{D}^f$; μ^z; *the trajectory solutions* $\mathscr{H}_{11}$, $\breve{\mathscr{D}}_{11}$ *and* $\mathscr{D}$ *of the dynamical equations (8.57)–(8.63). Then,* $\mathscr{W}(\varepsilon, \mathscr{H}_{11}(\varepsilon), \breve{\mathscr{D}}_{11}(\varepsilon), \mathscr{D}(\varepsilon))$ *is a time-backward increasing function of* ε. *If* (K_z^*, l_z^*) *is in* $\mathscr{K}^z_{t_f, \mathscr{H}_{11}^f, \mathscr{H}_{12}^f, \mathscr{H}_{21}^f, \mathscr{H}_{22}^f, \breve{\mathscr{D}}_{11}^f, \mathscr{D}^f;\mu} \times \mathscr{L}^z_{t_f, \mathscr{H}_{11}^f, \mathscr{H}_{12}^f, \mathscr{H}_{21}^f, \mathscr{H}_{22}^f, \breve{\mathscr{D}}_{11}^f, \mathscr{D}^f;\mu}$ *defined on* $[t_0, t_f]$ *with the corresponding solutions* $\mathscr{H}_{11}^*$, $\breve{\mathscr{D}}_{11}^*$ *and* $\mathscr{D}^*$ *of the dynamical equations (8.57)–(8.63) such that, for* $\tau \in [t_0, t_f]$

$$
\begin{aligned}
0 = &\frac{\partial}{\partial \varepsilon}\mathscr{W}(\tau, \mathscr{H}_{11}^*(\tau), \breve{\mathscr{D}}_{11}^*(\tau), \mathscr{D}^*(\tau)) + \frac{\partial}{\partial \operatorname{vec}(\mathscr{Y}_{11})}\mathscr{W}(\tau, \mathscr{H}_{11}^*(\tau), \breve{\mathscr{D}}_{11}^*(\tau), \mathscr{D}^*(\tau)) \\
&\cdot \operatorname{vec}(\mathscr{F}_{11}(\tau, \mathscr{H}_{11}^*(\tau), \mathscr{H}_{12}^*(\tau), \mathscr{H}_{21}^*(\tau), K_z^*(\tau))) \\
&+ \frac{\partial}{\partial \operatorname{vec}(\breve{\mathscr{Z}}_{11})}\mathscr{W}(\tau, \mathscr{H}_{11}^*(\tau), \breve{\mathscr{D}}_{11}^*(\tau), \mathscr{D}^*(\tau)) \\
&\cdot \operatorname{vec}(\breve{\mathscr{G}}_{11}(\tau, \breve{\mathscr{D}}_{11}^*(\tau), \mathscr{H}_{11}(\tau), K_z^*(\tau), l_z^*(\tau))) \\
&+ \frac{\partial}{\partial \operatorname{vec}(\mathscr{Z})}\mathscr{W}(\tau, \mathscr{H}_{11}^*(\tau), \breve{\mathscr{D}}_{11}^*(\tau), \mathscr{D}^*(\tau)) \\
&\cdot \operatorname{vec}(\mathscr{G}(\tau, \mathscr{H}_{11}^*(\tau), \mathscr{H}_{12}^*(\tau), \mathscr{H}_{21}^*(\tau), \mathscr{H}_{22}^*(\tau), l_z^*(\tau)))
\end{aligned}
\tag{8.66}
$$

then K_z^* *and* l_z^* *are optimal feedback and feedforward parameters. Moreover*

$$
\mathscr{W}(\varepsilon, \mathscr{Y}_{11}, \breve{\mathscr{Z}}_{11}, \mathscr{Z}) = \mathscr{V}(\varepsilon, \mathscr{Y}_{11}, \breve{\mathscr{Z}}_{11}, \mathscr{Z}) \tag{8.67}
$$

where $\mathscr{V}(\varepsilon, \mathscr{Y}_{11}, \breve{\mathscr{Z}}_{11}, \mathscr{Z})$ *is the value function.*

Proof. It is already contained in the rigorous proof of the sufficiency with the essential conditions aforementioned from [10].

8.4 The Quest for Risk-Averse Control Solutions

In this section, it can be seen at this early point in the algorithmic development that the Mayer-form verification theorem of dynamic programming given in [9] is used to approach the optimal control of Mayer form. In particular, it is required to denote the terminal time and states of a family of optimization problems by $(\varepsilon, \mathscr{Y}_{11}, \mathscr{Y}_{12}, \mathscr{Y}_{21}, \mathscr{Y}_{22}, \breve{\mathscr{Z}}_{11} \mathscr{Z})$ rather than $(t_f, \mathscr{H}_{11}^f, \mathscr{H}_{12}^f, \breve{\mathscr{H}}_{21}^f, \mathscr{H}_{22}^f, \breve{\mathscr{D}}_{11}^f, \mathscr{D}^f)$. As a result, this consideration also suggests that the value of the optimization problem depends on the terminal conditions. For instance, for any $\varepsilon \in [t_0, t_f]$, the state variables of the dynamical equations (8.57)–(8.63) are further expressed as $\mathscr{H}_{11}(\varepsilon) = \mathscr{Y}_{11}$, $\mathscr{H}_{12}(\varepsilon) = \mathscr{Y}_{12}$, $\mathscr{H}_{21}(\varepsilon) = \mathscr{Y}_{21}$, $\mathscr{H}_{22}(\varepsilon) = \mathscr{Y}_{22}$, $\breve{\mathscr{D}}_{11}(\varepsilon) = \breve{\mathscr{Z}}_{11}$, and $\mathscr{D}(\varepsilon) = \mathscr{Z}$.

As all would be expected by the quadratic-affine nature of (8.64), a candidate solution to the HJB equation (8.65) is developed as follows: For instance, let $k \in \mathbb{Z}^+$

and $\left(\varepsilon, \mathscr{Y}_{11}, \mathscr{Y}_{12}, \mathscr{Y}_{21}, \mathscr{Y}_{22}, \breve{\mathscr{Z}}_{11}, \mathscr{Z}\right)$ be any interior point of the reachable set $\mathscr{Q}$. A the real-valued function $\mathscr{W}\left(\varepsilon, \mathscr{Y}_{11}, \breve{\mathscr{Z}}_{11}, \mathscr{Z}\right)$ described by

$$\mathscr{W}(\varepsilon, \mathscr{Y}_{11}, \breve{\mathscr{Z}}_{11}, \mathscr{Z}) = z_0^T \sum_{r=1}^{k} \mu_r(\mathscr{Y}_{11}^r + \mathscr{E}_{11}^r(\varepsilon))z_0$$

$$+2z_0^T \sum_{r=1}^{k} \mu_r(\breve{\mathscr{Z}}_{11}^r + \breve{\mathscr{T}}_{11}^r(\varepsilon)) + \sum_{r=1}^{k} \mu_r(\mathscr{Z}^r + \mathscr{T}^r(\varepsilon)) \quad (8.68)$$

is differentiable. The time parametric functions $\mathscr{E}_{11}^r \in \mathscr{C}^1([t_0, t_f]; \mathbb{R}^{(n+m+c)\times(n+m+c)})$, $\breve{\mathscr{T}}_{11}^r \in \mathscr{C}^1([t_0, t_f]; \mathbb{R}^{n+m+c})$, and $\mathscr{T}^r \in \mathscr{C}^1([t_0, t_f]; \mathbb{R})$ are yet to be determined.

According to the result of [10], the derivative of $\mathscr{W}\left(\varepsilon, \mathscr{Y}_{11}, \breve{\mathscr{Z}}_{11}, \mathscr{Z}\right)$ with respect to ε is obtained as follows

$$\frac{d}{d\varepsilon}\mathscr{W}(\varepsilon, \mathscr{Y}_{11}, \mathscr{Y}_{12}, \mathscr{Y}_{22}, \mathscr{Z}) = z_0^T \sum_{r=1}^{k} \mu_r(\mathscr{F}_{11}^r(\varepsilon, \mathscr{Y}_{11}, \mathscr{Y}_{12}, \mathscr{Y}_{21}, K_z) + \frac{d}{d\varepsilon}\mathscr{E}_{11}^r(\varepsilon))z_0$$

$$+ 2z_0^T \sum_{r=1}^{k} \mu_r(\breve{\mathscr{G}}_{11}^r\left(\varepsilon, \mathscr{Y}_{11}, \breve{\mathscr{Z}}_{11}, K_z, l_z\right) + \frac{d}{d\varepsilon}\breve{\mathscr{T}}_{11}^r(\varepsilon))$$

$$+ \sum_{r=1}^{k} \mu_r(\mathscr{G}^r\left(\varepsilon, \mathscr{Y}_{11}, \mathscr{Y}_{12}, \mathscr{Y}_{21}, \mathscr{Y}_{22}, \breve{\mathscr{Z}}_{11}, l_z\right) + \frac{d}{d\varepsilon}\mathscr{T}^r(\varepsilon)) \quad (8.69)$$

provided that the admissible 2-tuple $(K_z, l_z) \in \overline{K}_z \times \overline{L}_z$.

The contribution of the results (8.68) and (8.69) to the HJB equation (8.65) is dominant in the following way; e.g.,

$$\min_{(K_z, l_z)\in\overline{K}_z\times\overline{L}_z} \Big\{ z_0^T \sum_{r=1}^{k} \mu_r(\mathscr{F}_{11}^r(\varepsilon, \mathscr{Y}_{11}, \mathscr{Y}_{12}, \mathscr{Y}_{21}, K_z) + \frac{d}{d\varepsilon}\mathscr{E}_{11}^r(\varepsilon))z_0$$

$$+ 2z_0^T \sum_{r=1}^{k} \mu_r(\breve{\mathscr{G}}_{11}^r(\varepsilon, \mathscr{Y}_{11}, \breve{\mathscr{Z}}_{11}, K_z, l_z) + \frac{d}{d\varepsilon}\breve{\mathscr{T}}_{11}^r(\varepsilon))$$

$$+ \sum_{r=1}^{k} \mu_r(\mathscr{G}^r(\varepsilon, \mathscr{Y}_{11}, \mathscr{Y}_{12}, \mathscr{Y}_{21}, \mathscr{Y}_{22}, \breve{\mathscr{Z}}_{11}, l_z) + \frac{d}{d\varepsilon}\mathscr{T}^r(\varepsilon))\Big\} \equiv 0. \quad (8.70)$$

In return, differentiating the expression within the bracket (8.70) with respect to K_z and l_z yields the necessary conditions for an interior extremum of the performance index with risk consequences (8.64) on $[0, t_f]$. In other words, the extremizing K_z and l_z must be

$$K_z(\varepsilon, \mathscr{Y}_{11}) = -R_z^{-1}(\varepsilon)B_z^T(\varepsilon)\sum_{r=1}^{k}\hat{\mu}_r\mathscr{Y}_{11}^r \tag{8.71}$$

$$l_z(\varepsilon, \breve{\mathscr{Z}}_{11}) = -R_z^{-1}(\varepsilon)B_z^T(\varepsilon)\sum_{r=1}^{k}\hat{\mu}_r\breve{\mathscr{Z}}_{11}^r, \qquad \hat{\mu}_r = \mu_r/\mu_1. \tag{8.72}$$

In view of (8.71) and (8.72), the zero value of the expression inside of the bracket of (8.70) that is pursued here for any $\varepsilon \in [0, t_f]$ when $\mathscr{Y}_{11}^r$, $\breve{\mathscr{Z}}_{11}^r$ and $\mathscr{Z}^r$ evaluated at the time-backward differential equations (8.57)–(8.63) requires

$$\frac{d}{d\varepsilon}\mathscr{E}_{11}^1(\varepsilon) = (A_z(\varepsilon) + B_z(\varepsilon)K_z(\varepsilon))^T\mathscr{H}_{11}^1(\varepsilon) + \mathscr{H}_{11}^1(\varepsilon)(A_z(\varepsilon) + B_z(\varepsilon)K_z(\varepsilon)) + K_z^T(\varepsilon)R_z(\varepsilon)K_z(\varepsilon) + Q_z(\varepsilon) \tag{8.73}$$

$$\begin{aligned}\frac{d}{d\varepsilon}\mathscr{E}_{11}^r(\varepsilon) &= (A_z(\varepsilon) + B_z(\varepsilon)K_z(\varepsilon))^T\mathscr{H}_{11}^r(\varepsilon) + \mathscr{H}_{11}^r(\varepsilon)(A_z(\varepsilon) + B_z(\varepsilon)K_z(\varepsilon))\\ &+ \sum_{\nu=1}^{r-1}\frac{2r!}{\nu!(r-\nu)!}(\mathscr{H}_{11}^\nu(\varepsilon)\Pi_1(\varepsilon) + \mathscr{H}_{12}^\nu(\varepsilon)\Pi_3(\varepsilon))\mathscr{H}_{11}^{r-\nu}(\varepsilon)\\ &+ \sum_{\nu=1}^{r-1}\frac{2r!}{\nu!(r-\nu)!}(\mathscr{H}_{11}^\nu(\varepsilon)\Pi_2(\varepsilon) + \mathscr{H}_{12}^\nu(\varepsilon)\Pi_4(\varepsilon))\mathscr{H}_{21}^{r-\nu}(\varepsilon)\end{aligned} \tag{8.74}$$

$$\frac{d}{d\varepsilon}\breve{\mathscr{T}}_{11}^1(\varepsilon) = (A_z(\varepsilon) + B_z(\varepsilon)K_z(\varepsilon))^T\breve{\mathscr{D}}_{11}^1(\varepsilon) + K_z^T(\varepsilon)R_z(\varepsilon)l_z(\varepsilon) + \mathscr{H}_{11}^1(\varepsilon)(B_z(\varepsilon)l_z(\varepsilon) + B_z(\varepsilon)u_r(\varepsilon) + E_z(\varepsilon)d(\varepsilon)) \tag{8.75}$$

$$\frac{d}{d\varepsilon}\breve{\mathscr{T}}_{11}^r(\varepsilon) = \mathscr{H}_{11}^r(\varepsilon)(B_z(\varepsilon)l_z(\varepsilon) + B_z(\varepsilon)u_r(\varepsilon) + E_z(\varepsilon)d(\varepsilon)) + (A_z(\varepsilon) + B_z(\varepsilon)K_z(\varepsilon))^T\breve{\mathscr{D}}_{11}^r(\varepsilon), \quad 2 \le r \le k \tag{8.76}$$

$$\begin{aligned}\frac{d}{d\epsilon}\mathscr{T}_1(\varepsilon) &= \mathrm{Tr}\{\mathscr{H}_{11}^1(\varepsilon)\Pi_1(\varepsilon) + \mathscr{H}_{12}^1(\varepsilon)\Pi_3(\varepsilon)\}\\ &+ \mathrm{Tr}\{\mathscr{H}_{21}^1(\varepsilon)\Pi_2(\varepsilon) + \mathscr{H}_{22}^1(\varepsilon)\Pi_4(\varepsilon)\} + l_z^T(\varepsilon)R_z(\varepsilon)l_z(\varepsilon)\\ &+ 2(\breve{\mathscr{D}}_{11}^1)^T(\varepsilon)(B_z(\varepsilon)l_z(\varepsilon) + B_z(\varepsilon)u_r(\varepsilon) + E_z(\varepsilon)d(\varepsilon))\end{aligned} \tag{8.77}$$

$$\begin{aligned}\frac{d}{d\varepsilon}\mathscr{T}_r(\varepsilon) &= \mathrm{Tr}\{\mathscr{H}_{11}^r(\varepsilon)\Pi_1(\varepsilon) + \mathscr{H}_{12}^r(\varepsilon)\Pi_3(\varepsilon)\}\\ &+ 2(\breve{\mathscr{D}}_{11}^r)^T(\varepsilon)(B_z(\varepsilon)l_z(\varepsilon) + B_z(\varepsilon)u_r(\varepsilon) + E_z(\varepsilon)d(\varepsilon))\\ &+ \mathrm{Tr}\{\mathscr{H}_{21}^r(\varepsilon)\Pi_2(\varepsilon) + \mathscr{H}_{22}^r(\varepsilon)\Pi_4(\varepsilon)\}, \quad 2 \le r \le k.\end{aligned} \tag{8.78}$$

At the same time, the boundary condition of $\mathscr{W}(\varepsilon, \mathscr{Y}_{11}, \breve{\mathscr{Z}}_{11}, \mathscr{Z})$ implies that the initial-value conditions $\mathscr{E}_{11}^r(0) = 0$, $\breve{\mathscr{T}}_{11}^r(0) = 0$, and $\mathscr{T}^r(0) = 0$ for the forward-in-time differential equations (8.73)–(8.78) and yields a value function

$$\mathscr{W}(\varepsilon, \mathscr{Y}_{11}, \breve{\mathscr{Z}}_{11}, \mathscr{Z}) = \mathscr{V}(\varepsilon, \mathscr{Y}_{11}, \breve{\mathscr{Z}}_{11}, \mathscr{Z})$$

$$= z_0^T \sum_{r=1}^{k} \mu_r \mathscr{H}_{11}^r(t_0) z_0 + 2z_0^T \sum_{r=1}^{k} \mu_r \breve{\mathscr{D}}_{11}^r(t_0) + \sum_{r=1}^{k} \mu_r \mathscr{D}^r(t_0)$$

for which the sufficient condition (8.66) of the verification theorem is satisfied so that the extremizing feedback parameters (8.71) and (8.72) become optimal

$$K_z^*(\varepsilon) = -R_z^{-1}(\varepsilon) B_z^T(\varepsilon) \sum_{r=1}^{k} \hat{\mu}_r \mathscr{H}_{11}^{*r}(\varepsilon)$$

$$l_z^*(\varepsilon) = -R_z^{-1}(\varepsilon) B_z^T(\varepsilon) \sum_{r=1}^{k} \hat{\mu}_r \breve{\mathscr{D}}_{11}^{*r}(\varepsilon).$$

Such encouraging results treat the development for resilient controls as based on the respect for preferential measures of risk in relation to effective uncertainty management that is now summarized for the class of time-delay networked stochastic systems with bandlimited control channel constraints as depicted in Fig. 8.6.

Theorem 8.4.1 (Resilient Control for Networked Stochastic Systems). *Under the assumptions of (A_z, B_z) uniformly stabilizable and (C_z, A_z) uniformly detectable, the networked control system governed by (8.19) is subject to the chi-squared measure of performance (8.20). Suppose $k \in \mathbb{Z}^+$ and the sequence $\mu = \{\mu_r \geq 0\}_{r=1}^k$ with $\mu_1 > 0$ are fixed. Then, the statistical optimal control law over $[t_0, t_f]$ is a two-degrees-of-freedom controller with a soft constraint pre-specified*

$$u_g^*(t) = K_z^*(t)\hat{z}^*(t) + l_z^*(t) + u_r(t), \quad t = t_f - \tau \tag{8.79}$$

$$K_z^*(\tau) = -R_z^{-1}(\tau) B_z^T(\tau) \sum_{r=1}^{k} \hat{\mu}_r \mathscr{H}_{11}^{r*}(\tau) \tag{8.80}$$

$$l_z^*(\tau) = -R_z^{-1}(\tau) B_z^T(\tau) \sum_{r=1}^{k} \hat{\mu}_r \breve{\mathscr{D}}_{11}^{r*}(\tau) \tag{8.81}$$

where $\hat{\mu}_r = \mu_r / \mu_1$ represent different levels of influence as they deem important to the performance distribution. Accordingly, $\{\mathscr{H}_{11}^{r*}(\tau)\}_{r=1}^k$ and $\left\{\breve{\mathscr{D}}_{11}^{r*}(\tau)\right\}_{r=1}^k$ are the optimal solutions of the backward-in-time differential equations

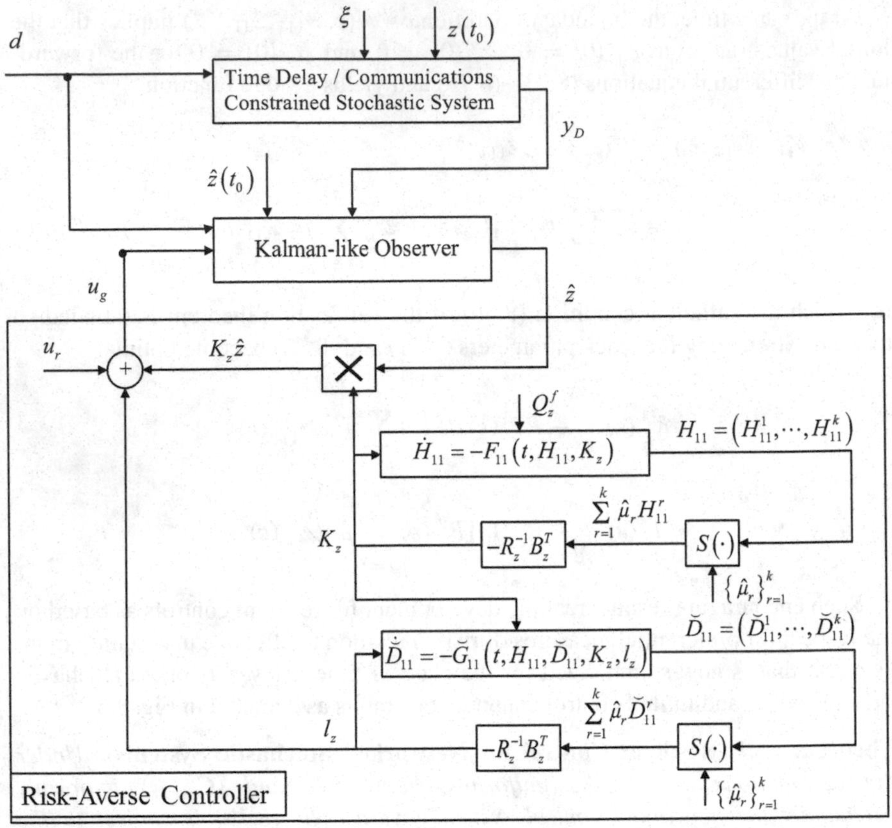

Fig. 8.6 Implementation by the resilient controlled system

$$
\begin{aligned}
\frac{d}{d\tau}\mathscr{H}_{11}^{1*}(\tau) = &- (A_z(\tau)+B_z(\tau)K_z^*(\tau))^T\,\mathscr{H}_{11}^{1*}(\tau) - \mathscr{H}_{11}^{1*}(\tau)(A_z(\tau) + B_z(\tau)K_z^*(\tau)) \\
&- K_z^{*T}(\tau)R_z(\tau)K_z^*(\tau) - Q_z(\tau)
\end{aligned} \tag{8.82}
$$

$$
\begin{aligned}
\frac{d}{d\tau}\mathscr{H}_{11}^{r*}(\tau) = &- (A_z(\tau)+B_z(\tau)K_z^*(\tau))^T\,\mathscr{H}_{11}^{r*}(\tau) - \mathscr{H}_{11}^{r*}(\tau)(A_z(\tau) + B_z(\tau)K_z^*(\tau)) \\
&- \sum_{\nu=1}^{r-1}\frac{2r!}{\nu!(r-\nu)!}(\mathscr{H}_{11}^{\nu*}(\tau)\Pi_1(\tau) + \mathscr{H}_{12}^{\nu*}(\tau)\Pi_3(\tau))\mathscr{H}_{11}^{r-\nu*}(\tau) \\
&- \sum_{\nu=1}^{r-1}\frac{2r!}{\nu!(r-\nu)!}(\mathscr{H}_{11}^{\nu*}(\tau)\Pi_2(\tau) + \mathscr{H}_{12}^{\nu*}(\tau)\Pi_4(\tau))\mathscr{H}_{21}^{r-\nu*}(\tau)
\end{aligned} \tag{8.83}
$$

and

$$
\begin{aligned}
\frac{d}{d\tau}\breve{\mathcal{D}}_{11}^{1*}(\tau) &= -(A_z(\tau)+B_z(\tau)K_z^*(\tau))^T\breve{\mathcal{D}}_{11}^{1*}(\tau) - K_z^{*T}(\tau)R_z(\tau)l_z^*(\tau) \\
&\quad - \mathcal{H}_{11}^{1*}(\tau)(B_z(\tau)l_z^*(\tau)+B_z(\tau)u_r(\tau)+E_z(\tau)d(\tau)) \qquad (8.84) \\
\frac{d}{d\tau}\breve{\mathcal{D}}_{11}^{r*}(\tau) &= -(A_z(\tau)+B_z(\tau)K_z^*(\tau))^T\breve{\mathcal{D}}_{11}^{r*}(\tau) \\
&\quad - \mathcal{H}_{11}^{r*}(\tau)(B_z(\tau)l_z^*(\tau)+B_z(\tau)u_r(\tau)+E_z(\tau)d(\tau)), \quad 2\le r\le k \qquad (8.85)
\end{aligned}
$$

where the terminal-value conditions $\mathcal{H}_{11}^{1*}(t_f)=Q_z^f$, $\mathcal{H}_{11}^{r*}(t_f)=0$ *for* $2\le r\le k$ *as well as* $\breve{\mathcal{D}}_{11}^{r*}(t_f)=0$ *for* $1\le r\le k$.

8.5 Chapter Summary

This chapter reviewed a number of still open issues pertaining to networked controls that can help characterize how hard, at least on theoretical grounds, it is to design an admissible set of output feedback control strategies. In the case of time-delay measurements and communication channel constraints, such an optimal control policy is feasible and thus, all the time and random events can order uncertain prospects of the chi-squared random cost. In accordance with preferential measures of performance risks, associated is the recently proposed mean-risk aware performance index, which then involves naturally two parameters, one of which is always the mean but the other parameter is the appropriate measures of performance dispersion. As a final remark, it is hoped that the results herein will provide a strong impetus for further research in the development of resilient controls for performance risk aversion while subject to networking effects.

References

1. Tatikonda, S.C.: Control under communication constraints. PhD thesis, Department of Electrical Engineering and Computer Science. MIT, Cambridge (2000)
2. Barrett, G., Lafortune, S.: On the synthesis of communicating controllers with decentralized information structures for discrete-event systems. In: Proceedings of IEEE Conference on Decision and Control, Tamba, pp. 3281–3286 (1998)
3. Barrett, G., Lafortune, S.: Decentralized supervisory control with communicating controllers. IEEE Trans. Autom. Control **45**, 1620–1638 (2000)
4. Mahmoud, M.S.: Resilient Control of Uncertain Dynamical Systems. Lecture Notes in Control and Information Sciences, vol. 303. Springer, Berlin/New York (2004)
5. Pham, K.D.: Linear-Quadratic Controls in Risk-Averse Decision Making: Performance-Measure Statistics and Control Decision Optimization. Springer Briefs in Optimization, ISBN 978-1-4614-5078-8 (2012)

6. Pham, K.D.: Statistical control paradigms for structural vibration suppression. Ph.D. dissertation, Department of Electrical Engineering, University of Notre Dame, Indiana (2004). Available via http://etd.nd.edu/ETD-db/theses/available/etd-04152004-121926/unrestricted/PhamKD052004.pdf.Cited15January2014
7. Hirshleifer, J.: Investment, Interest and Capital. Prentice Hall, Englewood Cliffs (1970)
8. Markowitz, H.: Portfolio Selection: Efficient Diversification of Investments. Wiley, New York (1970)
9. Fleming, W.H., Rishel, R.W.: Deterministic and Stochastic Optimal Control. Springer, New York (1975)
10. Pham, K.D.: Performance-reliability-aided decision-making in multiperson quadratic decision games against jamming and estimation confrontations. In: F. Giannessi (ed.), J. Optim. Theory Appl. **149**(1), 599–629 (2011)

Chapter 9
Risk-Averse Control of Networked Systems with Compensation of Measurement Delays and Control Rate Constraints

9.1 Introduction

With increased emphasis on large complex systems constituted by multi-level interconnections of stochastic dynamical systems, there is a growing and urgent need to automatically control of large-scale interconnected dynamical systems by output feedback with compensation of observation delays and control rate constraints both locally and remotely on a routine basis. The continued advance of processing, enabling tools, and basic technologies in areas including synchronization of networks, decentralized control with communications between controllers and controlled systems, etc. is facilitating remarkable leaps forward in unattended and attended sensors, command, control and communications. There have beeb, as noted, many publications proposing various solutions. By some reports [1], technical issues related to control under communications were investigated. In fact, other critical aspects of various communicating controllers for discrete-event systems were described in [2] and [3].

As of today, absent from discussion were investigations of the means by which stabilization analysis and control optimization for networked control systems were able to go beyond the universal condition of expected values and mean-variance rules in order to rank stochastic dominance for the certain class of performance cost distributions. As shall see in [4, 5], and [6], these traditional approaches are of limited generality and not realistic as they rule out asymmetry or skewness in the probability distributions of the non-normal random costs. Such an evaluation has generated a renewed demand for re-investigating the networked control problem of linear stochastic systems with feedback measurements and control actuation subject to network delays and control rate constraints to a depth of details not seen before; e.g., admissible sets of control laws for ordering uncertain prospects and thereby leveraging complete knowledge of the entire chi-squared distribution associated with the integral-quadratic-form cost over a finite horizon; not just means and/or variances as often can be seen from the existing literature.

K.D. Pham, *Resilient Controls for Ordering Uncertain Prospects*, Springer Optimization and Its Applications 98, DOI 10.1007/978-3-319-08705-4_9

The chapter is organized as follows. Section 9.2 is focused on a problem class of time-invariant linear stochastic systems together with time-delay observations and control rate constraints. In addition, the latest development in characterization and management of uncertain prospects pertaining to the restrictive family of finite-horizon integral quadratic costs is of particular interest. Moreover, Sect. 9.3 contains the problem statements toward the accompanying formulation of optimal rules for ordering uncertain prospects of the chi-squared random costs. Subsequently, the feasibility of time-varying linear memoryless control laws supported by Kalman-like estimator and subject to control rate constraints is put forward, as can be seen in Sect. 9.4. Some final remarks are further concluded in Sect. 9.5.

9.2 Relevance of Performance Statistics for Stochastic Dominance

Without a loss of generality, the development hereafter assumes a fixed probability space $(\Omega, \mathbb{F}, \{\mathbb{F}_{t_0,t} : t \in [t_0, t_f]\}, \mathbb{P})$ with filtration satisfying the usual conditions. Throughout the chapter, all filtrations are right continuous and complete and $\mathbb{F}_{t_f} \triangleq \{\mathbb{F}_{t_0,t} : t \in [t_0, t_f]\}$. In addition, let $\mathcal{L}^2_{\mathbb{F}_{t_f}}([t_0, t_f]; \mathbb{R}^n)$ denote the space of $\mathbb{F}_{t_f}$-adapted random processes $\{\hbar(t) : t \in [t_0, t_f]$ such that $E\{\int_{t_0}^{t_f} ||\hbar(t)||^2 dt\} < \infty\}$.

When modeling a controlled system, different elements; e.g., levels of detail, feedback measurements, time delay effects, control rate constraints, exogenous disturbances, etc. are considered by the control designer of that system. For instance, what follows is the class of stochastic systems to provide a central point for methods and approaches to the optimal control problem at hand

$$\dot{x}(t) = Ax(t) + Bu(t) + w(t), \quad x(t_0) \tag{9.1}$$

whereupon $x(t)$ is the controlled state process valued in $\mathbb{R}^n$; $u(t)$ is the control process valued in a closed convex subset of an appropriate metric space; and $w(t) \in \mathbb{R}^p$ is an exogenous state noise taking the form of a mutually uncorrelated stationary $\mathbb{F}_{t_f}$-adapted Gaussian process with its mean $E\{w(t)\} = m_w(t)$ and covariance $cov\{w(t_1), w(t_2)\} = W\delta(t_1 - t_2)$ for all t_1, $t_2 \in [t_0, t_f]$ and $W > 0$.

In addition, the feedback measurements being shared between the system and controller are governed by the following partial information structure and subject to a known network delay ζ, as shown in Fig. 9.1

$$y(t) = Cx(t - \zeta) + Du(t - \zeta) + v(t - \zeta) \tag{9.2}$$

whereas the exogenous measurement noise $v(t) \in \mathbb{R}^q$ is another mutually uncorrelated stationary $\mathbb{F}_{t_f}$-adapted Gaussian process with its mean $E\{v(t)\} = m_v(t)$ and covariance $cov\{v(t_1), v(t_2)\} = V\delta(t_1 - t_2)$ for all t_1, $t_2 \in [t_0, t_f]$ and $V > 0$.

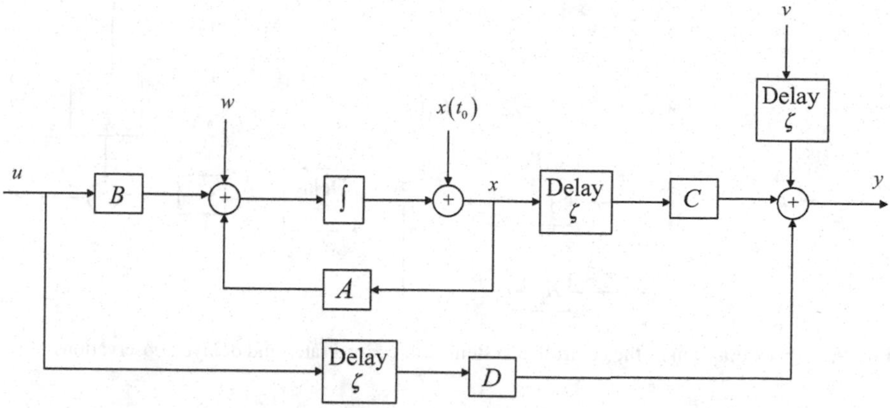

Fig. 9.1 Structure of stochastic controlled system with delayed observations

Not surprisingly, the restricted class of performance measures is of primarily interest, which varies only in a range bounded from below. This is so if, the controlled system cannot be better off without loosing anything. Henceforth, it is assumed that

$$J = x^T(t_f)Q_f x(t_f) + \int_{t_0}^{t_f} [x^T(\tau)Qx(\tau) + u^T(\tau)Ru(\tau) + \dot{u}^T(\tau)S\dot{u}(\tau)]d\tau \tag{9.3}$$

which is grew out of the need to trade offs among state regulation, control efforts and control rate constraints. Of note, the terminal penalty weighting $Q_f \in \mathbb{R}^{n\times n}$, the state regulation weighting $Q \in \mathbb{R}^{n\times n}$, the control effort weighting $R \in \mathbb{R}^{m\times m}$ and the control rate weighting $S \in \mathbb{R}^{m\times m}$ are the continuous-time matrix functions with the properties of symmetry and positive semi-definiteness. In addition, R and S are invertible.

Next the complexity involved in satisfying control and control rate constraints contributes to the development of the augmented system as depicted in Fig. 9.2

$$\dot{z}(t) = A_z z(t) + B_z \dot{u}(t) + w_z(t), \quad z(t_0) \tag{9.4}$$

$$y(t) = C_z z(t-\zeta) + v(t-\zeta) \tag{9.5}$$

subject to

$$J = z^T(t_f)Q_z^f z(t_f) + \int_{t_0}^{t_f} [z^T(\tau)Q_z z(\tau) + \dot{u}^T(\tau)S\dot{u}(\tau)]d\tau \tag{9.6}$$

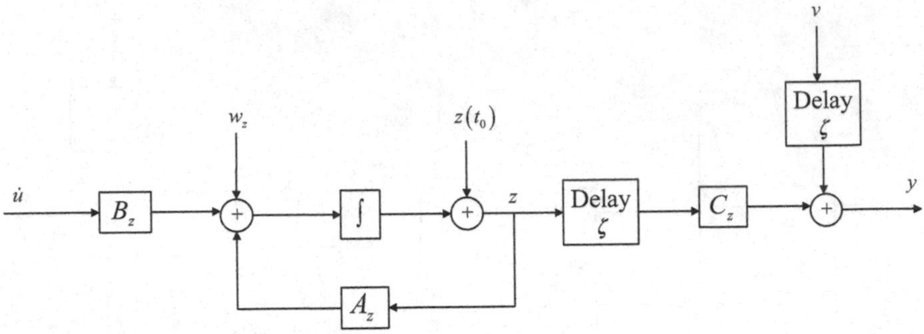

Fig. 9.2 Block diagram of the controlled system with control rates and delayed observations

from which the corresponding state variables and system coefficients are given by

$$z(t) \triangleq \begin{bmatrix} x(t) \\ u(t) \end{bmatrix}; z(t_0) \triangleq \begin{bmatrix} x(t_0) \\ 0_{m\times 1} \end{bmatrix}; w_z(t) \triangleq \begin{bmatrix} w(t) \\ 0_{m\times 1} \end{bmatrix}; m_{w_z}(t) \triangleq \begin{bmatrix} m_w(t) \\ 0 \end{bmatrix}; W_z \triangleq \begin{bmatrix} W & 0 \\ 0 & 0 \end{bmatrix}$$

$$A_z \triangleq \begin{bmatrix} A & B \\ 0 & 0 \end{bmatrix}; \ B_z \triangleq \begin{bmatrix} 0_{n\times m} \\ I_{m\times m} \end{bmatrix}; \ C_z \triangleq \begin{bmatrix} C & D \end{bmatrix}; \ Q_z^f \triangleq \begin{bmatrix} Q_f & 0 \\ 0 & 0 \end{bmatrix}; \ Q_z \triangleq \begin{bmatrix} Q & 0 \\ 0 & R \end{bmatrix}$$

whereas the aggregated process noise is characterized by $E\{w_z(t)\} = m_{w_z}(t)$ and $cov\{w_z(t_1), w_z(t_2)\} = W_z\delta(t_1 - t_2)$ for all $t_1, t_2 \in [t_0, t_f]$.

In view of (9.4)–(9.5), both mutually uncorrelated stationary white noises $\{w_z(t) : t \in [t_0, t_f]\}$ and $\{v(t) : t \in [t_0, t_f]\}$ are traditionally interpreted as the formal time derivatives of Brownian motions adapted to $\mathbb{F}_{t_f}$. In essence, one has a way to treat them more mathematically rigorous as the correspondent uncorrelated stationary Wiener random processes

$$dw_z(t) \triangleq \tilde{w}_z(t)dt = (w_z(t) - m_{w_z}(t))dt$$

$$dv(t) \triangleq \tilde{v}(t)dt = (v(t) - m_v(t))dt$$

whose the singular-value decompositions of $W_z \triangleq U_{W_z}\Lambda_{W_z}^{1/2}U_{W_z}^T$ and $V \triangleq U_V\Lambda_V^{1/2}U_V^T$.

In this Ito stochastic differential interpretation as illustrated in Fig. 9.3, the stochastic system dynamics (9.4)–(9.5) is equivalently described by

$$dz(t) = (A_z z(t) + B_z\dot{u}(t) + m_{w_z}(t))dt + U_{W_z}\Lambda_{W_z}^{1/2}dw_z(t), \quad z(t_0) \tag{9.7}$$

$$dy(t) = (C_z z(t-\zeta) + m_v(t-\zeta))dt + U_V\Lambda_V^{1/2}dv(t-\zeta), \tag{9.8}$$

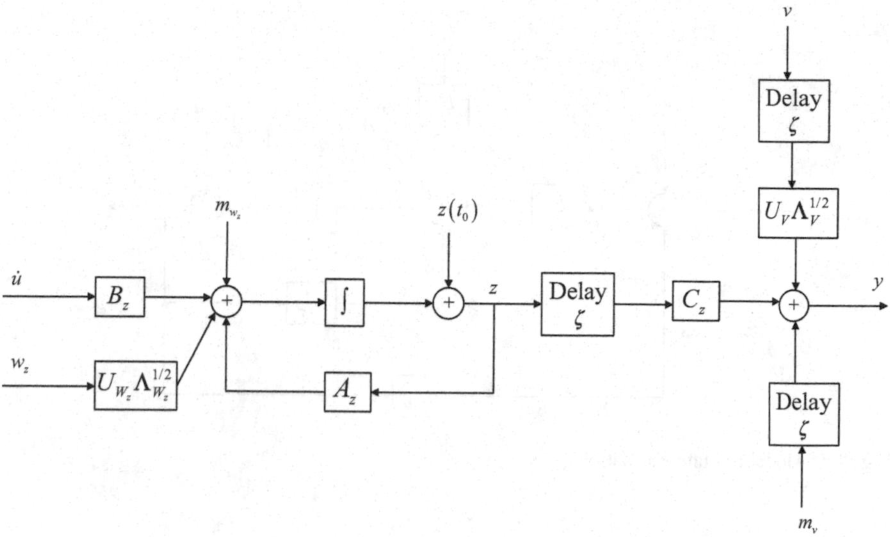

Fig. 9.3 Implementation diagram illustrating Brown motions of white noises

Besides the σ-algebras

$$\mathbb{F}_{t_0,t} \triangleq \sigma\{(w_z(\tau), v(\tau)) : t_0 \le \tau \le t\}$$

$$\mathcal{G}^y_{t_0,t} \triangleq \sigma\{y(\tau) : t_0 \le \tau \le t\}, \quad t \in [t_0, t_f]$$

defining the information available $\mathcal{G}^y_{t_f} \triangleq \{\mathcal{G}^y_{t_0,t} : t \in [t_0, t_f]\} \subset \{\mathbb{F}_{t_0,t} : t \in [t_0, t_f]\}$, there must be an appropriate delayed state estimator $\hat{z}(t-\zeta) \triangleq E\{z(t-\zeta)|\mathcal{G}^y_{t_f}\}$, which is the least mean-squared estimate of $z(t-\zeta)$ conditioned on the delayed observation $\mathcal{G}^y_{t_f}$ and is preserving the inherent linear Gaussian structure of (9.7)–(9.8) and the issue of network delay ζ. Of note, it is assumed that the estimate $\hat{z}(t-\zeta)$ is obtained directly from $y(t)$ by means of a Kalman filter, as shown in Fig. 9.4, which is now modified to include the network delay effect; e.g.,

$$d\hat{z}(t-\zeta) = (A_z\hat{z}(t-\zeta) + B_z\dot{u}(t-\zeta) + m_{w_z}(t-\zeta))dt + L_z(t-\zeta)(dy(t) - (C_z\hat{z}(t-\zeta) + m_v(t-\zeta)))dt, \quad \hat{z}(t_0) = z(t_0) \tag{9.9}$$

where the filtering gain, $L_z(t)$ and state-estimate error covariance, $\Sigma_z(t) \triangleq E\{[z(t)-\hat{z}(t)][z(t)-\hat{z}(t)]^T|\mathcal{G}^y_{t_f}\}$ are given by

$$L_z(t) = \Sigma_z(t)C_z^T V^{-1} \tag{9.10}$$

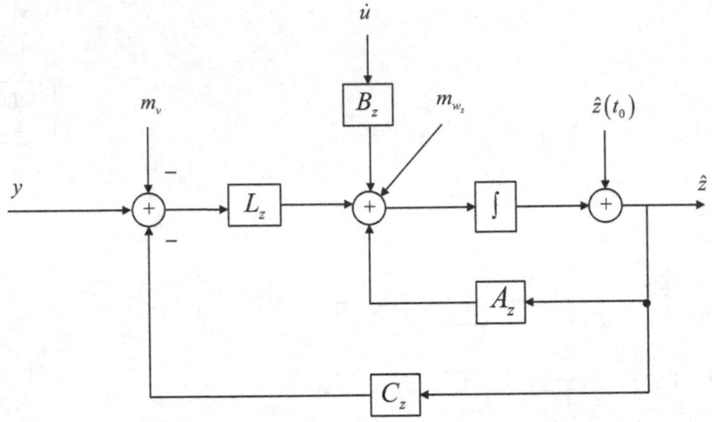

Fig. 9.4 Model of state estimation

$$\frac{d}{dt}\Sigma_z(t) = A_z\Sigma_z(t) + \Sigma_z(t)A_z^T - \Sigma_z(t)C_z^T V^{-1}C_z\Sigma_z(t) + W_z; \quad \Sigma_z(t_0) = 0. \tag{9.11}$$

As well known, in this case, the state-estimate errors $\tilde{z}(t) \triangleq z(t) - \hat{z}(t)$ are satisfying the stochastic differential equation with the initial-value condition $\tilde{z}(t_0) = 0$

$$d\tilde{z}(t) = (A_z - L_z(t)C_z)\tilde{z}(t)dt + U_{W_z}\Lambda_{W_z}^{1/2}dw_z(t) - L_z(t)U_V\Lambda_V^{1/2}dv(t), \tag{9.12}$$

Note, however, that at time t the Kalman filter (9.9) generates a least mean-squared estimate of $\hat{z}(t-\zeta)$ of the delayed state $z(t-\zeta)$. Therefore, the control rate process $\dot{u}(t)$ cannot be generated from $\hat{z}(t)$ but only from $\hat{z}(\sigma)$ for $\sigma \leq t - \zeta$. Consequently, the control optimization problem is thus seen to be that of selecting the control rate $\dot{u}(t)$ to optimize performance risks associated with the chi-squared random cost (9.6) and subject to the constraints (9.9) and (9.12).

In an associated work [7], it has been shown that the feedback control $u(t)$ and its rate $\dot{u}(t)$ for all $t \in [t_0, t_f]$ can be obtained from the current state estimate $\hat{z}(t)$ by the cascade combination of the Kalman estimator and a least mean-squared predictor. In particular, a predictor as illustrated in Fig. 9.5 that generates the best estimate $\hat{z}(t)$ of the current system state $\hat{z}(t-\zeta)$ (also known as best estimate of the delayed state), is now proposed of the form

$$\hat{z}(t) = \iota(t) + e^{A_z\zeta}[\hat{z}(t-\zeta) - \iota(t-\zeta)] \tag{9.13}$$

$$\dot{\iota}(t) = A_z\iota(t) + B_z\dot{u}(t) + m_{w_z}(t), \tag{9.14}$$

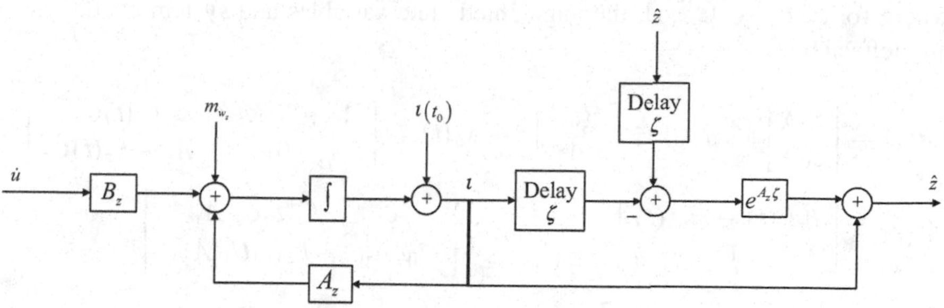

Fig. 9.5 Implementation by least mean-square predictor

As already mentioned, the information structure $\mathscr{G}_{t_f}^y$ is defined by a memory feedback via the stochastic differential equation (9.8). Consequently, the admissible set of feedback control rate laws therefore reduces to

$$\mathbb{U}^y[t_0, t_f] \triangleq \{\dot{u} \in \mathscr{L}^2_{\mathscr{G}^y_{t_f}}([t_0, t_f], \mathbb{R}^m), \mathbb{P} - a.s.\}$$

where $\mathbb{U}^y[t_0, t_f]$ is a closed convex subset of $\mathscr{L}^2_{\mathbb{F}_{t_f}}([t_0, t_f], \mathbb{R}^m)$.

Up to this point, the sequel development will adopt a slight abuse of notation for the time variable t, e.g., it is now meant for $t - \zeta$ for simplicity. Furthermore, to fulfil the expectations of such sufficient statistics for describing the conditional stochastic effects of control actions, the conditional probability density $p(z(t)|\mathscr{G}^y_{t_f})$ has to realize its complete knowledge simultaneously into two natural parameters: $\hat{z}(t)$ is to do with the conditional mean, the other $\Sigma_z(t)$ relates to the state-estimate error covariance and is independent with the control inputs and future feedback observations. With certain restrictions on the class of linear stochastic systems (9.7)–(9.8) and quadratic performance measure (9.9), many reasonable approximations and practical matters derive from the family of $\dot{u}(t) = \gamma(t, \hat{z}(t))$ which is further reduced to a simple manageable form; e.g., a linear time-varying feedback law generated from the accessible states $\hat{z}(t)$ by

$$\dot{u}(t) = K_z(t)\hat{z}(t) + l_z(t), \quad \forall\, t \in [t_0, t_f] \tag{9.15}$$

with the admissible gains $K_z \in \mathscr{C}([t_0, t_f]; \mathbb{R}^{m\times(n+m)})$ and $l_z \in \mathscr{C}([t_0, t_f]; \mathbb{R}^m)$ to which further defining properties will be defined subsequently.

Under the assumption of Lipschitz and linear growth conditions, uniformly in $t \in [t_0, t_f]$, the controlled system (9.7)–(9.8) has a unique $\mathbb{F}_{t_f}$-adapted continuous solution. In view of the control rate policy (9.15), the controlled system (9.7)–(9.8) is accordingly rewritten as

$$dq(t) = (A_q(t)q(t) + b_q(t))dt + G_q(t)dw_q(t), \quad q(t_0) \tag{9.16}$$

where for each $t \in [t_0, t_f]$, the augmented state variables and system coefficients are defined by

$$q(t) \triangleq \begin{bmatrix} \hat{z}(t) \\ \tilde{z}(t) \end{bmatrix}; \quad q(t_0) \triangleq \begin{bmatrix} z(t_0) \\ 0 \end{bmatrix}; \quad A_q(t) \triangleq \begin{bmatrix} A_z + B_z K_z(t) & L_z(t)C_z \\ 0 & A_z - L_z(t)C_z \end{bmatrix}$$

$$b_q(t) \triangleq \begin{bmatrix} B_z l_z(t) + m_{w_z}(t) \\ 0 \end{bmatrix}; \quad G_q(t) \triangleq \begin{bmatrix} 0 & L_z(t)U_V \Lambda_V^{1/2} \\ U_{W_z}\Lambda_{W_z}^{1/2} & -L_z(t)U_V \Lambda_V^{1/2} \end{bmatrix}$$

$$W_q \triangleq \begin{bmatrix} I_{(m+n)\times(m+n)} & 0 \\ 0 & I_{q\times q} \end{bmatrix}; E\{[w_q(t_1)-w_q(t_2)][w_q(t_1)-w_q(t_2)]^T\}=W_q|t_1-t_2|.$$

Similarly, the performance measure (9.9) which also incorporates new state variables, $q(t)$ is the quadratic utility function rewritten by

$$J = q^T(t_f)Q_q^f q(t_f) + \int_{t_0}^{t_f} [q^T(\tau)Q_q(\tau)q(\tau) + 2q^T(\tau)S_q(\tau) + l_z^T(\tau)Sl_z(\tau)]d\tau \tag{9.17}$$

where the continuous-time matrix-valued weightings are given by

$$Q_q(\tau) \triangleq \begin{bmatrix} Q_z + K_z^T(\tau)SK_z(\tau) & Q_z \\ Q_z & Q_z \end{bmatrix}; \quad Q_q^f \triangleq \begin{bmatrix} Q_z^f & Q_z^f \\ Q_z^f & Q_z^f \end{bmatrix}; \quad S_q(\tau) \triangleq \begin{bmatrix} K_z^T(\tau)Sl_z(\tau) \\ 0 \end{bmatrix}.$$

Till now, the closed-loop performance (9.17) pertaining to the problem of networked control with control rate constraints, is better understood in the generalized chi-squared nature of a random variable. Because asymmetry or skewness in the probability distribution is rarely considered to be a part of the stochastic control community, many respects of mean, variance, skewness, etc. and thereby representing closed-loop distributions of the chi-squared random cost will be advocated as effective statistical measures of performance risks. With this purpose in mind, the development of statistical optimal control theory herein continues to support the contention of performance risk aversion in feedback control synthesis.

Theorem 9.2.1 (Cumulant-Generating Function). *Let $q(\cdot)$ be a state variable of the controlled system (9.16) with the initial-value condition $q(\tau) \equiv q_\tau$ and $\tau \in [t_0, t_f]$. Further let the moment-generating function be*

$$\varphi(\tau, q_\tau, \theta) \triangleq \varrho(\tau, \theta) \exp\{q_\tau^T \Upsilon(\tau, \theta) q_\tau + 2q_\tau^T \eta(\tau, \theta)\} \tag{9.18}$$

$$\upsilon(\tau, \theta) \triangleq \ln\{\varrho(\tau, \theta)\}, \qquad \theta \in \mathbb{R}^+, \tag{9.19}$$

Then, the cumulant-generating function has the form of quadratic affine

$$\psi(\tau, q_\tau, \theta) = q_\tau^T \Upsilon(\tau, \theta) q_\tau + 2q_\tau^T \eta(\tau, \theta) + \upsilon(\tau, \theta) \tag{9.20}$$

where the scalar solution $\upsilon(\tau,\theta)$ solves the backward-in-time differential equation

$$\frac{d}{d\tau}\upsilon(\tau,\theta) = -\operatorname{Tr}\left\{\Upsilon(\tau,\theta)G_q(\tau)W_qG_q^T(\tau)\right\} - 2\eta^T(\tau,\theta)b_q(\tau) - \theta l_z^T(\tau)Sl_z(\tau), \quad \upsilon(t_f,\theta) = 0, \tag{9.21}$$

the matrix solution $\Upsilon(\tau,\theta)$ satisfies the backward-in-time differential equation

$$\frac{d}{d\tau}\Upsilon(\tau,\theta) = -A_q^T(\tau)\Upsilon(\tau,\theta) - \Upsilon(\tau,\theta)A_q(\tau) - 2\Upsilon(\tau,\theta)G_q(\tau)W_qG_q^T(\tau)\Upsilon(\tau,\theta) - \theta Q_q(\tau), \quad \Upsilon(t_f,\theta) = \theta Q_q^f, \tag{9.22}$$

and the vector solution $\eta(\tau,\theta)$ satisfies the backward-in-time differential equation

$$\frac{d}{d\tau}\eta(\tau,\theta) = -A_q^T(\tau)\eta(\tau,\theta) - \Upsilon(\tau,\theta)b_q(\tau) - \theta S_q(\tau), \quad \eta(t_f,\theta) = 0, \tag{9.23}$$

Also, the scalar solution $\varrho(\tau,\theta)$ satisfies the time-backward differential equation

$$\frac{d}{d\tau}\varrho(\tau,\theta) = -\varrho(\tau,\theta)[\operatorname{Tr}\left\{\Upsilon(\tau,\theta)G_q(\tau)W_qG_q^T(\tau)\right\} + 2\eta^T(\tau,\theta)b_q(\tau) + \theta l_z^T(\tau)Sl_z(\tau)], \quad \varrho(t_f,\theta) = 1, \tag{9.24}$$

Proof. Typically, the moment-generating function $\varpi(\tau,q_\tau,\theta) \triangleq e^{\{\theta J(\tau,q_\tau)\}}$ is accompanied by the cost-to-go function of (9.17) from an arbitrary state q_τ at a running time $\tau \in [t_0,t_f]$, that is,

$$J(\tau,q_\tau) = q^T(t_f)Q_q^f q(t_f) + \int_\tau^{t_f} [q^T(t)Q_q(t)q(t) + 2q^T(t)S_q(t) + l_z^T(t)Sl_z(t)]dt \tag{9.25}$$

subject to

$$dq(t) = (A_q(t)q(t) + b_q(t))dt + G_q(t)dw_q(t), \quad q(\tau) = q_\tau, \tag{9.26}$$

By definition, the moment-generating function is

$$\varphi(\tau,q_\tau,\theta) \triangleq E\{\varpi(\tau,q_\tau,\theta)\},$$

Subsequently, the total time derivative of $\varphi(\tau, q_\tau, \theta)$ is obtained as

$$dd\tau\varphi(\tau, q_\tau, \theta) = -\theta[q_\tau^T Q_q(\tau) q_\tau + 2q_\tau^T S_q(\tau) + l_z^T(\tau) S l_z(\tau)]\varphi(\tau, q_\tau, \theta),$$

Meanwhile, an application of the standard formula of Ito calculus results in

$$\begin{aligned}
d\varphi(\tau, q_\tau, \theta) &= E\{d\varpi(\tau, q_\tau, \theta)\} \\
&= E\Big\{\varpi_\tau(\tau, q_\tau, \theta)\,d\tau + \varpi_{q_\tau}(\tau, q_\tau, \theta)\,dq_\tau \\
&\quad + \frac{1}{2}\mathrm{Tr}\Big\{\varpi_{q_\tau q_\tau}(\tau, q_\tau, \theta) G_q(\tau) W_q G_q^T(\tau)\Big\}\,d\tau\Big\} \\
&= \varphi_\tau(\tau, q_\tau; \theta)d\tau + \varphi_{q_\tau}(\tau, q_\tau, \theta)(A_q(\tau)q_\tau + b_q(\tau))d\tau \\
&\quad + \frac{1}{2}\mathrm{Tr}\Big\{\varphi_{q_\tau q_\tau}(\tau, q_\tau, \theta) G_q(\tau) W_q G_q^T(\tau)\Big\}\,d\tau
\end{aligned}$$

which under the definition

$$\varphi(\tau, q_\tau, \theta) \triangleq \varrho(\tau, \theta)\exp\{q_\tau^T \Upsilon(\tau, \theta) q_\tau + 2q_\tau^T \eta(\tau, \theta)\}$$

and its partial derivatives leads to the result

$$\begin{aligned}
&-\theta[q_\tau^T Q_q(\tau) q_\tau + 2q_\tau^T S_q(\tau) + l_z^T(\tau) S l_z(\tau)]\varphi(\tau, q_\tau, \theta) \\
&= \Big\{\frac{\frac{d}{d\tau}\varrho(\tau, \theta)}{\varrho(\tau, \theta)} + 2q_\tau^T[\frac{d}{d\tau}\eta(\tau, \theta) + A_q^T(\tau)\eta(\tau, \theta) + \Upsilon(\tau, \theta) b_q(\tau)] \\
&+ \mathrm{Tr}\Big\{\Upsilon(\tau, \theta) G_q(\tau) W_q G_q^T(\tau)\Big\} + 2\eta^T(\tau, \theta) b_q(\tau) \\
&+ q_\tau^T[\frac{d}{d\tau}\Upsilon(\tau, \theta) + A_q^T(\tau)\Upsilon(\tau, \theta) + \Upsilon(\tau, \theta) A_q(\tau) \\
&+ 2\Upsilon(\tau, \theta) G_q(\tau) W_q G_q^T(\tau)\Upsilon(\tau, \theta)] q_\tau\Big\}\varphi(\tau, q_\tau, \theta),
\end{aligned}$$

At last, the nature of constant, linear and quadratic terms being independent of arbitrary initial values q_τ comes into light. Unsurprisingly, this insight leads to the results (9.21)–(9.24); and hereafter the proof is now complete.

What, precisely, are the uncertain prospects of (9.25)? The uncertain prospects of (9.25) are certainly contributing to the dispersion of distributions of (9.25). In fact, the statistical measures of risk pertaining to (9.25) are further transcribed through all performance-measure statistics. Therefore, a Maclaurin series expansion of (9.20) will carry the day

$$\psi(\tau, q_\tau, \theta) = \sum_{r=1}^{\infty} \frac{\partial^{(r)}}{\partial\theta^{(r)}}\psi(\tau, q_\tau, \theta)\Big|_{\theta=0}\frac{\theta^r}{r!} \tag{9.27}$$

in which all $\kappa_r \triangleq \frac{\partial^{(r)}}{\partial\theta^{(r)}}\psi(\tau,q_\tau,\theta)\Big|_{\theta=0}$ are called rth-order performance-measure statistics. Moreover, the series expansion coefficients are computed by using the cumulant-generating function (9.20)

$$\frac{\partial^{(r)}}{\partial\theta^{(r)}}\psi(\tau,q_\tau,\theta)\bigg|_{\theta=0} = q_\tau^T \frac{\partial^{(r)}}{\partial\theta^{(r)}}\Upsilon(\tau,\theta)\bigg|_{\theta=0} q_\tau + 2q_\tau^T \frac{\partial^{(r)}}{\partial\theta^{(r)}}\eta(\tau,\theta)\bigg|_{\theta=0} + \frac{\partial^{(r)}}{\partial\theta^{(r)}}\upsilon(\tau,\theta)\bigg|_{\theta=0}, \tag{9.28}$$

In view of (9.27), the rth- performance statistic subsequently takes the form of

$$\kappa_r = q_\tau^T \frac{\partial^{(r)}}{\partial\theta^{(r)}}\Upsilon(\tau,\theta)\bigg|_{\theta=0} q_\tau + 2q_\tau^T \frac{\partial^{(r)}}{\partial\theta^{(r)}}\eta(\tau,\theta)\bigg|_{\theta=0} + \frac{\partial^{(r)}}{\partial\theta^{(r)}}\upsilon(\tau,\theta)\bigg|_{\theta=0} \tag{9.29}$$

for any finite $1 \le r < \infty$. For notational convenience, there is, in turn, a need of changing notations as illustrated below

$$H_r(\tau) \triangleq \frac{\partial^{(r)}\Upsilon(\tau,\theta)}{\partial\theta^{(r)}}\bigg|_{\theta=0}, \quad \breve{D}_r(\tau) \triangleq \frac{\partial^{(r)}\eta(\tau,\theta)}{\partial\theta^{(r)}}\bigg|_{\theta=0} \tag{9.30}$$

$$D_r(\tau) \triangleq \frac{\partial^{(r)}\upsilon(\tau,\theta)}{\partial\theta^{(r)}}\bigg|_{\theta=0}, \tag{9.31}$$

At the close of the development, as the next result effectively forecasts the higher-order characteristics associated with (9.20), what are at stake for all the mathematical statistics of (9.20), come more clearly into focus. Appropriate choices of K_z and l_z are thereby capable of reshaping the closed-loop performance distributions in accordance with a priori criteria for stochastic dominance.

Theorem 9.2.2 (Performance-Measure Statistics). *Suppose that (A_q, B_q) is uniformly stabilizable and (C_q, A_q) is uniformly detectable. The kth-order statistic of the chi-squared performance measure (9.20) is given by*

$$\kappa_k = q^T(t_0)H_k(t_0)q(t_0) + 2q^T(t_0)\breve{D}_k(t_0) + D_k(t_0), \quad k \in \mathbb{Z}^+ \tag{9.32}$$

where the cumulant-generating components $H_k(\varsigma)$, $\breve{D}_k(\varsigma)$ and $D_k(\varsigma)$ evaluated at $\varsigma = 0$ satisfy the cumulant-generating equations (with the dependence of $H_k(\varsigma)$, $\breve{D}_k(\varsigma)$ and $D_k(\varsigma)$ upon K_z and l_z suppressed)

$$\frac{d}{d\varsigma}H_1(\varsigma) = -A_q^T(\varsigma)H_1(\varsigma) - H_1(\varsigma)A_q(\varsigma) - Q_q(\varsigma), \quad H_1(t_f) = Q_q^f \tag{9.33}$$

$$\frac{d}{d\varsigma}H_r(\varsigma) = -A_q^T(\varsigma)H_r(\varsigma) - H_r(\varsigma)A_q(\varsigma)$$
$$-\sum_{\nu=1}^{r-1}\frac{2r!}{\nu!(r-\nu)!}H_\nu(\varsigma)G_q(\varsigma)W_qG_q^T(\varsigma)H_{r-\nu}(\varsigma), \quad H_r(t_f) = 0, \quad 2 \le r \le k \tag{9.34}$$

and

$$\frac{d}{d\varsigma}\breve{D}_1(\varsigma) = -A_q^T(\varsigma)\breve{D}_1(\varsigma) - H_1(\varsigma)b_q(\varsigma) - S_q(\varsigma), \quad \breve{D}_1(t_f) = 0 \tag{9.35}$$

$$\frac{d}{d\varsigma}\breve{D}_r(\varsigma) = -A_q^T(\varsigma)\breve{D}_r(\varsigma) - H_r(\varsigma)b_q(\varsigma), \quad \breve{D}_r(t_f) = 0, \quad 2 \le r \le k \tag{9.36}$$

and, finally

$$\frac{d}{d\varsigma}D_1(\varsigma) = -\operatorname{Tr}\{H_1(\varsigma)G_q(\varsigma)W_qG_q^T(\varsigma)\} - 2\breve{D}_1^T(\varsigma)b_q(\varsigma) - l_z^T(\varsigma)Sl_z(\varsigma) \tag{9.37}$$

$$\frac{d}{d\varsigma}D_r(\varsigma) = -\operatorname{Tr}\{H_r(\varsigma)G_q(\varsigma)W_qG_q^T(\varsigma)\} - 2\breve{D}_r^T(\varsigma)b_q(\varsigma), \quad 2 \le r \le 2 \tag{9.38}$$

with the terminal-value conditions $D_r(t_f) = 0$ for $1 \le r \le k$.

The you-got-what you-had-coming explanations of the mathematical statistics of (9.20) were considered as measures of risk for non-symmetric distributions, a procedure proposed being tractably mathematical is further described in terms of the system description of (9.7)–(9.8)

$$\kappa_k = z_0^T H_{11}^k(t_0)z_0 + 2z_0^T\breve{D}_{11}^k(t_0) + D^k(t_0), \quad k \in \mathbb{Z}^+ \tag{9.39}$$

according to the matrix and vector partitions

$$H_r(\varsigma) \triangleq \begin{bmatrix} H_{11}^r(\varsigma) & H_{12}^r(\varsigma) \\ (H_{12}^r)^T(\varsigma) & H_{22}^r(\varsigma) \end{bmatrix}; \quad \breve{D}_r(\varsigma) \triangleq \begin{bmatrix} \breve{D}_{11}^r(\varsigma) \\ \breve{D}_{21}^r(\varsigma) \end{bmatrix}; \quad D^k(\varsigma) \triangleq D_k(\varsigma) \tag{9.40}$$

from which the sub-matrix and vector components depend on K_z and l_z and satisfy the backward-in-time matrix-valued differential equations

$$\frac{d}{d\varsigma}H_{11}^{1}(\varsigma) = -(A_z + B_z K_z(\varsigma))^T H_{11}^{1}(\varsigma) - H_{11}^{1}(\varsigma)(A_z + B_z K_z(\varsigma)) - K_z^T(\varsigma) S K_z(\varsigma) - Q_z, \quad H_{11}^{1}(t_f) = Q_z^f \tag{9.41}$$

$$\frac{d}{d\varsigma}H_{11}^{r}(\varsigma) = -(A_z + B_z K_z(\varsigma))^T H_{11}^{r}(\varsigma) - H_{11}^{r}(\varsigma)(A_z + B_z K_z(\varsigma)) - \sum_{\nu=1}^{r-1}\frac{2r!}{\nu!(r-\nu)!}(H_{11}^{\nu}(\varsigma)\Pi_1(\varsigma) + H_{12}^{\nu}(\varsigma)\Pi_3(\varsigma))H_{11}^{r-\nu}(\varsigma) - \sum_{\nu=1}^{r-1}\frac{2r!}{\nu!(r-\nu)!}(H_{11}^{\nu}(\varsigma)\Pi_2(\varsigma) + H_{12}^{\nu}(\varsigma)\Pi_4(\varsigma))H_{21}^{r-\nu}(\varsigma), \quad H_{11}^{r}(t_f) = 0 \tag{9.42}$$

$$\frac{d}{d\varsigma}H_{12}^{1}(\varsigma) = -(A_z + B_z K_z(\varsigma))^T H_{12}^{1}(\varsigma) - H_{11}^{1}(\varsigma)L_z(\varsigma)C_z - H_{12}^{1}(\varsigma)(A_z - L_z(\varsigma)C_z) - Q_z, \quad H_{12}^{1}(t_f) = Q_z^f \tag{9.43}$$

$$\frac{d}{d\varsigma}H_{12}^{r}(\varsigma) = -(A_z+B_z K_z(\varsigma))^T H_{12}^{r}(\varsigma) - H_{12}^{r}(\varsigma)(A_z - L_z(\varsigma)C_z) - H_{11}^{r}(\varsigma)L_z(\varsigma)C_z - \sum_{\nu=1}^{r-1}\frac{2r!}{\nu!(r-\nu)!}(H_{11}^{\nu}(\varsigma)\Pi_1(\varsigma) + H_{12}^{\nu}(\varsigma)\Pi_3(\varsigma))H_{12}^{r-\nu}(\varsigma) - \sum_{\nu=1}^{r-1}\frac{2r!}{\nu!(r-\nu)!}(H_{11}^{\nu}(\varsigma)\Pi_2(\varsigma) + H_{12}^{\nu}(\varsigma)\Pi_4(\varsigma))H_{22}^{r-\nu}(\varsigma), \quad H_{12}^{r}(t_f) = 0 \tag{9.44}$$

$$\frac{d}{d\varsigma}H_{21}^{1}(\varsigma) = -H_{21}^{1}(\varsigma)(A_z + B_z K_z(\varsigma)) - (L_z(\varsigma)C_z)^T H_{11}^{1}(\varsigma) - (A_z - L_z(\varsigma)C_z)^T H_{21}^{1}(\varsigma) - Q_z, \quad H_{21}^{1}(t_f) = Q_z^f \tag{9.45}$$

$$\frac{d}{d\varsigma}H_{21}^{r}(\varsigma) = -H_{21}^{r}(\varsigma)(A_z+B_z K_z(\varsigma)) - (A_z - L_z(\varsigma)C_z)^T H_{21}^{r}(\varsigma) - (L_z(\varsigma)C_z)^T H_{11}^{r}(\varsigma) - \sum_{\nu=1}^{r-1}\frac{2r!}{\nu!(r-\nu)!}(H_{21}^{\nu}(\varsigma)\Pi_1(\varsigma) + H_{22}^{\nu}(\varsigma)\Pi_3(\varsigma))H_{11}^{r-\nu}(\varsigma) - \sum_{\nu=1}^{r-1}\frac{2r!}{\nu!(r-\nu)!}(H_{21}^{\nu}(\varsigma)\Pi_2(\varsigma) + H_{22}^{\nu}(\varsigma)\Pi_4(\varsigma))H_{21}^{r-\nu}(\varsigma), \quad H_{21}^{r}(t_f) = 0 \tag{9.46}$$

$$\frac{d}{d\varsigma}H_{22}^1(\varsigma) = -H_{22}^1(\varsigma)(A_z - L_z(\varsigma)C_z) - H_{21}^1(\varsigma)L_z(\varsigma)C_z - Q_z$$
$$- (L_z(\varsigma)C_z)^T H_{12}^1(\varsigma) - (A_z - L_z(\varsigma)C_z)^T H_{22}^1(\varsigma), \quad H_{22}^1(t_f) = Q_z^f \tag{9.47}$$

$$\frac{d}{d\varsigma}H_{22}^r(\varsigma) = -H_{22}^r(\varsigma)(A_z - L_z(\varsigma)C_z) - (A_z - L_z(\varsigma)C_z)^T H_{22}^r(\varsigma) - H_{21}^r(\varsigma)L_z(\varsigma)C_z$$
$$- (L_z(\varsigma)C_z)^T H_{12}^r(\varsigma) - \sum_{\nu=1}^{r-1} \frac{2r!}{\nu!(r-\nu)!}(H_{21}^\nu(\varsigma)\Pi_1(\varsigma) + H_{22}^\nu(\varsigma)\Pi_3(\varsigma))H_{12}^{r-\nu}(\varsigma)$$
$$- \sum_{\nu=1}^{r-1} \frac{2r!}{\nu!(r-\nu)!}(H_{21}^\nu(\varsigma)\Pi_2(\varsigma) + H_{22}^\nu(\varsigma)\Pi_4(\varsigma))H_{22}^{r-\nu}(\varsigma), \quad H_{22}^r(t_f) = 0 \tag{9.48}$$

$$\frac{d}{d\varsigma}\breve{D}_{11}^1(\varsigma) = -(A_z + B_z K_z(\varsigma))^T \breve{D}_{11}^1(\varsigma) - K_z^T(\varsigma)S l_z(\varsigma)$$
$$- H_{11}^1(\varsigma)(B_z l_z(\varsigma) + m_{w_z}(\varsigma)), \quad \breve{D}_{11}^1(t_f) = 0 \tag{9.49}$$

$$\frac{d}{d\varsigma}\breve{D}_{11}^r(\varsigma) = -(A_z + B_z K_z(\varsigma))^T \breve{D}_{11}^r(\varsigma)$$
$$- H_{11}^r(\varsigma)(B_z l_z(\varsigma) + m_{w_z}(\varsigma)), \quad \breve{D}_{11}^r(t_f) = 0, \quad 2 \le r \le k \tag{9.50}$$

$$\frac{d}{d\varsigma}\breve{D}_{21}^r(\varsigma) = -(A_z - L_z(\varsigma)C_z)^T \breve{D}_{21}^r(\varsigma) - (L_z(\varsigma)C_z)^T \breve{D}_{11}^r(\varsigma)$$
$$- H_{21}^r(\varsigma)(B_z l_z(\varsigma) + m_{w_z}(\varsigma)), \quad \breve{D}_{21}^r(t_f) = 0, \quad 1 \le r \le k \tag{9.51}$$

$$\frac{d}{d\varsigma}D_1(\varsigma) = -\mathrm{Tr}\{H_{11}^1(\varsigma)\Pi_1(\varsigma) + H_{12}^1(\varsigma)\Pi_3(\varsigma) + H_{21}^1(\varsigma)\Pi_2(\varsigma) + H_{22}^1(\varsigma)\Pi_4(\varsigma)\}$$
$$- 2(\breve{D}_{11}^1)^T(\varsigma)(B_z l_z(\varsigma) + m_{w_z}(\varsigma)) - l_z^T(\varsigma)S l_z(\varsigma), \quad D_1(t_f) = 0 \tag{9.52}$$

$$\frac{d}{d\varsigma}D_r(\varsigma) = -\mathrm{Tr}\{H_{11}^r(\varsigma)\Pi_1(\varsigma) + H_{12}^r(\varsigma)\Pi_3(\varsigma) + H_{21}^r(\varsigma)\Pi_2(\varsigma) + H_{22}^r(\varsigma)\Pi_4(\varsigma)\}$$
$$- 2(\breve{D}_{11}^r)^T(\varsigma)(B_z l_z(\varsigma) + m_{w_z}(\varsigma)), \quad D_r(t_f) = 0, \quad 2 \le r \le k \tag{9.53}$$

whereupon $\Pi_1(\varsigma) \triangleq L_z(\varsigma)VL_z^T(\varsigma)$, $\Pi_2 = \Pi_3 \triangleq -L_z(\varsigma)VL_z^T(\varsigma)$, and $\Pi_4(\varsigma) \triangleq W_z + L_z(\varsigma)VL_z^T(\varsigma)$.

9.3 Toward the Optimal Decision Problem

Having recognized the dispersion measures (9.32) for the class of continuous distributions of (9.25), a risk-averse control can relate that any performance uncertainty the control designer may have anticipated is now addressed by selecting appropriate feedback control alternatives (K_z, l_z). Retrospective assessments of uncertain prospects regarding the restrictive family of (9.25) for the linear-quadratic class of stochastic systems often emphasize the work recently reported in [6]. In fact, there are certain technicalities that must be addressed at the information structures as governed by (9.41)–(9.53) together with the crucial state variables $H_{11}^r(\varsigma)$, $H_{12}^r(\varsigma)$, $H_{21}^r(\varsigma)$, $H_{22}^r(\tau)$, $\breve{D}_{11}^r(\varsigma)$, $\breve{D}_{21}^r(\varsigma)$ and $D^r(\varsigma)$. In order to have notational simplicity, it is required to denote the right members of (9.41)–(9.53)

$$\mathscr{F}_{11}^1(\varsigma, \mathscr{H}_{11}, \mathscr{H}_{12}, \mathscr{H}_{21}, K_z) \triangleq -(A_z + B_z K_z(\varsigma))^T \mathscr{H}_{11}^1(\varsigma)$$
$$- \mathscr{H}_{11}^1(\varsigma)(A_z + B_z K_z(\varsigma)) - K_z^T(\varsigma) S K_z(\varsigma) - Q_z$$

$$\mathscr{F}_{11}^r(\varsigma, \mathscr{H}_{11}, \mathscr{H}_{12}, \mathscr{H}_{21}, K_z) \triangleq -(A_z + B_z K_z(\varsigma))^T \mathscr{H}_{11}^r(\varsigma) - \mathscr{H}_{11}^r(\varsigma)(A_z + B_z K_z(\varsigma))$$
$$- \sum_{\nu=1}^{r-1} \frac{2r!}{\nu!(r-\nu)!} (\mathscr{H}_{11}^\nu(\varsigma)\Pi_1(\varsigma) + \mathscr{H}_{12}^\nu(\varsigma)\Pi_3(\varsigma))\mathscr{H}_{11}^{r-\nu}(\varsigma)$$
$$- \sum_{\nu=1}^{r-1} \frac{2r!}{\nu!(r-\nu)!} (\mathscr{H}_{11}^\nu(\varsigma)\Pi_2(\varsigma) + \mathscr{H}_{12}^\nu(\varsigma)\Pi_4(\varsigma))\mathscr{H}_{21}^{r-\nu}(\varsigma), \quad 2 \le r \le k$$

$$\mathscr{F}_{12}^1(\varsigma, \mathscr{H}_{11}, \mathscr{H}_{12}, \mathscr{H}_{22}, K_z) \triangleq -(A_z + B_z K_z(\varsigma))^T \mathscr{H}_{12}^1(\varsigma) - \mathscr{H}_{11}^1(\varsigma) L_z(\varsigma) C_z$$
$$- \mathscr{H}_{12}^1(\varsigma)(A_z - L_z(\varsigma)C_z) - Q_z$$

$$\mathscr{F}_{12}^r(\varsigma, \mathscr{H}_{11}, \mathscr{H}_{12}, \mathscr{H}_{22}, K_z) \triangleq -(A_z + B_z K_z(\varsigma))^T \mathscr{H}_{12}^r(\varsigma) - \mathscr{H}_{12}^r(\varsigma)(A_z - L_z(\varsigma)C_z)$$
$$- \mathscr{H}_{11}^r(\varsigma) L_z(\varsigma) C_z - \sum_{\nu=1}^{r-1} \frac{2r!}{\nu!(r-\nu)!} (\mathscr{H}_{11}^\nu(\varsigma)\Pi_1(\varsigma) + \mathscr{H}_{12}^\nu(\varsigma)\Pi_3(\varsigma))\mathscr{H}_{12}^{r-\nu}(\varsigma)$$
$$- \sum_{\nu=1}^{r-1} \frac{2r!}{\nu!(r-\nu)!} (\mathscr{H}_{11}^\nu(\varsigma)\Pi_2(\varsigma) + \mathscr{H}_{12}^\nu(\varsigma)\Pi_4(\varsigma))\mathscr{H}_{22}^{r-\nu}(\varsigma), \quad 2 \le r \le k$$

$$\mathscr{F}_{21}^1(\varsigma, \mathscr{H}_{11}, \mathscr{H}_{21}, \mathscr{H}_{22}, K_z) \triangleq -\mathscr{H}_{21}^1(\varsigma)(A_z + B_z K_z(\varsigma)) - (A_z - L_z(\varsigma)C_z)^T \mathscr{H}_{21}^1(\varsigma)$$
$$- (L_z(\varsigma)C_z)^T \mathscr{H}_{11}^1(\varsigma) - Q_z$$

$$\mathscr{F}_{21}^r(\varsigma, \mathscr{H}_{11}, \mathscr{H}_{21}, \mathscr{H}_{22}, K_z) \triangleq -\mathscr{H}_{21}^r(\varsigma)(A_z + B_z K_z(\varsigma)) - (A_z - L_z(\varsigma)C_z)^T \mathscr{H}_{21}^r(\varsigma)$$

$$
- (L_z(\varsigma)C_z)^T \mathscr{H}_{11}^r(\varsigma) - \sum_{\nu=1}^{r-1} \frac{2r!}{\nu!(r-\nu)!} (\mathscr{H}_{21}^\nu(\varsigma)\Pi_1(\varsigma) + \mathscr{H}_{22}^\nu(\varsigma)\Pi_3(\varsigma))\mathscr{H}_{11}^{r-\nu}(\varsigma)
$$

$$
- \sum_{\nu=1}^{r-1} \frac{2r!}{\nu!(r-\nu)!} (\mathscr{H}_{21}^\nu(\varsigma)\Pi_2(\varsigma) + \mathscr{H}_{22}^\nu(\varsigma)\Pi_4(\varsigma))\mathscr{H}_{21}^{r-\nu}(\varsigma), \quad 2 \le r \le k
$$

$$
\begin{aligned}
\mathscr{F}_{22}^1(\varsigma, \mathscr{H}_{22}, \mathscr{H}_{12}, \mathscr{H}_{21}) &\triangleq -\mathscr{H}_{22}^1(\varsigma)(A_z - L_z(\varsigma)C_z(\varsigma)) - \mathscr{H}_{21}^1(\varsigma)(L_z(\varsigma)C_z) \\
&- (L_z(\varsigma)C_z)^T \mathscr{H}_{12}^1(\varsigma) - (A_z - L_z(\varsigma)C_z)^T \mathscr{H}_{22}^1(\varsigma) - Q_z \\
\mathscr{F}_{22}^r(\varsigma, \mathscr{H}_{22}, \mathscr{H}_{12}, \mathscr{H}_{21}) &\triangleq -\mathscr{H}_{22}^r(\varsigma)(A_z - L_z(\varsigma)C_z) - \mathscr{H}_{21}^r(\varsigma)(L_z(\varsigma)C_z) \\
&- (L_z(\varsigma)C_z)^T \mathscr{H}_{12}^r(\varsigma) - (A_z - L_z(\varsigma)C_z)^T \mathscr{H}_{22}^r(\varsigma) \\
&- \sum_{\nu=1}^{r-1} \frac{2r!}{\nu!(r-\nu)!} (\mathscr{H}_{21}^\nu(\varsigma)\Pi_1(\varsigma) + \mathscr{H}_{22}^\nu(\varsigma)\Pi_3(\varsigma))\mathscr{H}_{12}^{r-\nu}(\varsigma) \\
&- \sum_{\nu=1}^{r-1} \frac{2r!}{\nu!(r-\nu)!} (\mathscr{H}_{21}^\nu(\varsigma)\Pi_2(\varsigma) + \mathscr{H}_{22}^\nu(\varsigma)\Pi_4(\varsigma))\mathscr{H}_{22}^{r-\nu}(\varsigma), \quad 2 \le r \le k
\end{aligned}
$$

$$
\begin{aligned}
\breve{\mathscr{G}}_{11}^1(\varsigma, \breve{\mathscr{D}}_{11}, \mathscr{H}_{11}, K_z, l_z) &\triangleq -(A_z + B_z K_z(\varsigma))^T \breve{\mathscr{D}}_{11}^1(\varsigma) - K_z^T(\varsigma) S l_z(\varsigma) \\
&- \mathscr{H}_{11}^1(\varsigma)(B_z l_z(\varsigma) + m_{w_z}(\varsigma)) \\
\breve{\mathscr{G}}_{11}^r(\varsigma, \breve{\mathscr{D}}_{11}, \mathscr{H}_{11}, K_z, l_z) &\triangleq -(A_z + B_z K_z(\varsigma))^T \breve{\mathscr{D}}_{11}^r(\varsigma) \\
&- \mathscr{H}_{11}^r(\varsigma)(B_z l_z(\varsigma) + m_{w_z}(\varsigma)), \quad 2 \le r \le k
\end{aligned}
$$

$$
\begin{aligned}
\breve{\mathscr{G}}_{21}^r(\varsigma, \breve{\mathscr{D}}_{11}, \breve{\mathscr{D}}_{21}, \mathscr{H}_{21}, l_z) &\triangleq (A_z - L_z(\varsigma)C_z)^T \breve{\mathscr{D}}_{21}^r(\varsigma) - (L_z(\varsigma)C_z)^T \breve{\mathscr{D}}_{11}^r(\varsigma) \\
&- \mathscr{H}_{21}^r(\varsigma)(B_z l_z(\varsigma) + m_{w_z}(\varsigma)), \quad 1 \le r \le k
\end{aligned}
$$

$$
\begin{aligned}
&\mathscr{G}_1(\varsigma, \mathscr{H}_{11}, \mathscr{H}_{12}, \mathscr{H}_{21}, \mathscr{H}_{22}, \breve{\mathscr{D}}_{11}, l_z) \triangleq \\
&\quad - \mathrm{Tr}\{\mathscr{H}_{11}^1(\varsigma)\Pi_1(\varsigma) + \mathscr{H}_{12}^1(\varsigma)\Pi_3(\varsigma) + \mathscr{H}_{21}^1(\varsigma)\Pi_2(\varsigma) + \mathscr{H}_{22}^1(\varsigma)\Pi_4(\varsigma)\} \\
&\quad - 2(\breve{\mathscr{D}}_{11}^1)^T(\varsigma)(B_z l_z(\varsigma) + m_{w_z}(\varsigma)) - l_z^T(\varsigma) S l_z(\varsigma) \\
&\mathscr{G}_r(\varsigma, \mathscr{H}_{11}, \mathscr{H}_{12}, \mathscr{H}_{21}, \mathscr{H}_{22}, \breve{\mathscr{D}}_{11}, l_z) \triangleq \\
&\quad - \mathrm{Tr}\{\mathscr{H}_{11}^r(\varsigma)\Pi_1(\varsigma) + \mathscr{H}_{12}^r(\varsigma)\Pi_3(\varsigma) + \mathscr{H}_{21}^r(\varsigma)\Pi_2(\varsigma) + \mathscr{H}_{22}^r(\varsigma)\Pi_4(\varsigma)\} \\
&\quad - 2(\breve{\mathscr{D}}_{11}^r)^T(\varsigma)(B_z l_z(\varsigma) + m_{w_z}(\varsigma)), \quad 2 \le r \le k
\end{aligned}
$$

where the k-tuple variables $\mathscr{H}_{11}, \mathscr{H}_{12}, \mathscr{H}_{21}, \mathscr{H}_{22}, \breve{\mathscr{D}}_{11}, \breve{\mathscr{D}}_{21}$ and $\mathscr{D}$ defined by

$$\mathscr{H}_{11}(\cdot) \triangleq (\mathscr{H}_{11}^1(\cdot), \ldots, \mathscr{H}_{11}^k(\cdot)), \qquad \mathscr{H}_{12}(\cdot) \triangleq (\mathscr{H}_{12}^1(\cdot), \ldots, \mathscr{H}_{12}^k(\cdot))$$

$$\mathscr{H}_{21}(\cdot) \triangleq (\mathscr{H}_{21}^1(\cdot), \ldots, \mathscr{H}_{21}^k(\cdot)), \qquad \mathscr{H}_{22}(\cdot) \triangleq (\mathscr{H}_{22}^1(\cdot), \ldots, \mathscr{H}_{22}^k(\cdot))$$

$$\breve{\mathscr{D}}_{11}(\cdot) \triangleq (\breve{\mathscr{D}}_{11}^1(\cdot), \ldots, \breve{\mathscr{D}}_{11}^k(\cdot)), \qquad \breve{\mathscr{D}}_{21}(\cdot) \triangleq (\breve{\mathscr{D}}_{21}^1(\cdot), \ldots, \breve{\mathscr{D}}_{21}^k(\cdot))$$

$$\mathscr{D}(\cdot) \triangleq (\mathscr{D}^1(\cdot), \ldots, \mathscr{D}^k(\cdot))$$

whereas all the continuous-time and matrix-valued states $\mathscr{H}_{11}^r, \mathscr{H}_{12}^r, \mathscr{H}_{21}^r, \mathscr{H}_{22}^r \in \mathscr{C}^1([t_0, t_f]; \mathbb{R}^{(n+m)\times(n+m)})$; vector-valued states $\breve{\mathscr{D}}_{11}^r, \breve{\mathscr{D}}_{21}^r \in \mathscr{C}^1([t_0, t_f]; \mathbb{R}^{n+m})$ and scalar-valued states $\mathscr{D}^r \in \mathscr{C}^1([t_0, t_f]; \mathbb{R})$ have the representations $\mathscr{H}_{11}^r(\cdot) \triangleq H_{11}^r(\cdot)$, $\mathscr{H}_{12}^r(\cdot) \triangleq H_{12}^r(\cdot)$, $\mathscr{H}_{21}^r(\cdot) \triangleq H_{21}^r(\cdot)$, $\mathscr{H}_{22}^r(\cdot) \triangleq H_{22}^r(\cdot)$, $\breve{\mathscr{D}}_{11}^r(\cdot) \triangleq \breve{D}_{11}^r(\cdot)$, $\breve{\mathscr{D}}_{21}^r(\cdot) \triangleq \breve{D}_{21}^r(\cdot)$, and $\mathscr{D}^r(\cdot) \triangleq D^r(\cdot)$.

In theory, there is always the possibility of showing the dynamical equations (9.41)–(9.53) to be bounded and Lipschitz continuous on $[t_0, t_f]$; see [8]. With these properties in mind, the precise product mappings of the dynamical equations (9.41)–(9.53) for the system control formulation can be sketched as follows

$$\mathscr{F}_{11} : [t_0, t_f] \times (\mathbb{R}^{(n+m)\times(n+m)})^{3k} \times \mathbb{R}^{m\times(n+m)} \mapsto (\mathbb{R}^{(n+m)\times(n+m)})^k$$

$$\mathscr{F}_{12} : [t_0, t_f] \times (\mathbb{R}^{(n+m)\times(n+m)})^{3k} \times \mathbb{R}^{m\times(n+m)} \mapsto (\mathbb{R}^{(n+m)\times(n+m)})^k$$

$$\mathscr{F}_{21} : [t_0, t_f] \times (\mathbb{R}^{(n+m)\times(n+m)})^{3k} \times \mathbb{R}^{m\times(n+m)} \mapsto (\mathbb{R}^{(n+m)\times(n+m)})^k$$

$$\mathscr{F}_{22} : [t_0, t_f] \times (\mathbb{R}^{(n+m)\times(n+m)})^{3k} \mapsto (\mathbb{R}^{(n+m)\times(n+m)})^k$$

$$\breve{\mathscr{G}}_{11} : [t_0, t_f] \times (\mathbb{R}^{(n+m)\times(n+m)})^{k} \times (\mathbb{R}^{n+m})^k \times \mathbb{R}^{m\times(n+m)} \times \mathbb{R}^m \mapsto (\mathbb{R}^{n+m})^k$$

$$\breve{\mathscr{G}}_{21} : [t_0, t_f] \times (\mathbb{R}^{(n+m)\times(n+m)})^{k} \times (\mathbb{R}^{n+m})^{2k} \times \mathbb{R}^m \mapsto (\mathbb{R}^{n+m})^k$$

$$\mathscr{G} : [t_0, t_f] \times (\mathbb{R}^{(n+m)\times(n+m)})^{4k} \times (\mathbb{R}^{n+m})^k \times \mathbb{R}^m \mapsto \mathbb{R}^k$$

whose the rules of action are governed by

$$\frac{d}{d\varsigma}\mathscr{H}_{11}(\varsigma) = \mathscr{F}_{11}(\varsigma, \mathscr{H}_{11}(\varsigma), \mathscr{H}_{12}(\varsigma), \mathscr{H}_{21}(\varsigma), K_z(\varsigma)), \quad \mathscr{H}_{11}(t_f) \tag{9.54}$$

$$\frac{d}{d\varsigma}\mathscr{H}_{12}(\varsigma) = \mathscr{F}_{12}(\varsigma, \mathscr{H}_{11}(\varsigma), \mathscr{H}_{12}(\varsigma), \mathscr{H}_{22}(\varsigma), K_z(\varsigma)), \quad \mathscr{H}_{12}(t_f) \tag{9.55}$$

$$\frac{d}{d\varsigma}\mathscr{H}_{21}(\varsigma) = \mathscr{F}_{21}(\varsigma, \mathscr{H}_{11}(\varsigma), \mathscr{H}_{12}(\varsigma), \mathscr{H}_{22}(\varsigma), K_z(\varsigma)), \quad \mathscr{H}_{21}(t_f) \tag{9.56}$$

$$\frac{d}{d\varsigma}\mathscr{H}_{22}(\varsigma) = \mathscr{F}_{22}(\varsigma, \mathscr{H}_{12}(\varsigma), \mathscr{H}_{21}(\varsigma), \mathscr{H}_{22}(\varsigma)), \quad \mathscr{H}_{22}(t_f) \tag{9.57}$$

$$\frac{d}{d\varsigma}\breve{\mathscr{D}}_{11}(\varsigma) = \breve{\mathscr{G}}_{11}(\varsigma, \breve{\mathscr{D}}_{11}, \mathscr{H}_{11}(\varsigma), K_z(\varsigma), l_z(\varsigma)), \quad \breve{\mathscr{D}}_{11}(t_f) \tag{9.58}$$

$$\frac{d}{d\varsigma}\breve{\mathscr{D}}_{21}(\varsigma) = \breve{\mathscr{G}}_{21}(\varsigma, \breve{\mathscr{D}}_{11}, \breve{\mathscr{D}}_{21}, \mathscr{H}_{21}(\varsigma), l_z(\varsigma)), \quad \breve{\mathscr{D}}_{21}(t_f) \tag{9.59}$$

$$\frac{d}{d\varsigma}\mathscr{D}(\varsigma) = \mathscr{G}(\varsigma, \mathscr{H}_{11}(\varsigma), \mathscr{H}_{12}(\varsigma), \mathscr{H}_{21}(\varsigma), \mathscr{H}_{22}(\varsigma), \breve{\mathscr{D}}_{11}(\varsigma), l_z(\varsigma)) \tag{9.60}$$

under the following definitions

$$\mathscr{F}_{11} \triangleq \mathscr{F}^1_{11} \times \cdots \times \mathscr{F}^k_{11}, \quad \mathscr{F}_{12} \triangleq \mathscr{F}^1_{12} \times \cdots \times \mathscr{F}^k_{12}, \quad \mathscr{F}_{21} \triangleq \mathscr{F}^1_{21} \times \cdots \times \mathscr{F}^k_{21}$$
$$\mathscr{F}_{22} \triangleq \mathscr{F}^1_{22} \times \cdots \times \mathscr{F}^k_{22}, \quad \breve{\mathscr{G}}_{11} \triangleq \breve{\mathscr{G}}^1_{11} \times \cdots \times \breve{\mathscr{G}}^k_{11}, \quad \breve{\mathscr{G}}_{21} \triangleq \breve{\mathscr{G}}^1_{21} \times \cdots \times \breve{\mathscr{G}}^k_{21}$$
$$\mathscr{G} \triangleq \mathscr{G}^1 \times \cdots \times \mathscr{G}^k$$

and the terminal-value conditions

$$\mathscr{H}_{11}(t_f) = \mathscr{H}_{12}(t_f) = \mathscr{H}_{21}(t_f) = \mathscr{H}_{22}(t_f) \triangleq (Q^f_z, 0, \ldots, 0)$$
$$\breve{\mathscr{D}}_{11}(t_f) = \breve{\mathscr{D}}_{21}(t_f) \triangleq (0, \ldots, 0), \quad \mathscr{D}(t_f) \triangleq (0, \ldots, 0),$$

Not surprisingly, the product system (9.54)–(9.60) uniquely determines $\mathscr{H}_{11}$, $\mathscr{H}_{12}$, $\mathscr{H}_{21}$, $\mathscr{H}_{22}$, $\breve{\mathscr{D}}_{11}$, $\breve{\mathscr{D}}_{21}$ and $\mathscr{D}$ once admissible feedback parameters K_z and l_z are specified. Thereby, $\mathscr{H}_{11} \equiv \mathscr{H}_{11}(\cdot, K_z)$, $\mathscr{H}_{12} \equiv \mathscr{H}_{12}(\cdot, K_z)$, $\mathscr{H}_{21} \equiv \mathscr{H}_{21}(\cdot, K_z)$, $\mathscr{H}_{22} \equiv \mathscr{H}_{22}(\cdot, K_z)$, $\breve{\mathscr{D}}_{11} \equiv \breve{\mathscr{D}}_{11}(\cdot, K_z, l_z)$, $\breve{\mathscr{D}}_{21} \equiv \breve{\mathscr{D}}_{21}(\cdot, K_z, l_z)$, and $\mathscr{D} \equiv \mathscr{D}(\cdot, K_z, l_z)$.

Throughout the work of [6], the author continued to press for substantive reform, optimal rules for ordering uncertain prospects of the chi-squared random cost (9.25), which have strong ties for non-symmetric distributions. Perhaps the most significance of this viewpoint involves potential necessary and sufficient conditions when ordering stochastic dominance for the class of probabilistic distributions with equal and unequal means herein. Certainly, it is encouraging to see the optimal control problem sustained by the feedback control parameters K_z and l_z is accustomed to adopting the mean-risk aware performance index as defined below.

Definition 9.3.1 (Mean-Risk Aware Performance Index). Fix $k \in \mathbb{Z}^+$ and the sequence $\mu = \{\mu_i \geq 0\}_{i=1}^k$ with $\mu_1 > 0$. The performance index with risk awareness for the stochastic system with control rate constraints (9.16)–(9.17) is defined by

$$\phi_0 : \{t_0\} \times (\mathbb{R}^{(n+m)\times(n+m)})^k \times (\mathbb{R}^{n+m})^k \times \mathbb{R}^k \mapsto \mathbb{R}^+$$

with the rule of action

$$\phi_0\left(t_0, \mathscr{H}_{11}(t_0), \breve{\mathscr{D}}_{11}(t_0), \mathscr{D}(t_0)\right) \triangleq \underbrace{\mu_1\kappa_1}_{\text{Mean Measure}} + \underbrace{\mu_2\kappa_2 + \cdots + \mu_k\kappa_k}_{\text{Risk Measures}}$$

$$= \sum_{r=1}^{k} \mu_r \left[z_0^T \mathscr{H}_{11}^r(t_0) z_0 + 2z_0^T \breve{\mathscr{D}}_{11}^r(t_0) + \mathscr{D}^r(t_0) \right] \qquad (9.61)$$

where additional parametric design of freedom μ_r, chosen by risk-averse controller designers, represent different levels of robustness prioritization according to the importance of the resulting performance-measure statistics to the probabilistic performance distribution. And the unique solutions $\{\mathscr{H}_{11}^r(\varsigma)\}_{r=1}^k$, $\left\{\breve{\mathscr{D}}_{11}^r(\varsigma)\right\}_{r=1}^k$, and $\{\mathscr{D}^r(\varsigma)\}_{r=1}^k$ evaluated at $\varsigma = t_0$ satisfy the dynamical equations (9.54)–(9.60).

In view of the performance-measure statistics (9.39) as the natural measures of dispersion, this result is of special significance in risk-averse control theory. It further implies that in order to generate the admissible set of feedback control laws, appropriate choices of parameters K_z and l_z should be further defined as follows.

Definition 9.3.2 (Admissible Feedback Parameters). For the given terminal data $(t_f, \mathscr{H}_{11}^f, \mathscr{H}_{12}^f, \mathscr{H}_{21}^f, \mathscr{H}_{22}^f, \breve{\mathscr{D}}_{11}^f, \mathscr{D}^f)$, the classes of admissible feedback gains and feedforward inputs are defined as follows. Let compact subsets $\overline{K}_z \subset \mathbb{R}^{m\times(n+m)}$ and $\overline{L}_z \subset \mathbb{R}^{n+m}$ be the sets of allowable gain and input values. With $k \in \mathbb{Z}^+$ and the sequence $\mu = \{\mu_r \geq 0\}_{r=1}^k$ and $\mu_1 > 0$ given, the sets of admissible $\mathscr{K}^z_{t_f,\mathscr{H}_{11}^f,\mathscr{H}_{12}^f,\mathscr{H}_{21}^f,\mathscr{H}_{22}^f,\breve{\mathscr{D}}_{11}^f,\mathscr{D}^f;\mu}$ and $\mathscr{L}^z_{t_f,\mathscr{H}_{11}^f,\mathscr{H}_{12}^f,\mathscr{H}_{21}^f,\mathscr{H}_{22}^f,\breve{\mathscr{D}}_{11}^f,\mathscr{D}^f;\mu}$ are the classes of time-continuous matrices $\mathscr{C}([t_0, t_f]; \mathbb{R}^{m\times(n+m)})$ and vectors $\mathscr{C}([t_0, t_f]; \mathbb{R}^{n+m})$ with values $K_z(\cdot) \in \overline{K}_z$ and $l_z(\cdot) \in \overline{L}_z$ for which the solutions to the dynamical equations (9.54)–(9.60) exist on the finite interval $[t_0, t_f]$.

Careful review of the excellent treatment in [9] is only now beginning, and gives motivation for the development hereafter. In a very real sense, the traditional end-point problem and the use of dynamic programming are intended to be kept. Appropriate modifications in the sequence of results are committed to conform with the unique terminologies of statistical optimal control as can been seen below.

Definition 9.3.3 (Optimization Problem of Mayer Type). Suppose $k \in \mathbb{Z}^+$ and the sequence $\mu = \{\mu_r \geq 0\}_{r=1}^k$ with $\mu_1 > 0$ are fixed. Then, the optimization problem over $[t_0, t_f]$ is defined as the minimization of the performance index (9.61) with respect to (K_z, l_z) in $\mathscr{K}^z_{t_f,\mathscr{H}_{11}^f,\mathscr{H}_{12}^f,\mathscr{H}_{21}^f,\mathscr{H}_{22}^f,\breve{\mathscr{D}}_{11}^f,\mathscr{D}^f;\mu} \times \mathscr{L}^z_{t_f,\mathscr{H}_{11}^f,\mathscr{H}_{12}^f,\mathscr{H}_{21}^f,\mathscr{H}_{22}^f,\breve{\mathscr{D}}_{11}^f,\mathscr{D}^f;\mu}$ and subject to the dynamical equations (9.54)–(9.60).

In an important sense, the parametrization of the terminal time and states; e.g., $(t_f, \mathscr{H}_{11}^f, \mathscr{H}_{12}^f, \mathscr{H}_{21}^f, \mathscr{H}_{22}^f, \breve{\mathscr{D}}_{11}^f, \mathscr{D}^f)$ as $(\varepsilon, \mathscr{Y}_{11}, \mathscr{Y}_{12}, \mathscr{Y}_{21}, \mathscr{Y}_{22}, \breve{\mathscr{Z}}_{11}, \mathscr{Z})$ represents a crucial move in the direction of embedding the aforementioned optimization into a larger optimal control problem.

Definition 9.3.4 (Reachable Set). Denote $\mathscr{Q} \triangleq \Big\{(\varepsilon, \mathscr{Y}_{11}, \mathscr{Y}_{12}, \mathscr{Y}_{21}, \mathscr{Y}_{22}, \breve{\mathscr{Z}}_{11}, \mathscr{Z}) \in [t_0, t_f] \times (\mathbb{R}^{(n+m)\times(n+m)})^{4k} \times (\mathbb{R}^{n+m})^k \times \mathbb{R}^k$ so that $\mathscr{K}^z_{t_f,\mathscr{H}^f_{11},\mathscr{H}^f_{12},\mathscr{H}^f_{21},\mathscr{H}^f_{22},\breve{\mathscr{D}}^f_{11},\mathscr{D}^f;\mu} \times \mathscr{L}^z_{t_f,\mathscr{H}^f_{11},\mathscr{H}^f_{12},\mathscr{H}^f_{21},\mathscr{H}^f_{22},\breve{\mathscr{D}}^f_{11},\mathscr{D}^f;\mu} \neq \emptyset\Big\}$.

How new all this may be related? Perhaps, the value function for this optimization problem is now depending on parameterizations of the terminal-value conditions.

Definition 9.3.5 (Value Function). Suppose that $(\varepsilon, \mathscr{Y}_{11}, \mathscr{Y}_{12}, \mathscr{Y}_{21}, \mathscr{Y}_{22}, \breve{\mathscr{Z}}_{11}, \mathscr{Z}) \in \mathscr{Q}$ be given. Then, the value function $\mathscr{V}(\varepsilon, \mathscr{Y}_{11}, \breve{\mathscr{Z}}_{11}, \mathscr{Z})$ is defined by

$$\mathscr{V}(\varepsilon, \mathscr{Y}_{11}, \breve{\mathscr{Z}}_{11}, \mathscr{Z}) = \inf_{K_z \in \overline{K}_z, l_z \in \overline{L}_z} \phi_0(t_0, \mathscr{H}_{11}(t_0), \breve{\mathscr{D}}_{11}(t_0), \mathscr{D}(t_0)),$$

By convention, $\mathscr{V}(\varepsilon, \mathscr{Y}_{11}, \breve{\mathscr{Z}}_{11}, \mathscr{Z}) = +\infty$ when the Cartesian product that follows $\mathscr{K}^z_{t_f,\mathscr{H}^f_{11},\mathscr{H}^f_{12},\mathscr{H}^f_{21},\mathscr{H}^f_{22},\breve{\mathscr{D}}^f_{11},\mathscr{D}^f;\mu} \times \mathscr{L}^z_{t_f,\mathscr{H}^f_{11},\mathscr{H}^f_{12},\mathscr{H}^f_{21},\mathscr{H}^f_{22},\breve{\mathscr{D}}^f_{11},\mathscr{D}^f;\mu}$ is empty.

Of note, the initial-cost problem is the necessary underpinning in statistical optimal control. Such an optimization of Mayer type cannot take hold unless the Hamilton-Jacobi-Bellman (HJB) equation satisfied by the value function. But moving beyond the unique terminologies, it is increasingly obvious that one has to resort to a comparative analysis as can be seen in [9].

Theorem 9.3.1 (HJB Equation for Mayer Problem). *Let an interior point of $\mathscr{Q}$; e.g., $(\varepsilon, \mathscr{Y}_{11}, \mathscr{Y}_{12}, \mathscr{Y}_{21}, \mathscr{Y}_{22}, \breve{\mathscr{Z}}_{11}, \mathscr{Z})$ at which the value function $\mathscr{V}(\varepsilon, \mathscr{Y}_{11}, \breve{\mathscr{Z}}_{11}, \mathscr{Z})$ is differentiable. If there exist optimal feedback gain $K_z^* \in \mathscr{K}^z_{t_f,\mathscr{H}^f_{11},\mathscr{H}^f_{12},\mathscr{H}^f_{21},\mathscr{H}^f_{22},\breve{\mathscr{D}}^f_{11},\mathscr{D}^f;\mu}$ and affine input $l_z^* \in \mathscr{L}^z_{t_f,\mathscr{H}^f_{11},\mathscr{H}^f_{12},\mathscr{H}^f_{21},\mathscr{H}^f_{22},\breve{\mathscr{D}}^f_{11},\mathscr{D}^f;\mu}$, then the partial differential equation of dynamic programming*

$$\begin{aligned}
0 = \min_{K_z \in \overline{K}_z, l_z \in \overline{L}_z} \Big\{ & \frac{\partial}{\partial \varepsilon} \mathscr{V}\left(\varepsilon, \mathscr{Y}_{11}, \breve{\mathscr{Z}}_{11}, \mathscr{Z}\right) \\
& + \frac{\partial}{\partial \operatorname{vec}(\mathscr{Y}_{11})} \mathscr{V}\left(\varepsilon, \mathscr{Y}_{11}, \breve{\mathscr{Z}}_{11}, \mathscr{Z}\right) \operatorname{vec}\left(\mathscr{F}_{11}\left(\varepsilon, \mathscr{Y}_{11}, \mathscr{Y}_{12}, \mathscr{Y}_{21}, K_z\right)\right) \\
& + \frac{\partial}{\partial \operatorname{vec}(\breve{\mathscr{Z}}_{11})} \mathscr{V}\left(\varepsilon, \mathscr{Y}_{11}, \breve{\mathscr{Z}}_{11}, \mathscr{Z}\right) \operatorname{vec}(\breve{\mathscr{G}}_{11}(\varepsilon, \breve{\mathscr{Z}}_{11}, \mathscr{Y}_{11}, K_z, l_z)) \\
& + \frac{\partial}{\partial \operatorname{vec}(\mathscr{Z})} \mathscr{V}\left(\varepsilon, \mathscr{Y}_{11}, \breve{\mathscr{Z}}_{11}, \mathscr{Z}\right) \operatorname{vec}(\mathscr{G}(\varepsilon, \mathscr{Y}_{11}, \mathscr{Y}_{12}, \mathscr{Y}_{21}, \mathscr{Y}_{22}, \breve{\mathscr{Z}}_{11}, l_z))\Big\}
\end{aligned} \tag{9.62}$$

is satisfied. The boundary condition of (9.62) is given by

$$\mathscr{V}\left(t_0, \mathscr{H}_{11}(t_0), \breve{\mathscr{D}}_{11}(t_0), \mathscr{D}(t_0)\right) = \phi_0\left(t_0, \mathscr{H}_{11}(t_0), \breve{\mathscr{D}}_{11}(t_0), \mathscr{D}(t_0)\right),$$

Proof. A rigorous proof of the necessary condition herein can be found in [10].

A side from the relevance of the necessary condition to the dynamic programming methodology, there is the further emergence of a framework for assessing a sufficient condition which gives rise to the result hereafter.

Theorem 9.3.2 (Verification Theorem). *Fix $k \in \mathbb{Z}^+$ and let $\mathscr{W}(\varepsilon, \mathscr{Y}_{11}, \breve{\mathscr{Z}}_{11}, \mathscr{Z})$ be a continuously differentiable solution of (9.62) satisfying the boundary condition*

$$\mathscr{W}(t_0, \mathscr{H}_{11}(t_0), \breve{\mathscr{D}}_{11}(t_0), \mathscr{D}(t_0)) = \phi_0\left(t_0, \mathscr{H}_{11}(t_0), \breve{\mathscr{D}}_{11}(t_0), \mathscr{D}(t_0)\right).$$

Let the terminal-value condition $(t_f, \mathscr{H}_{11}^f, \mathscr{H}_{12}^f, \mathscr{H}_{21}^f, \mathscr{H}_{22}^f, \breve{\mathscr{D}}_{11}^f, \mathscr{D}^f)$ be in $\mathscr{Q}$; the 2-tuple (K_z, l_z) in $\mathscr{K}^z_{t_f, \mathscr{H}_{11}^f, \mathscr{H}_{12}^f, \mathscr{H}_{21}^f, \mathscr{H}_{22}^f, \breve{\mathscr{D}}_{11}^f, \mathscr{D}^f;\mu} \times \mathscr{L}^z_{t_f, \mathscr{H}_{11}^f, \mathscr{H}_{12}^f, \mathscr{H}_{21}^f, \mathscr{H}_{22}^f, \breve{\mathscr{D}}_{11}^f, \mathscr{D}^f;\mu}$; the trajectory solutions $\mathscr{H}_{11}$, $\breve{\mathscr{D}}_{11}$ and $\mathscr{D}$ of the dynamical equations (9.54)–(9.60). Then, $\mathscr{W}(\varepsilon, \mathscr{H}_{11}(\varepsilon), \breve{\mathscr{D}}_{11}(\varepsilon), \mathscr{D}(\varepsilon))$ is a time-backward increasing function of ε. If (K_z^*, l_z^*) is in $\mathscr{K}^z_{t_f, \mathscr{H}_{11}^f, \mathscr{H}_{12}^f, \mathscr{H}_{21}^f, \mathscr{H}_{22}^f, \breve{\mathscr{D}}_{11}^f, \mathscr{D}^f;\mu} \times \mathscr{L}^z_{t_f, \mathscr{H}_{11}^f, \mathscr{H}_{12}^f, \mathscr{H}_{21}^f, \mathscr{H}_{22}^f, \breve{\mathscr{D}}_{11}^f, \mathscr{D}^f;\mu}$ defined on $[t_0, t_f]$ with the corresponding solutions $\mathscr{H}_{11}^*$, $\breve{\mathscr{D}}_{11}^*$ and $\mathscr{D}^*$ of the dynamical equations (9.54)–(9.60) such that, for $\tau \in [t_0, t_f]$

$$\begin{aligned}
0 = {} & \frac{\partial}{\partial \varepsilon}\mathscr{W}(\tau, \mathscr{H}_{11}^*(\tau), \breve{\mathscr{D}}_{11}^*(\tau), \mathscr{D}^*(\tau)) + \frac{\partial}{\partial\, \mathrm{vec}(\mathscr{Y}_{11})}\mathscr{W}(\tau, \mathscr{H}_{11}^*(\tau), \breve{\mathscr{D}}_{11}^*(\tau), \mathscr{D}^*(\tau)) \\
& \cdot \mathrm{vec}(\mathscr{F}_{11}(\tau, \mathscr{H}_{11}^*(\tau), \mathscr{H}_{12}^*(\tau), \mathscr{H}_{21}^*(\tau), K_z^*(\tau))) \\
& + \frac{\partial}{\partial\, \mathrm{vec}(\breve{\mathscr{Z}}_{11})}\mathscr{W}(\tau, \mathscr{H}_{11}^*(\tau), \breve{\mathscr{D}}_{11}^*(\tau), \mathscr{D}^*(\tau)) \\
& \cdot \mathrm{vec}(\breve{\mathscr{G}}_{11}(\tau, \breve{\mathscr{D}}_{11}^*(\tau), \mathscr{H}_{11}(\tau), K_z^*(\tau), l_z^*(\tau))) \\
& + \frac{\partial}{\partial\, \mathrm{vec}(\mathscr{Z})}\mathscr{W}(\tau, \mathscr{H}_{11}^*(\tau), \breve{\mathscr{D}}_{11}^*(\tau), \mathscr{D}^*(\tau)) \\
& \cdot \mathrm{vec}(\mathscr{G}(\tau, \mathscr{H}_{11}^*(\tau), \mathscr{H}_{12}^*(\tau), \mathscr{H}_{21}^*(\tau), \mathscr{H}_{22}^*(\tau), l_z^*(\tau)))
\end{aligned} \tag{9.63}$$

then K_z^* and l_z^* are optimal feedback and feedforward parameters. Moreover

$$\mathscr{W}(\varepsilon, \mathscr{Y}_{11}, \breve{\mathscr{Z}}_{11}, \mathscr{Z}) = \mathscr{V}(\varepsilon, \mathscr{Y}_{11}, \breve{\mathscr{Z}}_{11}, \mathscr{Z}) \tag{9.64}$$

where $\mathscr{V}(\varepsilon, \mathscr{Y}_{11}, \breve{\mathscr{Z}}_{11}, \mathscr{Z})$ is the value function.

Proof. It is already contained in the rigorous proof of the sufficiency with the essential conditions aforementioned from [10].

9.4 Persisting Statistical Optimal Control in Resiliency

The understanding of the Mayer-form verification theorem of dynamic programming is used to approach the optimal control of Mayer form at hand. And yet understanding the terminal time and states of a family of optimization problems by $(\varepsilon, \mathcal{Y}_{11}, \mathcal{Y}_{12}, \mathcal{Y}_{21}, \mathcal{Y}_{22}, \breve{\mathcal{Z}}_{11}\mathcal{Z})$ rather than $(t_f, \mathcal{H}_{11}^f, \mathcal{H}_{12}^f, \breve{\mathcal{H}}_{21}^f, \mathcal{H}_{22}^f, \breve{\mathcal{D}}_{11}^f, \mathcal{D}^f)$ is perhaps the most important now facing the control designer. In effect, this contending view also suggests that the value of the optimization problem depends on the terminal conditions. For instance, for any $\varepsilon \in [t_0, t_f]$, the state variables of the dynamical equations (9.54)–(9.60) are further expressed as $\mathcal{H}_{11}(\varepsilon) = \mathcal{Y}_{11}$, $\mathcal{H}_{12}(\varepsilon) = \mathcal{Y}_{12}$, $\mathcal{H}_{21}(\varepsilon) = \mathcal{Y}_{21}$, $\mathcal{H}_{22}(\varepsilon) = \mathcal{Y}_{22}$, $\breve{\mathcal{D}}_{11}(\varepsilon) = \breve{\mathcal{Z}}_{11}$, and $\mathcal{D}(\varepsilon) = \mathcal{Z}$.

Subsequently, with the quadratic-affine nature of (9.61) in mind, many of curious forms associated with potential solutions to the HJB equation (9.62) come through more clearly as follows: Fix $k \in \mathbb{Z}^+$ and let $\left(\varepsilon, \mathcal{Y}_{11}, \mathcal{Y}_{12}, \mathcal{Y}_{21}, \mathcal{Y}_{22}, \breve{\mathcal{Z}}_{11}, \mathcal{Z}\right)$ be any interior point of the reachable set $\mathcal{Q}$ at which the real-valued function $\mathcal{W}\left(\varepsilon, \mathcal{Y}_{11}, \breve{\mathcal{Z}}_{11}, \mathcal{Z}\right)$ described by

$$\mathcal{W}(\varepsilon, \mathcal{Y}_{11}, \breve{\mathcal{Z}}_{11}, \mathcal{Z}) = z_0^T \sum_{r=1}^{k} \mu_r(\mathcal{Y}_{11}^r + \mathcal{E}_{11}^r(\varepsilon)) z_0 + 2z_0^T \sum_{r=1}^{k} \mu_r(\breve{\mathcal{Z}}_{11}^r + \breve{\mathcal{T}}_{11}^r(\varepsilon)) + \sum_{r=1}^{k} \mu_r(\mathcal{Z}^r + \mathcal{T}^r(\varepsilon)) \tag{9.65}$$

is differentiable. The time parametric functions $\mathcal{E}_{11}^r \in \mathcal{C}^1([t_0, t_f]; \mathbb{R}^{(n+m)\times(n+m)})$, $\breve{\mathcal{T}}_{11}^r \in \mathcal{C}^1([t_0, t_f]; \mathbb{R}^{n+m})$, and $\mathcal{T}^r \in \mathcal{C}^1([t_0, t_f]; \mathbb{R})$ are yet to be determined.

As similarly illustrated in [10], the derivative of $\mathcal{W}\left(\varepsilon, \mathcal{Y}_{11}, \breve{\mathcal{Z}}_{11}, \mathcal{Z}\right)$ with respect to ε is obtained as follows

$$\begin{aligned}\frac{d}{d\varepsilon}\mathcal{W}(\varepsilon, \mathcal{Y}_{11}, \mathcal{Y}_{12}, \mathcal{Y}_{22}, \mathcal{Z}) = {} & z_0^T \sum_{r=1}^{k} \mu_r(\mathcal{F}_{11}^r(\varepsilon, \mathcal{Y}_{11}, \mathcal{Y}_{12}, \mathcal{Y}_{21}, K_z) + \frac{d}{d\varepsilon}\mathcal{E}_{11}^r(\varepsilon)) z_0 \\ & + 2z_0^T \sum_{r=1}^{k} \mu_r(\breve{\mathcal{G}}_{11}^r\left(\varepsilon, \mathcal{Y}_{11}, \breve{\mathcal{Z}}_{11}, K_z, l_z\right) + \frac{d}{d\varepsilon}\breve{\mathcal{T}}_{11}^r(\varepsilon)) \\ & + \sum_{r=1}^{k} \mu_r(\mathcal{G}^r\left(\varepsilon, \mathcal{Y}_{11}, \mathcal{Y}_{12}, \mathcal{Y}_{21}, \mathcal{Y}_{22}, \breve{\mathcal{Z}}_{11}, l_z\right) + \frac{d}{d\varepsilon}\mathcal{T}^r(\varepsilon))\end{aligned} \tag{9.66}$$

provided that the admissible 2-tuple $(K_z, l_z) \in \overline{K}_z \times \overline{L}_z$.

After the establishment of the candidate solution of (9.65) and the result (9.66) in the HJB equation (9.62), one obtains the following result

$$\min_{(K_z,l_z)\in\overline{K}_z\times\overline{L}_z} \Big\{ z_0^T \sum_{r=1}^{k} \mu_r (\mathcal{F}_{11}^r(\varepsilon,\mathcal{Y}_{11},\mathcal{Y}_{12},\mathcal{Y}_{21},K_z) + \frac{d}{d\varepsilon}\mathcal{E}_{11}^r(\varepsilon)) z_0$$

$$+2z_0^T \sum_{r=1}^{k} \mu_r (\breve{\mathcal{G}}_{11}^r(\varepsilon,\mathcal{Y}_{11},\breve{\mathcal{Z}}_{11},K_z,l_z) + \frac{d}{d\varepsilon}\breve{\mathcal{T}}_{11}^r(\varepsilon))$$

$$+\sum_{r=1}^{k} \mu_r (\mathcal{G}^r(\varepsilon,\mathcal{Y}_{11},\mathcal{Y}_{12},\mathcal{Y}_{21},\mathcal{Y}_{22},\breve{\mathcal{Z}}_{11},l_z) + \frac{d}{d\varepsilon}\mathcal{T}^r(\varepsilon)) \Big\} \equiv 0, \quad (9.67)$$

While the concern here is most directly with (9.67), the differentiation of the expression within the bracket of (9.67) with respect to K_z and l_z yields the necessary conditions for an interior extremum of the performance index with risk consequences (9.61) on $[t_0, t_f]$. In other words, the extremizing K_z and l_z must be

$$K_z(\varepsilon,\mathcal{Y}_{11}) = -S^{-1}B_z^T \sum_{r=1}^{k} \hat{\mu}_r \mathcal{Y}_{11}^r \quad (9.68)$$

$$l_z(\varepsilon,\breve{\mathcal{Z}}_{11}) = -S^{-1}B_z^T \sum_{r=1}^{k} \hat{\mu}_r \breve{\mathcal{Z}}_{11}^r, \qquad \hat{\mu}_r = \mu_r/\mu_1, \quad (9.69)$$

In view of (9.68) and (9.69), the zero value of the expression inside of the bracket of (9.67) that is pursued here for any $\varepsilon \in [t_0, t_f]$ when $\mathcal{Y}_{11}^r$, $\breve{\mathcal{Z}}_{11}^r$ and $\mathcal{Z}^r$ evaluated at the time-backward differential equations (9.54)–(9.60) requires

$$\frac{d}{d\varepsilon}\mathcal{E}_{11}^1(\varepsilon) = (A_z + B_z K_z(\varepsilon))^T \mathcal{H}_{11}^1(\varepsilon) + \mathcal{H}_{11}^1(\varepsilon)(A_z + B_z K_z(\varepsilon))$$
$$+ K_z^T(\varepsilon) S K_z(\varepsilon) + Q_z \quad (9.70)$$

$$\frac{d}{d\varepsilon}\mathcal{E}_{11}^r(\varepsilon) = (A_z + B_z K_z(\varepsilon))^T \mathcal{H}_{11}^r(\varepsilon) + \mathcal{H}_{11}^r(\varepsilon)(A_z + B_z K_z(\varepsilon))$$
$$+\sum_{v=1}^{r-1} \frac{2r!}{v!(r-v)!} (\mathcal{H}_{11}^v(\varepsilon)\Pi_1(\varepsilon) + \mathcal{H}_{12}^v(\varepsilon)\Pi_3(\varepsilon)) \mathcal{H}_{11}^{r-v}(\varepsilon)$$
$$+\sum_{v=1}^{r-1} \frac{2r!}{v!(r-v)!} (\mathcal{H}_{11}^v(\varepsilon)\Pi_2(\varepsilon) + \mathcal{H}_{12}^v(\varepsilon)\Pi_4(\varepsilon)) \mathcal{H}_{21}^{r-v}(\varepsilon) \quad (9.71)$$

$$\frac{d}{d\varepsilon}\breve{\mathcal{T}}_{11}^1(\varepsilon) = (A_z + B_z K_z(\varepsilon))^T \breve{\mathcal{D}}_{11}^1(\varepsilon) + K_z^T(\varepsilon) S l_z(\varepsilon)$$
$$+ \mathcal{H}_{11}^1(\varepsilon)(B_z l_z(\varepsilon) + m_{w_z}(\varepsilon)) \quad (9.72)$$

$$\frac{d}{d\varepsilon}\breve{\mathscr{T}}_{11}^{r}(\varepsilon) = (A_z + B_z K_z(\varepsilon))^T \breve{\mathscr{D}}_{11}^{r}(\varepsilon) + \mathscr{H}_{11}^{r}(\varepsilon)(B_z l_z(\varepsilon) + m_{w_z}(\varepsilon)) \tag{9.73}$$

$$\frac{d}{d\epsilon}\mathscr{T}_1(\varepsilon) = \text{Tr}\{\mathscr{H}_{11}^{1}(\varepsilon)\Pi_1(\varepsilon) + \mathscr{H}_{12}^{1}(\varepsilon)\Pi_3(\varepsilon) + \mathscr{H}_{21}^{1}(\varepsilon)\Pi_2(\varepsilon) + \mathscr{H}_{22}^{1}(\varepsilon)\Pi_4(\varepsilon)\}$$
$$+ 2(\breve{\mathscr{D}}_{11}^{1})^T(\varepsilon)(B_z l_z(\varepsilon) + m_{w_z}(\varepsilon)) + l_z^T(\varepsilon) S l_z(\varepsilon) \tag{9.74}$$

$$\frac{d}{d\varepsilon}\mathscr{T}_r(\varepsilon) = \text{Tr}\{\mathscr{H}_{11}^{r}(\varepsilon)\Pi_1(\varepsilon) + \mathscr{H}_{12}^{r}(\varepsilon)\Pi_3(\varepsilon) + \mathscr{H}_{21}^{r}(\varepsilon)\Pi_2(\varepsilon) + \mathscr{H}_{22}^{r}(\varepsilon)\Pi_4(\varepsilon)\}$$
$$+ 2(\breve{\mathscr{D}}_{11}^{r})^T(\varepsilon)(B_z l_z(\varepsilon) + m_{w_z}(\varepsilon)), \quad 2 \le r \le k, \tag{9.75}$$

The boundary condition of $\mathscr{W}(\varepsilon, \mathscr{Y}_{11}, \breve{\mathscr{Z}}_{11}, \mathscr{Z})$ implies that the initial-value conditions $\mathscr{E}_{11}^{r}(0) = 0$, $\breve{\mathscr{T}}_{11}^{r}(0) = 0$, and $\mathscr{T}^{r}(0) = 0$ for the forward-in-time differential equations (9.70)–(9.75) and yields a value function

$$\mathscr{W}(\varepsilon, \mathscr{Y}_{11}, \breve{\mathscr{Z}}_{11}, \mathscr{Z}) = \mathscr{V}(\varepsilon, \mathscr{Y}_{11}, \breve{\mathscr{Z}}_{11}, \mathscr{Z})$$
$$= z_0^T \sum_{r=1}^{k} \mu_r \mathscr{H}_{11}^{r}(t_0) z_0 + 2 z_0^T \sum_{r=1}^{k} \mu_r \breve{\mathscr{D}}_{11}^{r}(t_0) + \sum_{r=1}^{k} \mu_r \mathscr{D}^{r}(t_0)$$

for which the sufficient condition (9.63) of the verification theorem is satisfied so that the extremizing feedback parameters (9.68) and (9.69) become optimal

$$K_z^*(\varepsilon) = -\, S^{-1} B_z^T \sum_{r=1}^{k} \hat{\mu}_r \mathscr{H}_{11}^{*r}(\varepsilon)$$

$$l_z^*(\varepsilon) = -\, S^{-1} B_z^T \sum_{r=1}^{k} \hat{\mu}_r \breve{\mathscr{D}}_{11}^{*r}(\varepsilon),$$

In summary, promulgation of the analysis and its results is the procedural mechanism to compute the family of resilient controllers with performance risk aversion. As shown in Fig. 9.6, the design method and computational approach utilize the statistical measures of risk as well as the forward and backward-in-time differential equations.

Theorem 9.4.1 (Risk-Averse Control). *Under the assumptions of (A_z, B_z) stabilizable and (C_z, A_z) detectable, the network control system governed by (9.7)–(9.8) is subject to the chi-squared measure of performance (9.6). Suppose $k \in \mathbb{Z}^+$ and the sequence $\mu = \{\mu_r \ge 0\}_{r=1}^{k}$ with $\mu_1 > 0$ are fixed. Then, the risk-averse control strategy with control rate constraints over $[t_0, t_f]$ with two degrees of freedom is constructed as*

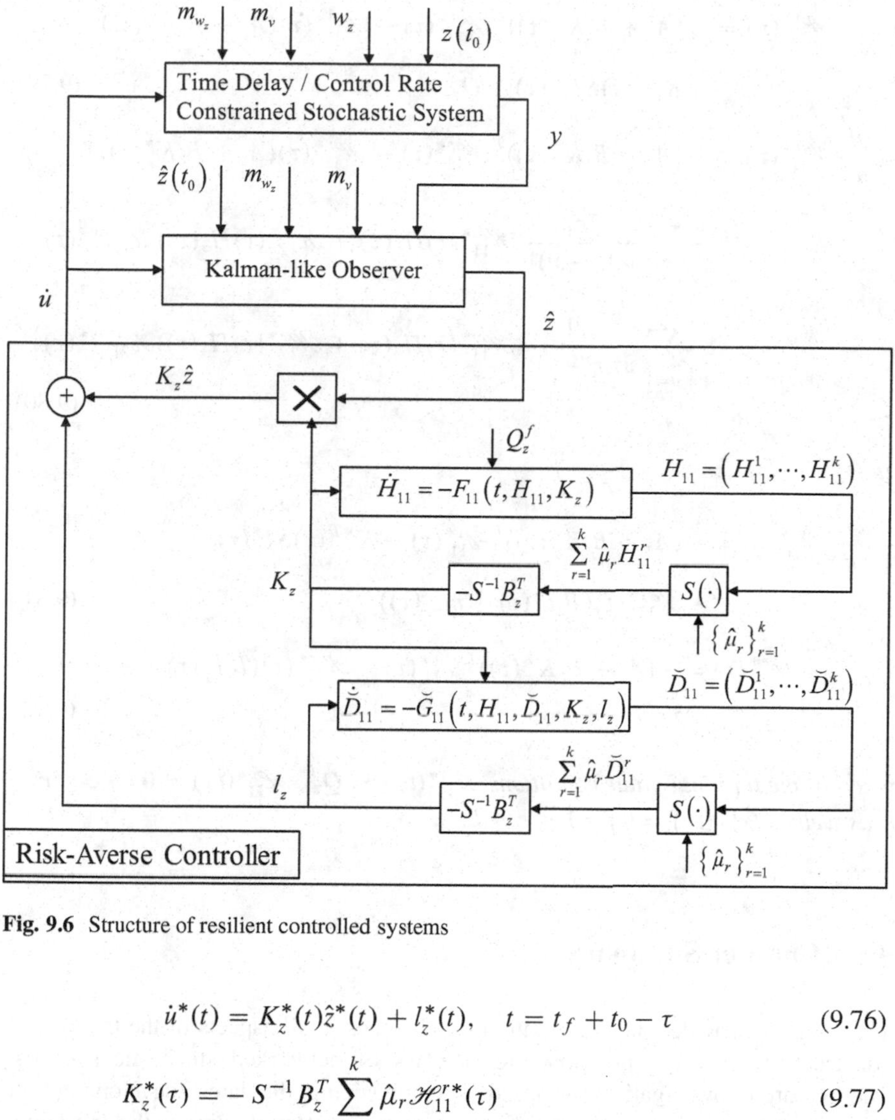

Fig. 9.6 Structure of resilient controlled systems

$$\dot{u}^*(t) = K_z^*(t)\hat{z}^*(t) + l_z^*(t), \quad t = t_f + t_0 - \tau \tag{9.76}$$

$$K_z^*(\tau) = -S^{-1}B_z^T \sum_{r=1}^{k} \hat{\mu}_r \mathscr{H}_{11}^{r*}(\tau) \tag{9.77}$$

$$l_z^*(\tau) = -S^{-1}B_z^T \sum_{r=1}^{k} \hat{\mu}_r \breve{\mathscr{D}}_{11}^{r*}(\tau) \tag{9.78}$$

where $\hat{\mu}_r = \mu_r/\mu_1$ represent different levels of influence as they deem important to the performance distribution. Finally, $\{\mathscr{H}_{11}^{r*}(\tau)\}_{r=1}^k$ and $\left\{\breve{\mathscr{D}}_{11}^{r*}(\tau)\right\}_{r=1}^k$ are the optimal solutions of the backward-in-time differential equations

$$\frac{d}{d\tau}\mathscr{H}_{11}^{1*}(\tau) = -(A_z + B_z K_z^*(\tau))^T \mathscr{H}_{11}^{1*}(\tau) - \mathscr{H}_{11}^{1*}(\tau)(A_z + B_z K_z^*(\tau)) - K_z^{*T}(\tau) S K_z^*(\tau) - Q_z \tag{9.79}$$

$$\frac{d}{d\tau}\mathscr{H}_{11}^{r*}(\tau) = -(A_z + B_z K_z^*(\tau))^T \mathscr{H}_{11}^{r*}(\tau) - \mathscr{H}_{11}^{r*}(\tau)(A_z + B_z K_z^*(\tau)) - \sum_{\nu=1}^{r-1} \frac{2r!}{\nu!(r-\nu)!}(\mathscr{H}_{11}^{\nu*}(\tau)\Pi_1(\tau) + \mathscr{H}_{12}^{\nu*}(\tau)\Pi_3(\tau))\mathscr{H}_{11}^{r-\nu*}(\tau) - \sum_{\nu=1}^{r-1} \frac{2r!}{\nu!(r-\nu)!}(\mathscr{H}_{11}^{\nu*}(\tau)\Pi_2(\tau) + \mathscr{H}_{12}^{\nu*}(\tau)\Pi_4(\tau))\mathscr{H}_{21}^{r-\nu*}(\tau) \tag{9.80}$$

and

$$\frac{d}{d\tau}\breve{\mathscr{D}}_{11}^{1*}(\tau) = -(A_z + B_z K_z^*(\tau))^T \breve{\mathscr{D}}_{11}^{1*}(\tau) - K_z^{*T}(\tau) S l_z^*(\tau) - \mathscr{H}_{11}^{1*}(\tau)(B_z l_z^*(\tau) + m_{w_z}(\tau)) \tag{9.81}$$

$$\frac{d}{d\tau}\breve{\mathscr{D}}_{11}^{r*}(\tau) = -(A_z + B_z K_z^*(\tau))^T \breve{\mathscr{D}}_{11}^{r*}(\tau) - \mathscr{H}_{11}^{r*}(\tau)(B_z l_z^*(\tau) + m_{w_z}(\tau)) \tag{9.82}$$

whereby the terminal-value conditions $\mathscr{H}_{11}^{1*}(t_f) = Q_z^f$, $\mathscr{H}_{11}^{r*}(t_f) = 0$ *for* $2 \le r \le k$ *as well as* $\breve{\mathscr{D}}_{11}^{r*}(t_f) = 0$ *for* $1 \le r \le k$.

9.5 Chapter Summary

This chapter provides the presentation of advances in all aspects of the theoretical constructs and design principles for a class of networked stochastic systems. The research investigation proposed is emphasizing the latest development in characterization and management of uncertain prospects pertaining to the restrictive family of finite-horizon integral quadratic costs associated with time-invariant linear stochastic systems, whereby the controlled systems have time-delay measurements and the incumbent controller is subject to control rate constraints. Associated performance index with mean and risk awareness is proven to be crucial when ordering uncertain prospects of the chi-squared random costs herein. As noted earlier, algorithms and procedures to guarantee the feasibility of output feedback control laws supported by state estimates in presence of time-delay observations, have also been put forward. Finally, applications and techniques that employ these research advances will be the drivers for future resilient controls with performance risk mitigation.

References

1. Tatikonda, S.C.: Control under communication constraints. PhD Thesis, Department of Electrical Engineering and Computer Science, MIT, Cambridge (2000)
2. Barrett, G., Lafortune, S.: On the synthesis of communicating controllers with decentralized information structures for discrete-event systems. In: Proceedings of IEEE Conference on Decision and Control, Tampa, pp. 3281–3286 (1998)
3. Barrett, G., Lafortune, S.: Decentralized supervisory control with communicating controllers. IEEE Trans. Autom. Control **45**, 1620–1638 (2000)
4. Hirshleifer, J.: Investment, interest and capital. Prentice Hall, Englewood Cliffs (1970)
5. Markowitz, H.: Portfolio Selection: Efficient Diversification of Investments. Wiley, New York (1970)
6. Pham, K.D.: Linear-Quadratic Controls in Risk-Averse Decision Making: Performance-Measure Statistics and Control Decision Optimization. Springer Briefs in Optimization. Springer, New York (2012). ISBN 978-1-4614-5078-8
7. Kleiman, D.L.: Optimal linear control for systems with time-delay and observation noise. IEEE Trans. Autom. Control **AC-14**, 524–527 (1969)
8. Pham, K.D.: Statistical control paradigms for structural vibration suppression. Ph.D. Dissertation, Department of Electrical Engineering. University of Notre Dame, Indiana, U.S.A. Available via http://etd.nd.edu/ETD-db/theses/available/etd-04152004-121926/unrestricted/PhamKD052004.pdf.Cited15January2014 (2004)
9. Fleming, W.H., Rishel, R.W.: Deterministic and Stochastic Optimal Control. Springer, New York (1975)
10. Pham, K.D.: Performance-reliability-aided decision-making in multiperson quadratic decision games against jamming and estimation confrontations. In: Giannessi, F. (ed.), J. Optim. Theory Appl. **149**(1), 599–629 (2011)

Chapter 10
Epilogue

10.1   Concern over Greater Uncertainty Quantification

Decision-making under uncertainty may be viewed as decisions between alternative probability distributions of performance costs, and the decision maker chooses between them in accordance to a consistent set of preferences. In view of the common acceptance of asymmetry or skewness among statisticians and engineers of the probability distributions, the popular expected value approach or the mean-variance preference rules may capture the first and second-order statistics associated with a certain class of probability distribution functions, but understanding of what is at stake here, remains, at best, persistently inadequate. The gulf between the mean-variance rules and almost universal reluctance to place emphases on higher-order statistics of the probability distributions, is indeed vast.

Among the topics of uncertainty analysis and quantification, the present work considers a certain restricted but important class of chi-squared distributions that include most integral-quadratic costs of practical interest in control engineering. A new model of ordering uncertain prospects has been proposed and is becoming more and more visible through different problem classes of resilient controls. In addition, characterization and management of performance-measure statistics in the process of reshaping closed-loop system performance are regarded as the enabling tools, which have brought about the mentioned results in statistical optimal control as evidenced enough in the monograph.

Nevertheless, these foundational results for resilient controls when ordering uncertain prospects, which have strong ties for non-symmetric distributions, are expected to experience further challenging obstacles. When ordering uncertain prospects with unequal means, no known selection rule uses both necessary and sufficient conditions for stochastic dominance. And thus, the proposed rules herein are merely reasonable approximations to the optimal rule, if any. In general, initial fruits of the progress and positive advancement constitute a strong foundation which

© Springer International Publishing Switzerland 2014   205
K.D. Pham, *Resilient Controls for Ordering Uncertain Prospects*, Springer
Optimization and Its Applications 98, DOI 10.1007/978-3-319-08705-4_10

represents a massive change in stochastic control research and actual practice. These changes have opened up many new possibilities and opportunities in resilient control designs to go forward.

10.2 Breakthrough for Resilient Controls

In the period reviewed, the issues of post-design performance analysis beyond the expected-value and mean-variance approaches are clear and numerous whereas, at present, available options to resolve these endemic problems are somewhat limited. Nevertheless *Resilient Controls for Ordering Uncertain Prospects* has been proposing various new model concepts which will modernize control sciences and guiding principles in response to emerging needs. Among the many positive developments that are being pursued in this monograph are initiatives to reform both control decisions with risk consequences and correct-by-design paradigms for performance reliability associated with the class of stochastic linear dynamical systems with integral quadratic costs and subject to network delays, control and communication constraints. In furtherance of these developments the methodology presented here is based on defining a mean-risk aware structure over the control problem of interest. This mean-risk aware structure is relied on the dual considerations of performance uncertainty and performance sensitivity. The methodology essentially encodes the control designer's knowledge about performance distributions with the generalized chi-squared behaviors, and uses this knowledge to order uncertain prospects. The feedback control strategy is then based on this priority ranking for performance risk aversion.

In whatever ways the ideas, concepts, and new understanding from this monograph may differ, a recurring theme supported background and contrast is the understanding of resilient controls in the face of uncertainty rather than in any specific models. It is no longer content with performance averages and thereby deciding to mitigate performance riskiness to ensure how much of the inherent or design-in reliability actually ends up in the developmental and operational phases.

10.3 The Door Opens

This monograph captures some of the winds of change that are blowing across uncertainty quantification and stochastic control communities and the creative efforts that are being made to increase the understanding of resilient controls in the face of uncertainty rather than in any specific models as an old order changes and new responses are demanded. Among obvious omissions are specific considerations of the application of decision theory to resilient controls of the linear-quadratic class of stochastic dynamical systems. Each of these topics is examined explicitly in several chapters; each proved to be elusive when the author tried to capture optimal rules for ordering uncertain prospects of more general families of probability distribution functions; and each merits more systematic analysis in a different kind of publication. No attempt has been made to write a concluding chapter. In opening the door of ordering uncertain prospects to a wider resilient control community, this monograph invites the control practitioners and theorists with reflective understanding of the distinctive context and of critical issues and options being faced.

Index

K.D. Pham, *Resilient Controls for Ordering Uncertain Prospects*, Springer Optimization and Its Applications 98, DOI 10.1007/978-3-319-08705-4

Printed in the United States
By Bookmasters